Biologically-Inspired Systems

Volume 16

Series Editor

Stanislav N. Gorb, Department of Functional Morphology and Biomechanics, Zoological Institute, Kiel University, Kiel, Germany

Motto: Structure and function of biological systems as inspiration for technical developments.

Throughout evolution, nature has constantly been called upon to act as an engineer in solving technical problems. Organisms have evolved an immense variety of shapes and structures from macro down to the nanoscale. Zoologists and botanists have collected a huge amount of information about the structure and functions of biological materials and systems. This information can be also utilized to mimic biological solutions in further technical developments. The most important feature of the evolution of biological systems is multiple origins of similar solutions in different lineages of living organisms. These examples should be the best candidates for biomimetics. This book series will deal with topics related to structure and function in biological systems and show how knowledge from biology can be used for technical developments in engineering and materials science. It is intended to accelerate interdisciplinary research on biological functional systems and to promote technical developments. Documenting of the advances in the field will be important for fellow scientists, students, public officials, and for the public in general. Each of the books in this series is expected to provide a comprehensive, authoritative synthesis of the topic.

More information about this series at http://www.springer.com/series/8430

Alexander E. Filippov • Stanislav N. Gorb

Combined Discrete and Continual Approaches in Biological Modeling

Springer

Alexander E. Filippov
Donetsk Institute for Physics
and Engineering
Donetsk, Ukraine

Stanislav N. Gorb
Zoological Institute
Kiel University
Kiel, Germany

ISSN 2211-0593 ISSN 2211-0607 (electronic)
Biologically-Inspired Systems
ISBN 978-3-030-41527-3 ISBN 978-3-030-41528-0 (eBook)
https://doi.org/10.1007/978-3-030-41528-0

This Springer imprint is published by the registered company Springer Nature Switzerland AG
The registered company address is: Gewerbestrasse 11, 6330 Cham, Switzerland

This book is dedicated to Prof. Dr. Werner Nachtigall (Saarbrücken), pioneer in the field of bionics/biomimetics in Germany, who by bringing together biology, physics, and engineering revolutionized emerging field of biomechanics.

Preface

Biological complexity is an obvious fact and is one of the various important outcomes of the process of evolution. Evolution has produced some remarkably complex organisms and systems (cells, tissues, organs, surfaces, limbs, etc.), although the actual level of complexity is very hard to define or measure in biology, especially with regard to properties such as gene content, the number of cell types, or morphology. Some biologists believe that evolution is progressive and had a direction that led toward so-called higher organisms. However, there is no evidence for this statement. Natural selection has no intrinsic direction, and organisms are selected for either increased or decreased complexity in response to local environmental conditions. Although there has been an increase in the maximum level of complexity over the history of life, there have always existed a high number of rather "simple" organisms. If we consider species distribution in the multidimensional space of the phenotype, mutations in such a system provide an isotropic swelling of the system without any preferences for "simple" or "complex" characters and organisms, whereas selection cuts off some forms (independently of the degree of complexity), leaving others to survive. The most challenging goal of biology is to understand why selection seems to "prefer" some particular forms and not the other ones.

The complexity of biological systems is tremendous at all levels of organization, from the genetic and molecular to the cellular and organismic level. Modern experimental biology is continuously making huge leaps from the macroscale down to the nanoscale, but the major problem of modern biology is the lack of a general methodology for dealing with the huge amount of data and the lack of concepts that can use the data for something meaningful. In other words, we need more appropriate modeling to organize our data and to extract novel knowledge from it.

Another problem of experimental biology is that we are dealing with systems which are very fragile and sensitive and behave statistically, which means that often no kind of experiment is possible and, if so, only at extremely high costs and usually without proper control samples/experiments. This is one of the most important fundamental differences between experimental biology and other experimental disciplines dealing with non-biological matter. Again, the most helpful approach in this context would be modeling, which can then be used as an experimental platform "in silico" in order to predict outcomes in real biological experiments. In some cases, a "virtual" experiment may even be the only viable possibility.

These considerations lead to the insight that in many instances, experimental biology can profit from new methods made possible by novel modeling techniques that take advantage of the recent developments in modeling and visualization. This book is not necessarily a coherent story in itself but a potpourri of different examples, which are united by the use of the same methodology and inspired by this insight. It is the result of a long-standing collaboration between a theoretical physicist who studied field theory and its applications to phase transitions and critical phenomena in physical kinetics, magnetism, superconductivity, and tribology and a field biologist who became interested in the functional mechanisms of certain biological structures and certain behavioral and ecological processes. This combination of expertise and our excitement about the beauty of the studied structures and processes resulted in this volume. The first main concept of the book is that the basic mechanisms behind biological systems are expected to be based on some kind of *self-organization.* This idea is not particularly new, and there are a vast number of sources describing models of pattern formations in development, ecology, and behavior.

The second main concept is that in our modeling, we apply a combination of *continual and discrete modeling approaches*, each of which is well established in biological modeling but which are not often used in combination.

The third main concept is the visualization of the results. Using modern computational techniques and visualization tools, we are able to visualize every single step of the process and to generate sort of a *virtual behavior* that enables the direct visualization and comparison of our models with real biological systems. This approach provides an excellent bridge for the collaboration between biologists and theoretical physicists, which would be impossible with the use of traditional biological vocabulary and mathematical language. Applying these concepts to various relevant problems of modern biology has led to some useful results, which has prompted us to compile and publish this book, in the hope of providing valuable insights and inspiring more research in a similar vein.

Chapters 1 and 2 describe our philosophy and general approach to modeling and to the kinds of biological systems under consideration.

Chapter 3 discusses some problems of fibrillar adhesion in biological and biomimetic systems (clusterization and functional significance of material gradients along adhesive setae). It also presents spatial models of the gecko foot hair and the functional significance of highly specialized nonuniform geometry.

Chapter 4, we takes a look at spatula-shaped contact geometries. Dimension-dependent adhesive properties and shear-induced adhesion of biological spatulate attachment devices form the central problem of this chapter. Questions about the role of fluids in insect feet adhesion and the role of self-aligning geometry in the formation of adhesive contact in a plant fruit are also considered.

Chapter 5 is devoted to the problem of anisotropic friction in snake skin. We start with the structural diversity of the snake skin microstructure and a generalized numerical model of the frictional behavior of the skin which is covered by anisotropic surface nanostructures. Finally, we discuss the role of these surface features in snake locomotion.

Chapter 6, we discusses specific contact systems based on mechanical interlocking, such as biological "micro-Velcro" and bird feather microhooks.

Chapter 7 discusses two further biomechanical problems at the microscale, namely, penile propulsion into the spiraled spermathecal ducts of female beetles and the role of viscoelasticity in resilin biomechanics.

Chapter 8 is devoted to pattern formation in biology. A two-dimensional (2D) problem is examined by using correlation analysis of symmetry breaking in a surface nanostructure ordering. Three-dimensional (3D) problems are discussed by means of two independent examples: pattern formation of colloid spheres in the water-repellent cerotegument of whip spiders and pattern formation in the superhydrophobic cuticle of springtails.

Finally, Chap. 9 deals with ecology and evolution and pattern formation in animal behavior. The first example is the attempt to forecast the long-term ant-species-dependent dynamics of a myrmecochorous plant community, while the second one concerns predator–prey behavior in a cylindrical world where we demonstrate that the influence of this particular space on such behavior is quite nontrivial.

Various parts of the original studies were carried out in the Department of Functional Morphology and Biomechanics, Kiel University, Germany; the Institute of Mechanics, Technical University of Berlin, Germany; and the Institute for Physics and Engineering, Donetsk, Ukraine. We wish to thank our coauthors of the original papers: Elena Gorb, Lars Heepe, Alexander Kovalev, Yoko Matsumura, and Clemens Schaber (Department of Functional Morphology and Biomechanics, Kiel University, Germany), Valentin Popov (Institute of Mechanics, Technical University of Berlin, Germany), Guido Westhoff (Hagenbeck Tiergarten, Hamburg, Germany), Jonas O. Wolff (Macquarie University, Department of Biological Sciences, Sydney, Australia), and Rhainer Guillermo-Ferreira (Universidade Federal de São Carlos (UFSCar), Departamento de Hidrobiologia (DHb), São Carlos, Brazil). We thank Renate Schilling for linguistic correction and Elena Gorb for final proofreading.

The main part of this book was written when Alexander Filippov was visiting Stanislav Gorb in the Department of Functional Morphology and Biomechanics of Kiel University, Germany, in 2017–2019. These visits were financially supported by the Georg Forster Research Award (Alexander von Humboldt Foundation, Germany) UKR 1118826 GFPR. We hereby gratefully acknowledge this generous support.

Kiel and Donetsk, 2020

Alexander Filippov
Stanislav Gorb

Contents

About the Authors

Alexander E. Filippov is Principal Scientist of Donetsk Institute for Physics and Engineering. He received his PhD degree in Solid-State Physics, Phase Transitions, and Critical Phenomenon in 1985. In 1990, he was habilitated as Doctor of Science in former USSR. He is Member of the Scientific Council of Donetsk Institute for Physics and Engineering of the National Academy of Sciences of Ukraine. He is a known specialist in the fields of theoretical and applied physics. His interest ranges from atomic scale friction and planetary systems dynamics to the questions of biological evolution and ecology. He discovered for the first time so-called physical branch of solution of the exact local renormalization group equation and developed new variant of the perturbation theory based on naturally small Fisher exponent. He performed a number of outstanding studies of critical phenomena in magnetic, traditional, and high-temperature superconducting systems. Most of these works are published in *Nature*, *PNAS*, *Physical Review Letters*, *Scientific Reports*, and other top journals. Using ideas of dynamic chaos and self-organization in nonlinear systems, he published a number of studies related to the general theory of biological evolution and some applications to the bacterial growth, ecology of seed dispersal, animal behavior, biomimetics, etc. He also has numerous publications in tribology on mesoscopic fractal and nanoscale surfaces, which are extremely well recognized and widely cited by the scientific community. He was invited by numerous working groups worldwide to theoretically solve specific experimental problems: Université Pierre et Marie Curie (France), University of Nijmegen (the Netherlands), Hong Kong Baptist University (Hong Kong), Tel Aviv University (Israel), Berlin Technical University (Germany), International School for Advanced Studies (SISSA, Trieste, Italy), University of Kiel (Germany), and Université de Montréal (Canada). In 2016, he received the Georg Forster Research Award from Alexander von Humboldt Foundation, Germany. He is Author of the book *Phase Transitions in Systems with Competitive Interactions* (1989) and has around 250 papers in peer-reviewed journals.

Stanislav Gorb is Professor and Director at the Zoological Institute of Kiel University, Germany. He received his PhD degree in Zoology and Entomology at the I.I. Schmalhausen Institute of Zoology of the Ukrainian Academy of Sciences in Kiev. He was a Postdoctoral Researcher at the University of Vienna (Austria), a Research Assistant at the University of Jena, and a Group Leader at the Max Planck Institutes for Developmental Biology in Tübingen and for Metals Research in Stuttgart (Germany). His research focuses on morphology, structure, biomechanics, physiology, and evolution of surface-related functional systems in animals and plants, as well as the development of biologically inspired technological surfaces and systems. He received the Schlossmann Award in Biology and Materials Science in 1995, International Forum Design Gold Award in 2011, Materialica "Best of" Award in 2011, Karl Ritter von Frisch Medal of German Zoological Society in 2018, and Outstanding Contribution Award of the International Society of Bionic Engineering in 2019. He is Corresponding Member of the Academy of the Sciences and Literature in Mainz, Germany (since 2010) and Member of the National Academy of Sciences Leopoldina, Germany (since 2011). He has authored and coauthored several books, more than 500 papers in peer-reviewed journals, and 4 patents.

Chapter 1
Introduction

Abstract In this chapter, we introduce general idea of the attraction, which is very common phenomenon for the processes in the Nature. We start from rather general biological example of the natural selection, where adaptation to the environmental conditions can be described as an attraction of some population distribution in the phenotype space to a center of ecological niche. The niche is mathematically represented as the "survival coefficient" which in turn can be linked to a kind of energy potential. This link allows establishing a very useful connection between biological and physical approaches to a wide range of problems. In particular, we discuss an evolution in complex potential with a lot of valleys in a multidimensional space accompanied by so-called "large river" effect, which corresponds to an extremely slow evolution of some, normally close to final, stages of the adaptation. This effect is related to the practically important states of the "frozen kinetics" which accompanies extremely wide spectrum of phenomena and allows understanding different physical and biological processes.

1.1 The 'Attracting Nature' of Nature

The most amazing feature of Nature is that it is extremely primitive and fantastically complex at the same time. This means that its basic laws are very simple, even primitive, but the observed structures and their dynamic behaviors are extremely complex. In some sense this book is devoted to a study of this strange combination of simplicity and complexity. Of course, we do not plan to solve or even discuss all the possible problems related to this issue. Rather, we will address some interesting phenomena related to this quite common observation which are relevant for a number of concrete problems in biology.

Electronic supplementary material The online version of this chapter (https://doi.org/10.1007/978-3-030-41528-0_1) contains supplementary material, which is available to authorized users.

A. E. Filippov, S. N. Gorb, *Combined Discrete and Continual Approaches in Biological Modeling*, Biologically-Inspired Systems 16,
https://doi.org/10.1007/978-3-030-41528-0_1

Since it can be ruled out that biological processes are controlled by a "super-brain" which calculates every single step, basic mechanisms may be expected to be very simple and even primitive. It is well accepted now that they should be determined by some kind of *self-organization*. For example, in biology general ideas of natural selection claim only that organisms basically transfer their genetically coded structure to the next generations with some deviations (mutations) and new forms being selected according to their advantages (or disadvantages) to survive and continue this transfer further.

While many philosophical questions may arise in this context, in this book we will concentrate on two main practical questions:

1. what does self-organization mean mathematically and
2. how can we apply this knowledge to generate new insights?

1.2 The Mathematics of Self-Organization

Let us start with a very simple first step using a biological example. We will select just one particular characteristic of some species which can be described quantitatively, for example, the length of the legs, the speed of running or any other parameter which can be directly measured. From the very beginning, it is obvious that there are limits to the value of any such parameter. It cannot exceed some upper and some lower boundary, and very likely has an optimum which is defined and limited by the species' living conditions.

Now we can show that this clear supposition immediately leads to important mathematical and practical consequences. In particular, from the mathematical point of view, it means that some function $S(x)$ exists in the one-dimensional space $\{x\}$ of the given parameter which represents the probability to survive with a particular value x of this parameter.

As a rule, we do not know the correct shape of the function $S(x)$, but it is commonly supposed to present some kind of normal (Gaussian) distribution around some optimal value x_0 which corresponds to the maximal probability of survival, with some width w_S describing the more or less acceptable deviations from this value where the probability of survival sufficiently differs from zero:

$$S(x, x_S) \sim \exp\left[-\left(\frac{x - x_S}{w_S}\right)^2\right]. \tag{1.1}$$

However, instant distribution of the same parameter of species also varies with some deviations, and the probability of finding its particular value at any given moment is proportional to another function $F(x)$. Typically, this function is also supposed to take on the shape of a simple Gaussian curve. However, generally speaking this does not necessarily have to be the case, and even if it is so, its center, width and amplitude do not need to coincide with the corresponding values for the probability to survive, as given by the function $S(x, x_S)$.

To calculate the values for a new generation, one has to account for the fact that due to mutations they will vary around every value $x \to x_G$ as well as around the new center, $F(x) \to F(x_G)$. These deviations are defined by the corresponding function $G(x, x_G) \sim \exp\left[-\left(\frac{x-x_G}{w_G}\right)^2\right]$.

The newly born generation can be described as a product of the factor $G(x, x_G)$ and the previous distribution $F(x_G)$ integrated over all the realizations of the parameter $x_G : F_1(x) = \int dx_G G(x, x_G) F(x_G)$.

Finally, the surviving population is given by the product of all the three functions:

$$F_2(x, x_S) = S(x, x_S) \int dx_G G(x, x_G) F(x_G). \tag{1.2}$$

The apparent complexity of this final formula is illusory. We simply multiplied three factors: heredity, variability and selection. Even intuitively one can predict that at every iteration $k = 1, 2, \ldots$ the survival coefficient $S(x, x_S)$ will cut off those parts of the next population $F_{k+1}(x, x_S)$ whose $S(x, x_S)$ is small, and will support those ones which are closer to its center x_S. So, the distribution $F_{k+1}(x, x_S)$ will gradually be "attracted" to the center. Of course, this can easily be checked quantitatively.

This observation may at first seem to be very specific and only valid for examples taken from biological evolution. In fact, however, the survival coefficient reflects an interaction between every individual x of the array $F(x)$ with the environment. It was written in the form of the collective mean-field factor $S(x, x_S)$ for the sake of simplicity only.

Indeed, any interaction with the environment leads completely different objects of nature to "be attracted" to "their natural places" and, finally, to converge and stabilize there. *Attraction* is the key word here and we will apply this concept to many different studies in the course of this book.

Of course, we are usually not able to account exactly for all the interactions. But we can make some suppositions about them and their mutual relationships. We can select and combine the strongest interactions into simpler combinations and play with them theoretically in order to model the real behavior of the objects as closely as possible.

It is important to note in this context that Nature is globally stable and that the majority of particular processes (as well as the mathematical problems which we are solving) are stable as well. In the simple example discussed above, such stability was provided by the exponential decrease of the Gaussian curve $S(x, x_S)$ which automatically suppressed all phenotypes with large deviations of the parameter x from x_S (Fig. 1.1).

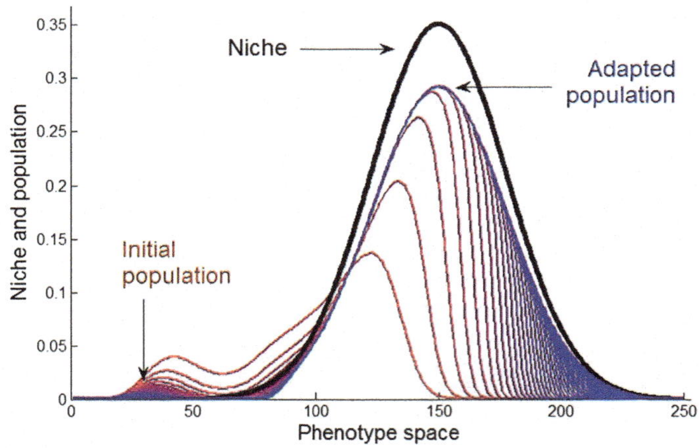

Fig. 1.1 Attraction of the population to the "ecological niche". The niche or "survival coefficient" is represented by the bold black line. Successive stages of evolution are shown by the parallel curves with the color changing from red to blue. See also supplementary Movie 1.1

Of course, the Gaussian curve is a strong simplification of reality. However, it can be considered an approximation – and it leads to a very pragmatic idea: that it does not really matter what happens to the values of the parameters in the central attraction zone. In fact, it is not even important how many subsystems and interactions between subsystems are included. The only feature that has to be controlled is the system's stability against large deviations from the center, where physically or biologically reasonable values of the parameters are expected. But if the factors, such as $S(x, x_S)$, are chosen properly, the system will stabilize itself almost automatically (Filippov 1993, 1998).

1.3 Frozen Kinetics or the Large River Effect

Now we will introduce another, more physical point of view, instead of the "survival coefficient" $S(x, x_S)$. Even in "purely biological" matters the probability of finding a particular realization of a variable x is defined by an energy U which is necessary to create a "niche" $U(x)$ for it. In physics, a coefficient that is analogous to $S(x, x_S)$ is derived from the so-called Boltzmann energy distribution:

$$S(x) \sim \exp\left[-U(x)\right] \tag{1.3}$$

where, as we see, this energy $U(x)$ also stays in the exponent (Landau and Lifshitz 1980; Lifshitz and Pitaevskii 1981).

Regardless of the branch of science concerned, this calculation of energy tells the system "what it costs" to create each single state. If we can calculate or even just

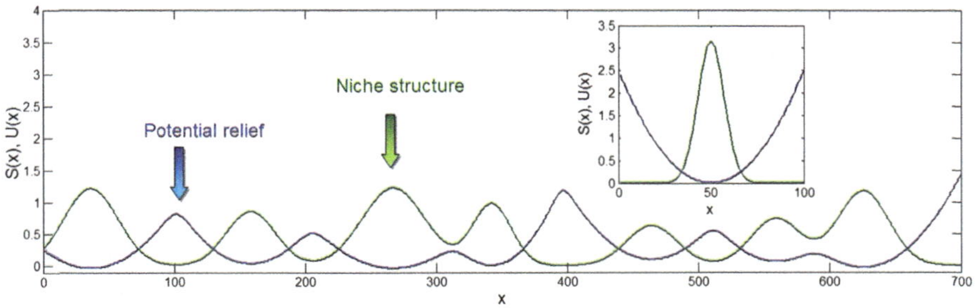

Fig. 1.2 Example of a relationship between the niche structure (green curve) and the effective potential relief (blue curve). A single niche and the corresponding potential valley are shown in the insert

estimate this cost from experiments or observations, we will be able to write down such an energy as a potential relief curve $U(x)$ in the parameter space $\{x\}$. Again, it is important to note that the number of parameters in space $\{x\}$ is not limited. If this number is large, a potential relief $U(x)$ will only be found in multi-dimensional space. This may be difficult to imagine visually, but it is no problem to calculate it with modern computers (Fig. 1.2).

Since all systems tend to minimize the expended energy, any system will relax to one or more minima of the potential $U(x)$. If we know the correct $U(x)$, then our work is reduced to a simple task of watching how the system is doing this according to the equation of motion. In a very simple and reduced manner this can be expressed by the following form:

$$\partial F(x,t)/\partial t = A \exp\left(-U(x)\right) - F(x,t). \tag{1.4}$$

From this equation it is immediately obvious that with time the initial distribution of density $F(x, t = 0)$ will be attracted to the structure described by the potential: $F(x, t) \to A \exp(-U(x))$.

At this point, the reader might interject that such an approach is a great oversimplification. In such an approach, we actually do not need any mathematics or any simulation since we already know from actual experiments everything we need to know and the calculation only masks the things which we have already found out without it.

The answer to this argument is: yes and no! The system is in fact attracted to the states which we somehow know in advance. But this attraction is a process. By watching intermediate stages of the process we can garner a lot of knowledge which we would otherwise never come by.

First of all, every process takes time. If this time period is very long, we will never see the final configuration. There are at least two reasons for this:

(1) The process may be so long that neither we personally nor even the whole of humanity will ever be able to see the final stages. For example, all the galaxies

develop and tend to some well predictable stages (e.g., spherical or elliptic ones). But this process takes billions of years. As a result, we can see many diverse views of spiral galaxies, including our own Milky Way. For us they are visually "frozen" and all we can do is to compare different galaxies which we see at different stages of their evolution and make some conclusions about typical scenarios of their motion.

Now, the everyday application of the proposed ***attractor approach*** is not as exotic as the study of galaxies. But the absolute majority of the phenomena around us can actually only be observed at intermediate stages of their evolution, even though they look static if we do not pay attention to the complete process. Take rocks and stones, for example, which are at an intermediate stage of their evolution and will slowly change with time. This phenomenon may be regarded as "***frozen kinetics***", but actually it is a particular form of a general phenomenon, the so-called "***large river effect***" (Zumbach 1993; Bagnuls and Bervillier 1994; Filippov 1994, 1995).

This effect refers to the course of a river on the Earth's landscape. Starting from the top of a mountain, the water quickly moves down in the form of waterfalls and small but very fast rivers; it collects in slower rivers moving between hills and further accumulates in wide and slowly moving large rivers which ultimately flow into the sea. This simple picture from school handbooks perfectly illustrates the sophisticated concept of nonlinear dynamics, discussed above in the context of effective potential relief. It probably cannot be proven mathematically, but common observation confirms that the same phenomenon takes place in the phase trajectories of nonlinear equations, caused by effective potential relief.

In the context of this particular discussion it may not be important that many small rivers merge into a single one (although it is important for this chapter), but it is definitely important here that close to the sea the slope of the landscape is usually very gentle and the river flows extremely slowly (Fig. 1.3).

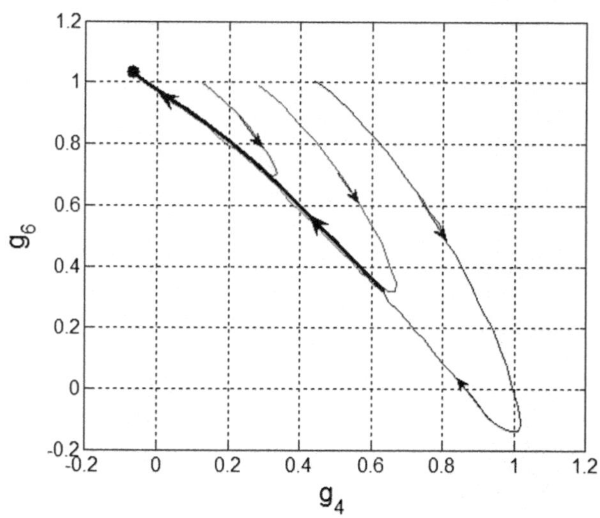

Fig. 1.3 "Large river effect" in nonlinear differential equations. Various flows with different initial conditions in abstract phase space $\{g_4, g_6\}$ quickly tend to a common trajectory (shown by the bold black line) and slowly move along it to a fixed point

The naïve assumption of traditional (linear) mathematics of previous centuries was that the attraction to a fixed point $\vec{r}*$ with time $t \to \infty$ can be described by the exponential function $|\vec{r} - \vec{r}*| \sim \exp(-\alpha t)$, where the characteristic rate α can be relatively small, but still assumes a constant nonzero value. However, the study of systems described by nonlinear realistic models shows us that in many cases such an attraction appears to be "logarithmically slow".

Formally it corresponds to variable exponent α which decreases with time and tends to zero $\alpha = \alpha(t) \to 0$ at $t \to \infty$. However, in the context of research this usually means that the evolution of the system parameters is so slow that they practically do not change at all during the time typically used for studying the problem. For this reason, this state may also be called ***frozen kinetics***.

In fact, the scientist's job here is similar to that of painters, sculptors, musicians and poets, who produce frozen motions of reality. Now positivistic science also has the instruments for joining this club.

1.4 Variable Potentials

All systems are open to the external world. While they relax into the stationary state, something may happen in their environment that partially or even completely changes the conditions which prevailed when the system started its evolution. So, even if we had enough time to wait for the final stages of evolution, in some cases we would still never see them, because they will have become impossible before they will be reached.

Mathematically this means that the "static" potential relief varies as well. It cannot be assumed to be static, but only "quasi-static". Thus, we have to change the function $U(x) \to U(x, t)$ (of course, only if we know how it varies, or at least if we can make some guesses about its variation and compare later the results with the real observations). This completely modifies our research strategy. Of course, in many cases we do not need to know the final (maybe dead) stages of evolution, but are much more interested in the intermediate (living) stages. Life is motion!

From a technical point of view, it is sometimes not very convenient to vary the potential relief directly. In many cases, it is easier instead to examine how external forces form and change an effective potential with time. If these changes are known, one can simply insert the forces into the right side of the equation of motion and solve the modified version:

$$\partial F(x,t)/\partial t = A \exp(-U(x)) - F(x,t) + \ldots External\_forces.$$

However, we need to remember that these forces are the result of the interaction between the system under consideration and other ones, which were originally separated from the study for simplicity's sake and only for this reason treated as "external" subsystems. So, in many cases it is easier to directly include additional subsystems into the study from the very beginning than to speculate theoretically about the forces caused by them. In a certain sense, this procedure returns the problem to its original form. However, this transfers the problem and the parameters into higher dimensionality.

By doing this, we sort of "oscillate" between these approaches. Remember that we started with the simplest possible presentation of the problem, with only one system moving into a potential valley and being attracted to a final configuration there. As a rule, such an oversimplification produces only a caricature of reality, which is much more complex. But, as with every good caricature, it allows us to see the most essential features of the object, many of which can even be predicted without numerical simulation.

And such a caricature gives us a picture which is easy to imagine and provides all the important details: where an initial configuration starts, how it moves and how it is attracted to the (one or few) valleys of the potential (Wilson and Kogut 1974; Ma 1976; Patashinskii and Pokrovskii 1979). For a start, it is helpful to have a very primitive and simplistic model of reality. This is called a "toy model" and we will be using many of them in this book.

The huge advantage of toy models is that we can quickly solve the equations and then visually compare our assumptions with the simulation results. If they contradict, something is wrong either in our understanding of the problem or in the modeling solution. In this case, we will have to modify something in order to harmonize both sides at this preliminary stage. It would be very difficult to do this harmonization later on, when the model has become very complicated and we are no longer able to envisage the complete picture. Something analogous happens in everyday life when we compare a caricature with a realistic painting. When we see a beautiful portrait by a great painter, we can feel the internal harmony of all the different sides of the portrayed person, but we will not be able to extract all their characteristics.

Yet in many cases we need a correct portrait of reality, especially if we return to applied sciences. By adding some "external forces" we can make the model more and more realistic, but at the same time we lose visual control of the problem (the ability to imagine it).

Now let us imagine that we would like to determine a "potential" in which the system relaxes to its natural configuration. For the complex form of the model this configuration will definitely appear as a multi-dimensional one. It is extremely difficult to model such a potential 'a priori' or even to imagine it. Fortunately, due to the increasing capability of computers we can now do something different. We can perform numerical simulations with more or less intuitively understandable forces operating between the subsystems and simultaneously accumulate different probability distributions. This gives us an *a posteriori* representative and realistic portrait of the system. Mathematicians call this a *phase portrait.*

Despite the expected complexity and multidimensionality, this approach has a number of practical advantages. We can explore the system's projections to different hyper-planes of parametric space step by step and accumulate at least some static representations of the complex mutable reality. So, in principle, we can perform the next "oscillation" in our description to the observable attraction to the minimums of the "potential" in the phase space. However, it is very easy to overestimate the potential of such an approach, because our capabilities are still limited.

Nature is generally simple, but combinatorically complex. The best description of it is somewhere in a middle of complexity. Now, this statement as well as our previous discussion of the potential relief may either seem purely philosophical or too sophisticated to you. But we are confident that this approach will open up a new road to an intuitive understanding of many different biological or related phenomena and at the same time will allow us to be braver in the formulation and treatment of various biologically inspired mathematical models.

1.5 Lessons to Be Learned from the Dynamics of a Myrmecochorous Plant Community

Let us take the model of a complex dynamic system (which will be more extensively described in Chap. 9): the long-term ant-species-dependent dynamics of a myrmecochorous plant community. Seed dispersal by ants is a widely spread phenomenon and myrmecochorous plants constitute a large portion of species in many ecosystems. For this discussion it is important that the ant species complex in the ecosystem is continuously changing in time and space, and the long-term effects of such ant–plant interactions on the plant community are quite difficult to study in a standard experimental field approach.

Seeds of myrmecochorous plants bear specialized lipid-rich appendages, elaiosomes, for attracting ants. Ant workers collect the seeds and usually carry them to their nests. From the point of view of our approach this means that the space of the forest becomes non-uniform and forms a kind of potential relief where some positions are more preferable for the seeds to appear than other places. The formation of an effective "potential" is very complex and involves many different factors so we will only mention the main ones here. Some seeds reach the nests, whereas others are dropped during transport. In the nests, the energy-laden elaiosomes are removed and consumed, whereas the intact and viable seeds are commonly deposited either in underground nest chambers or in "waste piles" outside the nest. The ants benefit by receiving high-quality food. In turn, myrmecochory provides plants with several selective advantages, such as protection of the seeds against predators and fire, avoidance of interspecific competition and reduction of the competition between the parental plant and its seedlings, and so on (Gorb and Gorb 2003).

It is important to note that for many plant species seed dispersal by ants is the only dispersal method used. These seeds, depending on their dimensions and elaiosome sizes, are attractive to different ant species to a different extent and have different dropping rates during transport to/from the nest. In the context of a potential approach this means that the potential relief is different for different species. But the art of model construction consists of an appropriate choice of complexity once it

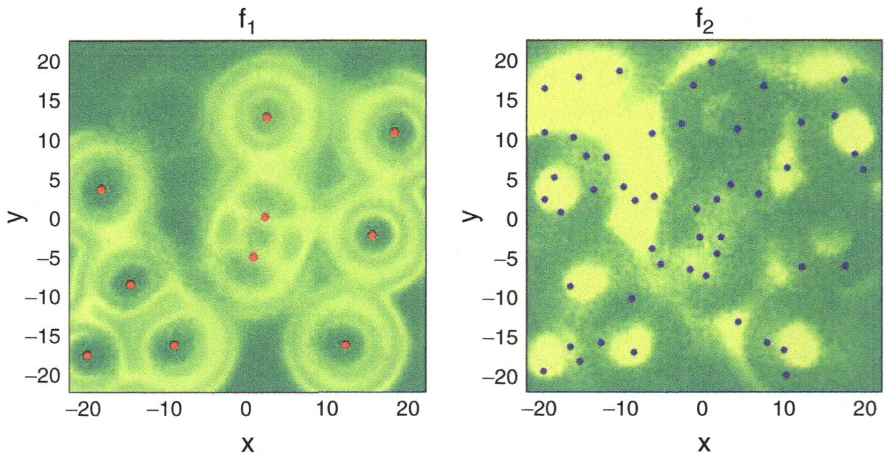

Fig. 1.4 Correspondences between the distributions of anthills and seeds. The positions of two different kinds of anthills are shown by red and blue dots, respectively. The densities f_1 and f_2 of two different kinds of seeds are shown in two subplots represented by the color maps, with brighter colors corresponding to higher densities

is developed enough to account for the most important properties of the system and at the same time to remain "somewhere in the middle of the complexity".

This example is also especially suitable for understanding the general phenomenon of "frozen kinetics" (or, as is mathematically more correct, the "large river effect"), because the ant species complex in the ecosystem is continuously changing in time and space, leading to long-term effects on ant–plant interactions and, ultimately, on the plant community (Fig. 1.4).

1.6 Adiabatic Approximation

The distribution of anthills in a particular forest is relatively stable. So, for a few years the seeds will experience the multi-valley effective potential caused by the anthills as practically static. Seed distributions and resulting distributions of plants are gradually attracted to a pattern corresponding to the quasi-static densities of the anthills. However, from time to time an ant family will move and build its nest (anthill) at another site. When this happens, the density distribution of the plants in this area will likely be at some intermediate stage. Now, the system will use this intermediate distribution as the initial condition, and the marathon of attraction to new potential minima resumes. This process never comes to an end; at least not while all the components of the system exist.

This observation completely changes the main goal of the study, which will now be to determine the correlation between the number of nests of different ant species and the stability of the ecosystem under consideration, but not by simply calculating the final distribution. This example thus illustrates another idea, which is derived from physics and is extremely important for practical applications of the attractor approach: the so-called ***adiabatic approximation***.

Such an approximation was applied for the first time to calculate systems consisting of electrons and ions (e.g., in crystals). Ions are much heavier and move much more slowly than electrons. In a very good approximation one can treat their distribution in space (e.g., in a crystal lattice) as static and study electrons moving in a practically static potential created by the ions (Tolpygo 1950; Born and Huang 1954; Baryakhtar et al. 1999). However, positively charged ions repulse one another (by the way, anthills repulse one another as well!) and need the attraction from the electrons to build a stable crystal lattice. This means they can only stay in a more or less static condition by being attracted by a collective potential of the "electron sea". But, in turn, this "sea" is self-consistently formed by the ion lattice. It is exactly this kind of interaction between ions and electrons which can also be seen between ants and seeds.

The effective "adiabatic behavior" is well illustrated in Movie 1.2. The description of the model generating this behavior is presented in Filippov et al. (2000), where the collective motion of two groups of predator–prey pairs with essentially different mobilities is studied. It can be seen that both the fast and slow subgroups treat either the instant or the time-averaged configuration of another group (represented by small and big circles, respectively) as the effective potentials.

It is exactly the kind of interaction which we have between ants and seeds, if we are ready to overcome visual difference between them and electrons with ions, of course. Furthermore, let us remember that both the density of the plants and the distribution of anthills tend to their frozen kinetics only as long as the forest exists in its present state. However, the forest is also continuously changing due to various reasons: climate, geology, biological evolution, anthropological factors. Here we consider these examples in a very particular context and it is helpful to remember that in reality we are dealing with a hierarchy of time scales. Thus, for every specific study a proper positioning of the model between different time scales, somewhere in the middle, can be a very fruitful approach.

1.6.1 Continuous or Discrete Modeling

The analogy between the system of ions and electrons on the one hand and of anthills and seeds on the other hand also shows that the approaches to physical and biological problems are basically quite similar. In both cases we are faced with the two alternatives of using either continuous or discrete descriptions. Thus, we can describe electrons as discrete particles moving in space or alternatively as a continuous electron density that adapts to the discrete crystal lattice of ions. Each of these two approaches has its own mathematical advantages and disadvantages and both are used in physics for solving a range of different problems.

The same two options exist in biology, where we can also choose either ***discrete or continuous models*** depending on the situation, or we may even combine both of them. For example, such a combination may be used for modeling the seed–ant interactions. It is convenient to treat the distribution of seeds as a continual density and to construct the effective potential for it starting from the discrete set of the anthills.

In this case, we approach the problem as follows. A subsystem of ant nests of two different species is treated as two separate arrays of discrete points $\{x_{kj}, y_{kj}\}$. The nests are assumed to initially be randomly placed inside a fixed area in two-dimensional space $[x, y]$. Here, the indices $\{k\}$ and $\{j\}$ denote two different species of ants $k = 1, 2$ and a particular ant nest $j = 1, \ldots, N_k$ in each subset of ant nests.

Index $k = 1$ corresponds to the nests of larger ants with larger foraging territories and $k = 2$ corresponds to the nests of smaller ants with smaller territories. Initial numbers of the nests at $t = 0$ are always fixed and, in this particular example, equal to $N_1 = 10$ and $N_2 = 50$, respectively. To simplify the simulations, we selected a square forest area of the size $L_x \times L_y$, with $L_x = L_y = 50$ m. The initial positions of the ant nests at $t = 0$ are given by the formulae: $x_{kj} = L_x\zeta_{kj}$, $y_{kj} = L_y\zeta_{kj}$, where $\zeta_{kj}(x, y)$ are δ-correlated random numbers $< \zeta_{kj}\zeta_{k'j'} >= \delta_{kk'}\delta_{jj'}$ uniformly distributed in the interval $[-1/2 \div 1/2]$ along the spatial coordinates x and y. Briefly, this means that the original nests are distributed uniformly and independently in the square forest area.

To construct the effective potential for the seeds, or in other words to model the ability of the ants to disperse the plant seeds, each ant nest is assumed to have, around its current position, a region with a positive impact on the probability $z_k^{k'}(x, y)$ of plants to survive and produce new seeds for the next generation. Here, we use the subscripts $k = 1, 2$ to denote the nests of different ant species, and the superscripts $k' = 1, 2$ to denote different plant species.

Large and small ants collect the seeds of different plant species (with different sizes) and dispose them either within a ring close to the borders of their territories (large ants) or in small circles within the entire territory of the ant colony (small ants). In the model, this is accounted for by a different spatial structure of the *preference coefficients*:

$$z_1(x, y) = \sum_{k=1}^{2} B_k^1 \sum_{j=1}^{N_1} \exp\left[-\left(\frac{r - R_1}{R_2}\right)^2\right]; \quad z_2(x, y) = \sum_{k=1}^{2} B_k^2 \sum_{j=1}^{N_1} \exp\left[-\left(\frac{r}{R_2}\right)^2\right] \tag{1.5}$$

where $r = \sqrt{(x - x_{kj})^2 + (y - y_{kj})^2}$ is the distance from an arbitrary point in area $L_x \times L_y$ to the position of the ant nest $\{x_{kj}, y_{kj}\}$. The values of the coefficients $B_k^{k'}$ should be chosen to reflect the experimentally observed preferences in the choice of different seeds by different ant species.

The coefficients of Eq. (1.5) form a non-uniform surface which plays a role with regard to the desired effective potential. The space-time evolution of the continual seed densities $f_{1,\,2}(x, y)$ is determined by their adaptation to this surface.

Exactly this approach was applied in our original study (Gorb et al. 2013) which will be further detailed in Chap. 9. However, it is an obvious simplification and, in principle, if it is required for a particular study, the description can be further refined. Remember that in reality the seeds are discrete objects. Due to the activity of the ants, which interact with them, they are moved in space and as a result are attracted to the preferred positions described by the coefficients of Eq. (1.5).

Besides, the seeds also interact with each other because they cannot merge with each other. From the mathematical point of view this means that there is an effective potential between them with a strong repulsing core which prevents mutual penetration. One could say that there is a strong short-range repulsion between them. If we return to the electrons, they also mutually repulse each other, but with a different, long-range Coulomb potential.

From a modeling perspective this is the only difference between the physical and the biological example. The electrons are attracted to the potential relief caused by the ions discretely distributed in some limited space, as the seeds are attracted to the anthills in the forest. If the electrons did not repulse each other, they would quickly collapse into one or very few super-particles in the most preferable place(s) of the lattice potential. The lattice would also become unstable and transform into something different.

So, the appropriate density, which is not too small and not too large, is assured by the proper choice of the mutual interaction. This example illustrates two things: (1) there are close parallels between physical and biological studies, and (2) the correct choice of the kind of interaction is important for the global stability of the model.

In our study on the myrmecochorous plant community presented later on in this book (cf. Chap. 9), we did not go deeply into a discussion of the discrete nature of seeds and their mutual interactions and simply assumed a continuous density. However, discrete modeling can be used whenever it is deemed necessary. Below (cf. Chap. 8) we will discuss such examples in detail.

1.6.2 Continuous and Discrete Modeling in Multidimensional Space

Let us continue the discussion of the advantages and disadvantages of continuous and discrete approaches. It is very easy to represent or imagine a one-dimensional distribution of some density as a simple curve along an axis corresponding to some solitary parameter. By the way, this is exactly what all university students of different specializations (from physics to biology and economics) are asked to do right from the beginning. So, this is absolutely natural for our imagination. This is why we used such a presentation to start the discussion of the attraction of "something to something". It is a little bit more sophisticated to do this in a two-dimensional space. But this is nothing in comparison to what we get in three-dimensional space. How to represent and to read cases like these? One of the tricks to do so, discovered by physics many years ago, is the so-called *Fermi surface*. The basic idea is rather simple: let us find a surface with a constant value of density (constant energy in the case of the Fermi surface) and plot its projection onto a planar computer screen. After such a procedure we will start to see the surface!

Actually, this is not extremely sophisticated. It is absolutely natural and it is exactly what we can see in nature. When ordinary day light falls onto any surface, this is in fact an interaction between the electromagnetic field of light and the electronic density of matter. The light is reflected by a surface of a certain electronic density (where it is large enough to support the reflection). So, each time in everyday life we perform the procedure of visualization of 3D density. Despite of complexity behind it, this is easy to percept for us. We certainly enjoy it and even call everything we can easily accept as "obvious" (in the sense that it is visible).

People often refuse to see such "obvious" pictures of densities found by numerical calculations. Even for people from the field of theoretical physics in the middle of the twentieth century (the time of the strongest progress in the field) it was sometimes quite difficult. Some even called the Fermi surfaces obtained from this kind of calculations "monsters". But fortunately we live in the twenty-first century, we are better equipped for processing huge amounts of data, and we can now apply such a representation to biological structures. This is how we are going to address various problems in this book later on.

1.7 Disadvantages of the Continuous Approach

However, let us return to the disadvantages of the continuous approach. We are able to visualize 3D distributions if we are not afraid to see the obvious. However, what if a further dimension is added? This additional dimensionality is time! Albert Einstein solved this question for scientists, and the cinematograph solved it for everybody else. The majority of people may not understand Einstein's relativistic theory, but all of them understand movies. So, when dealing with four-dimensional densities we can apply the same method: visualize a 3D surface of constant density for three of the parameters, and fly along the last one of the parameters in "next dimensionality". This parameter may be real time, but may also be something else. Actually, in any plot we are exploring something moving along one of the axes. Why not use the same idea in this case as well? For example, we can simply rotate three-dimensional surfaces and watch them move.

However, with the continuous progress in computing, the number of additional dimensionalities $d = 5, 6, \ldots$ is growing. We will not be able to imagine all of them. But this is not the only problem. Another important limitation is that the number of necessary computer operations will grow dramatically. For example, if the number of calculation cells (N) along one side of a box is $N = 100$, the size of the array in d-dimensional space will be N^d. So, for $d = 5$ it will already be $N^d = 10, 000, 000, 000$ (ten billions!).

The more parameters we use, the less advantage a continuous description seems to have in comparison to a discrete description. And this is not the only problem. Up to this point we hardly said anything about long-range interactions. It was only briefly mentioned in the above discussion of the adaptation of the seeds to the adiabatically slow potential of anthills and of electrons inside the crystal lattice of slowly moving ions. In both cases, the interaction between the seeds and between the particles was mutual repulsion. But the seeds repulse each other only at very short distances (due to their hard cores which do not allow them to merge with each other. The Coulomb interaction between the electrons, however, is a long-range repulsion). Its potential $U_{Coulomb}(r) \sim 1/r$ very slowly decreases with the distance r. In the majority of cases, the interaction between different substances is screened at some characteristic distance r_0, as given by $U_{Screen}(r) \sim \exp(-r/r_0)/r$, but it is not immediately going to zero.

Let us imagine (for definiteness and simplicity) that we have only two continuous probability densities, $\rho_1(r_1)$ and $\rho_2(r_2)$, interacting at a distance. From the mathematical point of view, this means that every point of the density along the first coordinate r_1 interacts with each point r_2 along the second coordinate with a long-range potential $U(r_1 - r_2)$. This causes a non-local impact on the total energy of the system:

$$U_{interaction} = \int dr_1 dr_2 \rho_1(r_1) U(r_1 - r_2) \rho_2(r_2). \tag{1.6}$$

This impact immediately complicates the equations of motion of both densities:

$$\begin{aligned} \partial\rho_1(r_1)/\partial t &= \delta U_{interaction}/\delta r_1 + \ldots = \int dr_2 U(r_1 - r_2)\rho_2(r_2) + \ldots \\ \partial\rho_2(r_2)/\partial t &= \delta U_{interaction}/\delta r_2 + \ldots = \int dr_1 U(r_2 - r_1)\rho_1(r_1) + \ldots \end{aligned} \tag{1.7}$$

As a result, in order to solve these equations numerically one has to integrate over the complete volume of both densities at each step of the calculation! Since in many cases we are dealing with much larger numbers of densities $\rho_j(r_j)$, where $j = 1, 2, \ldots N$, the procedure will quickly turn into a mathematical nightmare.

The reader might now ask why we use the continuous approach at all. The answer is that there are many cases in which we do need long-range interactions to model a particular system under consideration.

Despite the fact that in reality "everything interacts with everything", in many systems long-range interactions can be combined priorly (before the simulation) to a reduced set of interactions, which compensate for (and thus screen) one another, thus allowing us to narrow down the model to a much simpler phenomenology with local interactions only:

$$U_{interaction} = \int dr\rho_1(r)U_{12}(r)\rho_2(r).$$

The observable world consists of *phenomena*, and it is absolutely natural for us to operate on a phenomenological level. For example, we normally ignore the long-range interaction of electromagnetic waves with electronic density and simply accept that a ray of light is reflected from a surface in the local point r. This is why we prefer such a description whenever it is feasible. Nevertheless, in some cases we simply have to calculate long-range interactions. If we cannot do this in a continual approach, we should return to the discrete one or at least to a combination of both approaches.

Instead of Eqs. (1.6 and 1.7), we can also write down direct interactions with a potential:

$$U_{interact}\left(r_{jk}\right) = U\left(\left|\vec{r}_j - \vec{r}_k\right|\right), \tag{1.8}$$

depending on the absolute distance between the particles $\left|\vec{r}_j - \vec{r}_k\right|$ in the positions $\vec{r}_j$ and $\vec{r}_k$. Here, the indexes j and k denote N particles of the system $j = 1, 2, \ldots N$ and $k = 1, 2, \ldots N$. Formally, the equation of motion is very simple:

$$\partial\vec{r}_j/\partial t = -\sum_k \partial U_{interact}\left(\vec{r}_{jk}\right)/\partial\vec{r}_j + F_{external}\left(\vec{r}_j\right), \tag{1.9}$$

where $F_{external}\left(\vec{r}_j\right)$ is the sum of all the external forces which are not directly included in the interaction $U_{interact}(r_{jk})$. But the simplicity of this equation is illusory. Supplementary Movie 1.3 illustrates how complex a trajectory can be even in the case when only one particle (red) pursues another one in a relatively simple static potential relief.

This is for the following reasons:

1. If every single one of N particles interacts with all the others, the number of operations at each time step grows with increasing N according to $N \cdot N = N^2$. So from the very beginning we have to consider how to minimize this number (maybe by restricting the interactions to only those with properly defined close neighbors).
2. Even in the reduced form, the compact Eqs. (1.8 and 1.9) generate a giant amount of information about the system. To be able to apply the equations practically one has to specify the spatial dependence of the interactions $U_{interact}(r_{jk})$. For different kinds of particles in the same system, the interactions can be different. Besides, it is necessary to describe some form of boundary of the system and introduce external interactions $F_{external}\left(\vec{r}_j\right)$ with impacts originating beyond the boundary of the system but directly included in the description. The complex behavior of such a system is illustrated in the supplementary Movie 1.4, where a certain number of "predators" (N_1) pursuing a certain number of "prey items" (N_2) is

shown, with an effective "potential" caused by a redistribution of prey density $\rho\left(\vec{r}\right)$ in space.

1.8 Lessons to Be Learned from the Adhesive System of Insects

Some of the additional forces can play an important role for the boundary conditions, some of them may be entered into the equation of motion as direct interactions with external subsystems, etc. Because of this uncertainty we cannot proceed without any further specification of the system and the model. To illustrate some important aspects of this process we will use one of the problems studied later on in the book – the adhesive system of insect feet. For this purpose we will concentrate on some particular questions which are convenient for exploring some basic ideas of our modeling approach.

It can be shown that the thin tape-like contact tips (spatulae) of insect hairs (setae) in combination with applied shear forces lead to the formation of a maximal real contact area without slippage (Popov 2010). Since only flexible materials can generate a large contact area between the pad and the substrate at minimal normal load, the flexibility of the material is important for the contact formation of these adhesive pads. On the other hand, elongated structures made of very flexible materials have low mechanical stability: insect setae consisting of very soft material may buckle and collapse, resulting in so-called clusterisation/condensation. In this case, the functional advantage of multiple adhesive contacts would strongly decrease.

From a mathematical point of view, we are dealing with two kinds of attraction. The first one is trivial attraction to a surface and the second one is the attraction between the setae. So this is a typical problem where internal and external interactions compete with each other and none of them can be excluded from the observation.

That is why the material properties of insect adhesive setae represent an *optimization problem*, which was solved in the course of biological evolution by developing gradients of thickness and mechanical properties. As a result, the problem involves one more degree of freedom, which makes the system an even better example for our current questions.

It is known from experimental studies that there is a gradient of material composition between the setal tip and the base. This gradient is hypothesized to be an evolutionary optimization enhancing the adaptation of adhesive pads to rough surfaces, while simultaneously preventing setal clusterisation. Such an optimization presumably increases the performance of the adhesive system in general. However, this hypothesis is difficult to prove experimentally using biological specimens. That is why we decided to test it with a numerical simulation.

According to the general idea described above, we applied to this problem a minimalistic but quite realistic model which is a compromise between numerical limitations caused by the power of the computers. We adapted the model to our particular problem by including the gradient of material properties of the modeled insect setae. This particular model includes the majority of the elements common for all the models mentioned above. In principle, it is constructed as follows: an array of initially parallel fibers is attached to a hard planar base. Stiffness of the fibers $F_{elastic}$ is continuously varied along their length and can be changed from very soft to much stiffer or even almost rigid (but still with some degree of flexibility).

This step is very suitable for explaining an important advantage of the discrete approach which we will use many times in the following chapters. If the discrete elements of the array (or "nodes" of a filament) are ordered and during the whole process belong to the same "fiber", one can always define the nearest neighbors. In some sense, this means that the problem which originally involves 3D multiple bodies can be partially (for the neighbors) transformed into effectively a 1D one-body problem.

This fantastic simplification practically does not reduce the generality of the problem. In some sense it even makes the model better suited for our goals, because it perfectly illustrates that a good caricature allows us to understand and imagine the system better than a perfect but too complex image of reality.

Of course, some memory of the 3D nature of the problem remains in this approximation as well. In particular, in this model it is reflected by the existence of both longitudinal $\vec{F}_{jk}^{\parallel}$ and transversal $\vec{F}_{j}^{\perp}$ stiffnesses of the fibers. Mathematically they can be simulated by the following interaction between the segments:

$$\vec{F}_{jk}^{\parallel} = K^{\parallel}\left(\vec{r}_j - \vec{r}_k\right)\left[1 - \left(\vec{r}_j - \vec{r}_k\right)^2/dr^2\right],\; k = j \pm 1 \text{ and } \vec{F}_{j}^{\perp} = K^{\perp}\left(2\vec{r}_j - \vec{r}_{j+1} - \vec{r}_{j-1}\right).$$

Here a couple of tricks are used to simplify the model further. The longitudinal force, $\vec{F}_{jk}^{\parallel}$, is described as a force caused by two-minima potential which tends to keep the distance between the points $\vec{r}_j$ and $\vec{r}_{j\pm1}$ close to the equilibrium length of the segment dr. The transversal force, $\vec{F}_{j}^{\perp}$, keeps $\vec{r}_j$ close to the mean value $\left(\vec{r}_{j+1} + \vec{r}_{j-1}\right)/2$ between its nearest neighbors and tends to maintain the angle between the neighboring segments close to 180° (Fig. 1.5).

Now let us turn to the external forces. In this example, such a force appears due to the adhesive nature of the setal surface. The ends of the fibers are attracted to the surface by molecular and capillary forces. For the sake of simplicity, we simulate this force by the gradient of the Morse potential, $U_{VdW}(r) = U_o(1 - \exp(-r/r_0))^2$, where r is the distance between the end of the fibers and the external surface. This potential has a form which is typical of such attraction forces which have a minimum at a characteristic distance from the surface r_0.

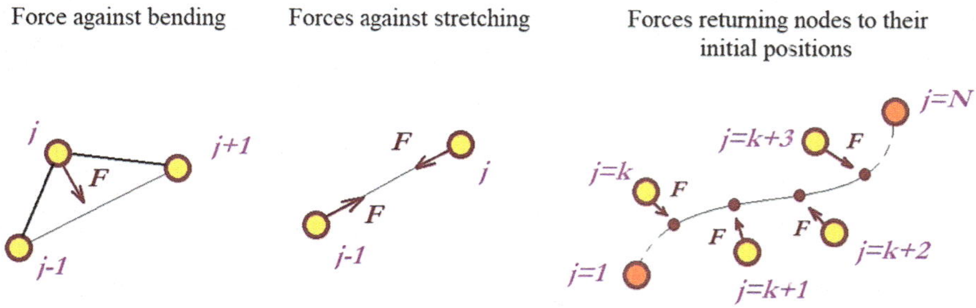

Fig. 1.5 Forces acting within one fiber in a discrete approach. The force acting against bending tends to return an intermediate "node" to the position between the two nearest ones (left). The force acting against stretching conserves initial distance between the two nearest neighbors. Forces conserving the original form return the positions of all the internal "nodes" to their initial places

Besides the interaction between the nearest neighbors along the individual fiber, there is an attraction between the different fibers as well. The fibers have elasticity gradient. The soft part of every fiber, which normally is physically thin and flexible, interacts with corresponding regions of other fibers of the array. This interaction force has the same origin (van der Waals or capillary forces) as the fiber ends' attraction to the external surface (e.g. the wall). Due to this, it naturally takes the same form: $U_{interact}(r_{jk}) = U_o(1 - \exp(-r_{jk}/r_0))^2$, with comparable characteristic parameters U_0, r_0. To keep the model simple, one can also reduce the mutual interaction between the fibers to the interaction between the closest neighbors in the array: $r_{nn+1} = \left|\vec{r}_n - \vec{r}_{n\pm1}\right|$ (not to be confounded with the interaction along an individual fiber!). Thus, we simultaneously have a simplification in both the interaction along the fibers and between them.

For the problem under consideration here, one can neglect the effects of inertia and treat the system as an over-damped one. In this approximation, the differential equation of motion does not contain the second time derivative and can be formally written in the general form of Eq. (1.9), $\partial\vec{r}/\partial t = \vec{F}/\gamma$, as above.

The only (but important) difference is that in this case the model includes physical parameters and interactions. For example, here γ is a dissipative constant, and force is accumulated due to the above-mentioned interactions $\vec{F} = \vec{F}_{elastic} + \vec{F}_{VdW} + \vec{F}_{interact}$. But if we normalize γ^{-1} to a typical relaxation time of the system, the equation of motion is formally reduced to its simplest form: $\partial\vec{r}/\partial t = \vec{F}$, as formally written above (Eq. 1.4) for a very general case.

Despite its seeming simplicity, this function already includes a large number of interactions, but it still needs some further additions. In this particular problem, which involves an elasticity gradient as well, the stiffness of fibers continuously varies along the vertical coordinate. To define this variation, we can apply the analytical smooth step function $\Theta(y) = 1/[1 + \exp(-(y - y_0)/\Delta)]$ with regulated position y_0 of the place starting from which the fiber can be easily bent and width Δ of the interval where the elasticity essentially modifies. This function tends to

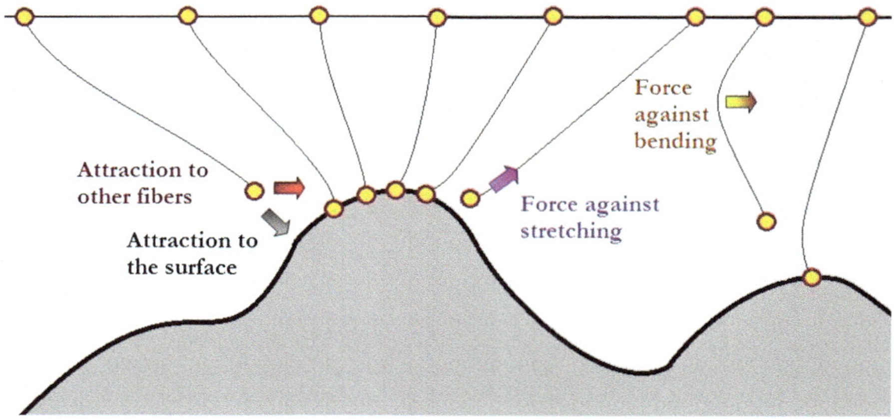

Fig. 1.6 Forces acting in the system of fibers in the presence of an external rough surface (below). Each fiber is attached to the common root system (schematically shown by horizontal line). It is subjected to at least two external forces (attraction to the surface and to the nearest fibers) and two internal forces acting against its deformation (by stretching and by bending)

1 when $y \ll y_0$, and gradually goes to zero in the opposite direction, $y \gg y_0$. At first glance, this seems to be an absolutely trivial modification. However, it shows up the strong advantage of the discrete approach, especially in combination with modern matrix-based computer calculations. In general, one can associate an individual interaction (or a variation of it) with every member of the discrete array, solve numerically the same equation with the only different physical meaning of different lines of the array and quickly extract the necessary information from universally organized visualization procedure (Fig. 1.6).

1.9 Lessons to Be Learned from Hairy Spatulate Contact Structures

The gradient of elasticity in setal fibers is one solution among many which have evolved in nature to enlarge the contact area with an external surface. Another solution is represented by some hairy adhesive systems of different animals which do not rely on flat punches on their tips, but rather on spatulate structures. Experimental observations allow numerous explanations of this particular contact geometry, e.g., enhanced adaptability to rough substrates and contact formation by shear forces rather than by normal load. It is very difficult, however, to infer the exact mechanism just from experimental observations, especially if it comes to understanding the role of spatulate terminal elements in biological fibrillar adhesion.

Fortunately, we can now apply a numerical approach to study the dynamics of such spatulate tips during contact formation on rough substrates. It is expected that such a model will be able to demonstrate clearly that the contact area increases with the applied shear force. Again, the numerical model here deals with the problem of optimization. It is expected to show that the applied shear force has an optimum which corresponds to the situation when maximal contact is formed but no slip occurs. In other words, some equilibrium of forces should exist which generates the strongest possible adhesion.

It is important that this kind of dynamic model can be constructed in 3D space because this allows us to study how the contact on a rough substrate can be generated by shear forces, especially if the spatulae are initially not aligned in the plane of the substrate (Filippov et al. 2011). In Chap. 4 we will use such a numerical approach to study the dynamics of spatulate tips during contact formation on rough substrates. Combining the different possibilities discussed above, one can quite realistically reproduce the system and ask at the same time formally different, but related questions. For example: what is the role of the thickness (elasticity) gradient and does the applied shear force contribute to the enhancement of the contact area on a rough substrate? And, of course, the most interesting question of all: is there an optimal shear distance/force for the single spatula?

The situation in this case is relatively simple and complex at the same time and as a result its description involves a number of different independent features already discussed above. First of all, we are dealing with a van der Waals attraction to the surface which competes with the resistance of the spatula to bending. According to the theory of elasticity, the elastic energy of a flexible plate in 3D space is given by a quite complex integral (Landau and Lifshitz 1976):

$$W_{elastic} = \frac{E}{24(1-\nu^2)} \iint dxdyh^3(x,y) \left\{ \left(\frac{\partial^2 z}{\partial x^2} + \frac{\partial^2 z}{\partial y^2} \right)^2 + 2(1-\nu) \left[\left(\frac{\partial^2 z}{\partial x \partial y} \right)^2 - \frac{\partial^2 z}{\partial x^2} \frac{\partial^2 z}{\partial y^2} \right] \right\}, \tag{1.10}$$

where E is Young's modulus of the plate material and ν is the Poisson ratio which is typically equal to $\nu = 1/3$. The competition of the forces leads to (over-damped) dynamics in the system which can be described as before by the equation of motion in the vertical direction:

$$\gamma \frac{\partial z(x,y)}{\partial t} = -\frac{\delta W_{elastic}[z]}{\delta z} - \frac{\delta U_{VdW}[z]}{\delta z}, \tag{1.11}$$

van der Waals interaction also produces a horizontal force on the rough surface:

$$F^x_{VdW} = -\delta U_{VdW}[z(x)]/\delta x.$$

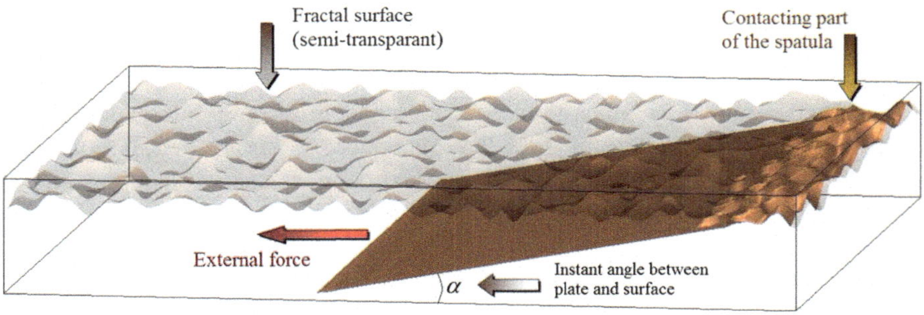

Fig. 1.7 Conceptual schematic of the model representing the development of the contact between a single spatula and the fractal substrate surface. Initially attached by a negligible part only, but pulled by the external horizontal force, the thin, two-dimensional plate (spatula) gradually rotates to a smaller angle between it and the surface. With time, this dynamic process favors much better (almost perfect) contact between the spatula and the substrate surface

This force competes with an external shear forceF^x. When F^x exceeds the total resistance of all instantly bonded segments$\int dxdyF^x_{VdW} > |F^x|$, the whole spatula moves along the x-direction according to the equation:

$$\gamma \partial x/\partial t = F^x - \int dxdyF^x_{VdW}. \tag{1.12}$$

Thus, a typical configuration of the moving foot can be described as follows: the plate of spatula is initially attached to the surface at one of its end segments by the van der Waals force $F_{VdW} = -\frac{\delta U_{VdW}[z]}{\delta z}$. In the course of time, it relaxes into an equilibrium state in which it adheres to the surface by additional segments. The rate of attachment depends on the angle α between the spatula and the surface and is normally faster for smaller angles α.

If the external force shown by red arrow in the figure is nonzero ($F \neq 0$), it pulls the spatula (plate) to the left, competing with the van der Waals attachment F_{VdW} of the adhering segments. If the total attachment force is stronger than the horizontal component of the external force, $\int dxdyF^x_{VdW} > |F^x|$, the spatula does not slide along the x-axis. However, the part which remains unattached can rotate and approach the surface $z(x, y) \to <Z>$ due to the action of the vertical component of the force $F^z > 0$; $F^z \sim z$ (Fig. 1.7).

This rotation reduces the distance between the hard and the flexible surface and can greatly enhance total adhesion. Generally speaking, one can expect that stronger shear forces will cause faster attachment. However, if the shear force is too strong, it can exceed the van der Waals locking and even lead to the detachment of previously attached segments. In this case, the spatula will start to slip along the surface, its rotation stops and additional segments do not adhere to the substrate.

These qualitative considerations give rise to the optimization problem already mentioned above: to what extent can the shear force be varied to stimulate the attachment without rupturing the contact? To answer this question, we performed two sets of numerical simulations: (1) with a fixed initial inclination angle α and varying force F and (2) with a fixed force F and a varying angle α. The complete results of these numerical experiments will be presented in Chap. 4. Here we will only comment on the main results which are of general interest.

As expected, a higher force F leads to a faster decrease of the inclination angle. If the force is smaller than some critical value for detachment, $F < F_{crit}$, then the plate gradually tends to a horizontal orientation for $t \to \infty$. When the force approaches the critical value F_{crit}, it becomes capable of breaking some of the already attached bonds and causing a slight shift of the spatula in the horizontal direction. Finally, if the force exceeds a threshold F_{crit}, it completely breaks any initial attachment and causes a permanent sliding of the plate. In this case, the spatula does not rotate and does not further approach the horizontal orientation, thus not leading to any increase in adhesion.

In the biological attachment system, the animal naturally cannot control the state (attachment and orientation) of each individual spatula, but presumably it can monitor the total resistance force of the entire array of spatulae, keeping it close to, but not exceeding, the critical shear force value.

This is a very interesting and non-trivial result from both the biological and mathematical points of view. It shows that complex and essentially nonlinear living systems can solve an optimization problem without balancing all the forces in a static state, but rather by very slow dynamics at conditions close to the critical threshold. Thus, it is another practical application of our approach which we call *large river effect* or *frozen kinetics*. In the following chapters we will discuss its numerous other applications.

References

Bagnuls C, Bervillier C (1994) Describing actual critical behaviour from field theory: a delicate matter. Phys Lett A 195:163–170

Baryakhtar VG, Zarotchentsev EV, Troitskaya EP (1999) Theory of adiabatic potential and atomic properties of simple metals. Taylor and Francis, Boca Raton

Born M, Huang K (1954) Dynamical theory of crystal lattices. Clarendon Press, Oxford

Filippov AE (1993) Phenomenological approach to construction of attractors. Theor Math Phys 94:325–338

Filippov AÉ (1994) Mimicry of phase transitions and the large-river effect. JETP Lett 60:141–144

Filippov AE (1995) Some applications of the large river effect to fluctuation theory. JETP 81:784–792

Filippov AE (1998) Niche structure as a dissipative attractor of evolution. J Obshej Biol (J Gen Biol) 58:81–98

Filippov AE, Soboleva TK, Wedderburn ME (2000) A mathematical model of competitive self-organization. Complex Int 8:1–9

Filippov A, Popov VL, Gorb SN (2011) Shear induced adhesion: contact mechanics of biological spatula-like attachment devices. J Theor Biol 276:126–131

Gorb EV, Gorb SN (2003) Seed dispersal by ants in a deciduous forest ecosystem. Kluwer Academic Publishers, Dordrecht

Gorb EV, Filippov AE, Gorb SN (2013) Long-term ant-species-dependent dynamics of a myrmecochorous plant community. Arthropod Plant Interact 703:277–286
Landau LD, Lifshitz EM (1976) Mechanics, vol 1, 3rd edn. Butterworth-Heinemann, Oxford
Landau LD, Lifshitz EM (1980) Statistical physics, vol 5, 3rd edn. Butterworth-Heinemann, Oxford
Lifshitz EM, Pitaevskii (1981) Physical kinetics, vol 10, 1st edn. Pergamon Press, Oxford
Ma SK (1976) Modern theory of critical phenomena. W. A. Benjamin, Inc., Reading
Patashinskii AZ, Pokrovskii VL (1979) Fluctuation theory of phase transitions. Pergamon Press, Oxford
Popov VL (2010) Contact mechanics and friction. Springer, Berlin/Heidelberg
Tolpygo KB (1950) Physical properties of a rock salt lattice made up of deformable ions. (In Russian. English translation: Ukrainian Journal of Physics, vol. 53, special issue, 2008). Zh Eksp Teor Fiz 20:497–509
Wilson KG, Kogut J (1974) The renormalization group and the ϵ expansion. Phys Rep 12:75–199
Zumbach G (1993) Almost second order phase transitions. Phys Rev Lett 71:2421–2424

Chapter 2
Various Methods of Pattern Formation

Abstract Evolution of different systems can be described in terms of their relaxation to the minimums of some effective potential relief. This observation faces us with a question how to generate corresponding potential patterns which describe adequately various physical and biological systems. In this chapter, we present a number of different ways to generate such potentials demanded by the problems of different kinds. For example, we reproduce such a generation in the framework of a simple theory of phase transitions, automatic blocking of the growing phase nucleation and universal large scale structure. Being frozen at late stages of evolution, they form majority of meta-stable structures which we observe in real world. Counting on mentioned above universality of naturally-generated fractal structures and their further utilization in next chapters of this book, we reproduce also formal algorithms of generation of such structures based on random deposition technique and Fourier-transform approaches.

In many examples considered in this book we will use the numerical generation of an effective potential (surface, pattern, density distribution, etc.). As mentioned in the Chap. 1 modeled systems will be automatically attracted to the correct configurations if the effective potential is known and the dynamic equations are appropriate. The potential can arise in real space or in some imaginary space of parameters. In a certain sense, this does not matter. For particular problems where the potential has some specific meaning, however, it does matter from either the physical or the biological point of view.

Electronic supplementary material The online version of this chapter (https://doi.org/10.1007/978-3-030-41528-0_2) contains supplementary material, which is available to authorized users.

A. E. Filippov, S. N. Gorb, *Combined Discrete and Continual Approaches in Biological Modeling*, Biologically-Inspired Systems 16,
https://doi.org/10.1007/978-3-030-41528-0_2

2.1 A Simple Theory of Phase Transitions and Pattern Formation

In this chapter, we will present some practically useful and technically convenient examples of effective potentials (or patterns, if we are talking about their formation in 2D or 3D space). Recently we also published this information in the paper (Filippov and Gorb 2019). These potentials or patterns may be generally divided into two categories:

- abstract, numerically generated patterns which are simple and convenient for theoretical numerical simulations.
- potentials that are provoked by the study of specific, experimentally observed systems.

In the latter case, the simplicity of the potential or the procedure of its generation are not so important, because in such a case we are more interested in the potential itself at the final stage of the study, or maybe even as a goal of the particular research, for example, if we would like to know how a particular pattern could appear self-consistently.

In the following, we will continuously switch between these two alternatives. Sometimes, we will use very abstract models to illustrate some general idea, or simply to generate some particular equations that reproduce what we have observed. But sometimes almost the complete study will be devoted to extracting the equations or procedures that reflect the particular nature of the problem under consideration.

The same distinction applies to the dimensionality of the surface (potential, line, etc.) generated by a procedure. In some cases it is enough to create a 1D line with specific properties. But even in such "trivial" cases our motivation can vary. For example, the reduction to one-dimensionality may be used just for the simplicity of further applications of the generated object. However, in many cases, the use of one-dimensionality may be more correct than generating and applying 2D patterns.

For many problems of contact mechanics, a real 3D problem can be transformed into an exact 1D model by the so-called Method of Reduced Dimensionality (MRD) discovered and actively developed during the last decades by Valentin Popov and co-workers (Popov and Heß 2015). We will cite and discuss several examples of MRD applications in later sections of this book (Chaps. 2, 3 and 4).

It should also be mentioned that the absolute majority of the patterns we used in the following studies of biological problems were generated by simulations of physical or chemical kinetic processes. Below, we basically report on methods of physical kinetics. For example, ideas from fluctuation theory, such as phase separation and phase transition, can be applied in pattern generation. Since it is practically impossible to reproduce all the mathematical foundations of this field, which have been published in hundreds of handbooks starting from the nineteenth century, in many cases we will simply assume that these are either known or can be proven.

Independently of the accuracy and correctness of the terminology, however, one can start with a very formal and methodically transparent approach. Let us assume that there is a fluctuating density $\rho(x)$ which is originally, at $t = 0$, randomly distributed in one-dimensional space $\{x\}$. Random initial distribution means that, on average, the fluctuations $\delta\rho(x, t)$ are equal to zero, $<\delta\rho(x, t)> = 0$, and originally independent at every point and time of the system: $<\delta\rho(x, t)\delta\rho(x', t')> = D\delta(x - x')\delta(t - t')$, so-called δ-correlated noise.

In the majority of real systems the densities of the fluctuations interact. These interactions can cause mutual attraction and reinforcement of the fluctuations leading to a nucleation of the regions with nonzero density $\rho(x)$ and their expansion in space. One of the most accepted theories of this process is the Landau theory of phase transitions (Landau 1937). The main idea of the Landau theory is that the energy of the system can be expressed as a function of the so-called "order parameter" density. In particular, it can be a functional of the spatially distributed real density $\rho(x) : U(\rho(x))$. The simplest variant of the theory corresponds to the case when the local form $U(\rho(x)) \rightarrow U(\rho)$ of the energy (depending on the uniform variableρ) has two minima corresponding to "ordered" ($\rho = \rho_0 \neq 0$) and "disordered" ($\rho = 0$) states.

Depending on the relationship between the energy minima $U(\rho_0)$ and $U(\rho = 0)$, one or the other state is energetically preferable. For example, if $U(\rho_0) < U(\rho = 0)$ and if the system originally was in another state, $\rho(t = 0) = 0$, it can be transferred to the energetically preferable state with $\rho = \rho_0 \neq 0$.

The typical form of the local energy $U(\rho)$ that corresponds to this quantitative discussion can be correctly derived from a microscopic description of the system, but this is beyond our scope here. Let us just assume that we know the correct function of the energy and can represent this in as simple form as possible. Starting from the beginning of the twentieth century, when physicists discovered modern quantum mechanics, they recommended: "if you do not know what to do with a function, expand it into to a series of the variables". We will follow this wise rule and assume that the local energy, depending on its density, can be approximated by the following form:

$$U(\rho) = \tau\rho^2 + b\rho^3 + g\rho^4 + o(\rho). \tag{2.1}$$

The symbol $o(\rho)$ means that some part of the energy, which is relatively small in comparison to the rest, can be neglected. With a proper choice of the parameters τ, b and g in the expansion $U(\rho) = \tau\rho^2 + b\rho^3 + g\rho^4 + o(\rho)$, the function will have two different minima and barrier between them in accordance with the above discussion.

Thus, one can expect that if we start with an initially disordered fluctuating field $\rho(x)$ and find the proper kinetic equation, we will get a process of its ordering: $\rho(x) \rightarrow \rho_0$. Unfortunately, this is not as easy as it sounds when the function $\rho(x)$ is not uniform. Generally, the densities $\rho(x)$ in different points are different and interact with each other. Thus, the total energy has some nonlocal addition due to the interaction:

$$U_{nonlocal}[\rho(x)] = \int dx_1 dx_2 \rho_1(x_1) V(|x_1 - x_2|) \rho_2(x_2). \quad (2.2)$$

Nobody likes such nonlocal forms in numerical modeling because they lead to time-consuming calculations. This is especially true if $V(|x_1 - x_2|)$ represents long-range interaction. However, another simplification is possible if the interaction is short-ranged enough. We will not present the complete proof of this step here but only ascertain that it is possible to expand the energy $U_{nonlocal}[\rho(x)]$ over the gradients $\nabla\rho = \partial\rho/\partial x$ of density $\rho(x)$. In a standard case, this can be transformed into the following form:

$$U_{nonlocal}[\rho(x)] \approx \int dx\Big[a\rho(x)^2 + c(\nabla\rho(x))^2 + \ldots\Big]. \quad (2.3)$$

The first term here joins the analogous local term $\tau\rho^2$ in Eq. (2.1) and simply renormalizes it: $(\tau + a) \to \tau$. The gradient term $c(\nabla\rho)^2$ is qualitatively more important. It conserves information about the interaction between the densities in different points of space. Normally, it tends to equalize the densities in different points and make the total density distribution $\rho(x)$ smoother. And it is the main reason why originally uncorrelated fluctuations of density, $<\delta\rho(x, t)\delta\rho(x', t')> = D\delta(x - x')\delta(t - t')$, tend to the ordered state with $<\delta\rho(x, t)> = 0$ and $<\rho(x, t \to \infty)> = \rho_0 = const$.

At this point, the reader may ask why we need such a complex description if ultimately everything tends to the simple constant $\rho_0 = const$. The reason is that we are not interested in the trivial final state of the kinetic process, but in its intermediate states. The absolute majority of real surfaces and substances are "frozen" in intermediate states of the kinetic process, and in these stages they produce practically important patterns with non-uniform $\rho(x)$ and with nonzero gradients $\nabla\rho$. Moreover, in many cases (in more complex systems than we are studying here) even the final stage of the kinetic process is represented by domains and other structures with non-uniform distributions $\rho(x)$. Below, we will study some of them, but for now we will return to the simplest (!) case.

Combining Eqs. (2.1) and (2.3), one can write the energy function in the following form (Patashinskii and Pokrovskii 1979):

$$\begin{aligned}\Phi[\rho(x)] &\approx \int dx\left[\frac{c}{2}(\nabla\rho(x))^2 + \frac{\tau}{2}\rho(x)^2 + \frac{b}{3}\rho(x)^3 + \frac{g}{4}\rho(x)^4 + \ldots\right] = \\ &= \int dx\left[\frac{c}{2}(\nabla\rho(x))^2 + F(\rho(x))\right].\end{aligned} \quad (2.4)$$

However, it is still not enough to write down the complete equation for the fluctuating density (Landau and Khalatnikov 1965):

$$\partial\rho(x,t)/\partial t = -\gamma\delta\Phi[\rho(x,t)]/\delta\rho(x,t). \qquad (2.5)$$

If $\rho(x,t)$ really fluctuates, there is not only a random value $<\delta\rho(x,t)\delta\rho(x', t')> = D\delta(x - x')\delta(t - t')$ at the beginning of the process, but it will continue to fluctuate further. This means that Eq. (2.5) should also include a source of random fluctuations $\zeta(x,t)$. Mathematically, this has the same properties as the initial condition for the fluctuations of density:

$$< \zeta(x,t) >= 0, \quad < \zeta(x,t)\zeta(x',t') >= D_\zeta\delta(x - x')\delta(t - t'). \qquad (2.6)$$

This random source stays on the right-hand side of the equation of motion:

$$\partial\rho(x,t)/\partial t = -\gamma\,\delta\Phi[\rho(x,t)]/\delta\rho(x,t) + \zeta(x,t) \qquad (2.7)$$

and causes new fluctuations during the whole process of the density nucleation. Equation (2.7) represents the complete equation for this simplest (!) model. Taking into account the explicit form of the function $\Phi[\rho(x)]$ in Eq. (2.4) and performing the variation $\delta\Phi[\rho(x,t)]/\delta\rho(x,t)$, one can write down another form of the kinetic equation that is more convenient for solving it:

$$\partial\rho(x,t)/\partial t = -\gamma\left[c\Delta\rho(x,t) + \tau\rho(x,t)^2 + b\rho(x,t)^3 + g\rho(x,t)^4\right] + \zeta(x,t) \qquad (2.8)$$

where $\Delta\rho(x)$ is the Laplasian operator which, in the 1D case, is simply equal to the second derivative $\Delta\rho(x) = \partial^2\rho(x)/\partial x^2$.

Starting from a random distribution, the density evolves with time according to Eq. (2.8). The barrier in the local part of the energy $F(\rho(x))$ does not allow the majority of the density fluctuations to grow immediately to the minimum corresponding to the nonzero equilibrium density $\rho_0 = const.$ Only relatively large fluctuations can pass the barrier and grow. These will grow in both senses: their amplitude at maximum tends to ρ_0 and they expand in space (Kuzovlev et al. 1993a, b).

A typical intermediate configuration with different nuclei of nonzero density is shown in Fig. 2.1. If there is a physical cause that freezes this configuration, we will get a 1D representation of the non-uniform pattern of density. The simplest possible reason for this may be a very small damping constant γ which controls the rate of the kinetic process in Eq. 2.8. If the characteristic time $1/\gamma$ of evolution of the distribution $\rho(x,t)$ to the equilibrium ρ_0 is extremely long in comparison with other time spans involved in a problem, the evolution of the density may be ignored and the intermediate distribution can be treated as practically static, similar to the galaxies developing over billions of years. Galactic patterns are much longer than our life span and the majority of processes on our planet, so they can be treated as practically static.

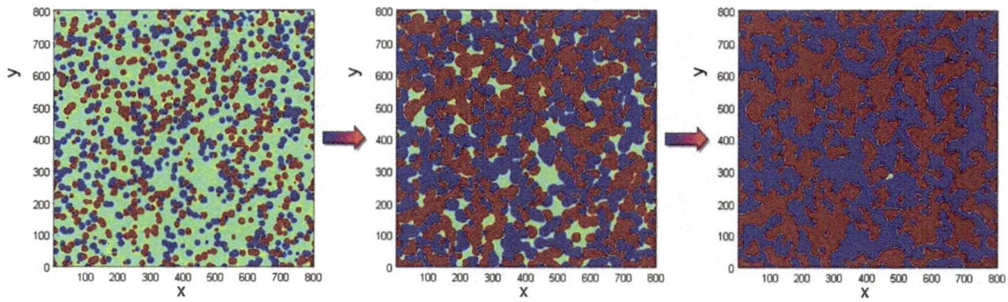

Fig. 2.1 Formation of the domain structure as a kinetic process. Two different kinds of domains are represented by red and blue colors, respectively. See also supplementary movie 2.1

However, in many cases the kinetic process may be stopped by some natural causes at an intermediate stage. Normally this happens when the system is open to external influences or consists of a number of interacting subsystems. In the first case, external forces can cause a change of the coefficients of energy expansion (cf. Eq. (2.4)) in the course of the density evolution. As a result, the system starts its development to some equilibrium $F(\rho)$ and finishes at a completely different equilibrium corresponding to the modified function $\widetilde{F}(\rho)$. In the second case, varying interactions between the subsystems can lead to the growth of different densities in different domains of the space. If such growing domains come into contact with each other, they will mutually block their further expansion and form a static (or maybe very slowly shifting) domain wall between them. Such blocking leads to the fixation of the domain structure and, on a large scale, will form a pattern that includes many domains of different sizes. The simplest way to obtain such a domain structure is to use the expansion $F(\rho)$ with even terms only:

$$\begin{aligned}\Phi[\rho(x)] &\approx \int dx\Big[\frac{c}{2}(\nabla\rho(x))^2 + \frac{\tau}{2}\rho(x)^2 + \frac{g}{4}\rho(x)^4 + \frac{u}{6}\rho(x)^6 + \ldots\Big] = \\ &= \int dx\Big[\frac{c}{2}(\nabla\rho(x))^2 + F(\rho(x))\Big].\end{aligned} \tag{2.9}$$

The equation of motion that corresponds to the functional Eq. (2.9) leads to the typical domain structure shown in Fig. 2.1. The same process in motion is illustrated in supplementary movie 2.1.

2.2 Automatic Blocking of the Nucleation and Freezing of the Process

In the above cases of "almost frozen" kinetics, we assumed that this process is much slower than other processes in a particular model. However, it has been shown more than 20 years ago that since the formation of new phase nuclei includes processes that prevent their appearance and growth in other regions in space, it should result in the autostabilization of an intermediate mixed state. This may be called "automatic blocking of the nucleation".

Normally this is caused by the nonlocal (long-range) interactions which occurred in an ordered process by itself. We know that numerical modeling does not like nonlocal interactions because these calculations are extremely time-consuming. However, it is possible to show that the effect of blocking long-range interactions can be approximated by including some local additions to the above models.

There are various mechanisms for the formation of an effective long-range interaction in such systems. For example, formation of the substance (and the surface, which is needed here) includes the reaction of a crystal lattice (striction) to a change in magnitude of the order parameter $\rho(x, t)$ during a phase transition. One can show that in the relatively simple case of an isotropic medium and quadratic striction, the local energy function of the free energy is modified in the following manner (Patashinskii and Pokrovskii 1979):

$$\Phi[\rho(x)] = \int dx\Big[\frac{C}{2}(\nabla\rho(x))^2 + F(\rho(x)) + g_u\rho(x)^2 u_{ii} +$$

$$+\frac{k^2}{2}u^2{}_{ii} + \mu\Big(u_{1k} - \frac{1}{3}\delta_{1k}u_{ll}\Big)^2\Big]. \tag{2.10}$$

If the lattice vibrations follow the variations of $\rho(x, t)$, we can utilize the condition $\delta\Phi[\rho, u]/\delta u_{ik} = 0$ to eliminate the variable u_{ik} and we will get, after some standard mathematical transformations, an effective function solely in terms of the field $\rho(x, t)$:

$$\Phi[\rho(x)] = \int dx\left[c\Big(\frac{C}{2}\nabla\rho(x)\Big)^2 + \widetilde{F}(\rho(x)) + \frac{\kappa}{2V}\rho(x)^2\int dx'\rho(x\prime)^2\right]. \tag{2.11}$$

Here, $\widetilde{F}(\rho(x))$ is the renormalized local form of $\widetilde{F}(\rho(x))$ with the same structure as the original function$F(\rho(x))$ (we shall henceforth omit the tilde), and the constant κ is defined by the expression $\kappa = \frac{g^2}{2V}\left[(k/2 + 2\mu/3)^{-1} - (k/2 + 2P/3)^{-1}\right]$. The constant P is determined by the external pressure or other constraints (twins, defects, etc.) which prevent the free expansion of the crystal and, in turn, determine the sign of κ. When $P > \mu$, $\kappa > 0$; otherwise $\kappa < 0$.

The nonlocal construct $\int dx[\rho(x)^2 \int dx'\rho(x')^2]$ in the function $\Phi[\rho(x)]$ generates a term with a long-range effect in the equation of motion for the field variable

$$\partial\rho(x,t)/\partial t = -\gamma\left[c\Delta\rho + \partial F/\partial\rho - \frac{\kappa}{2V}\rho(x,t)\int dx'\rho(x',t)^2\right] + \zeta(x,t), \tag{2.12}$$

whose presence significantly accelerates or slows down (or even totally stops) the ordering process, depending on both the magnitude and the sign of κ (plus/minus). A

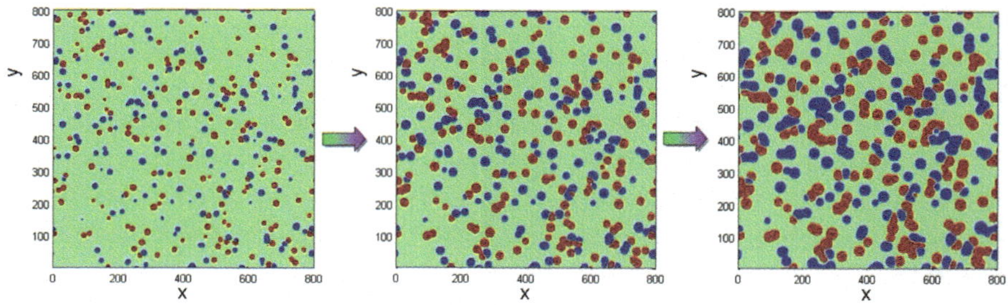

Fig. 2.2 Formation of the domain structure in kinetics when self-blocking is present. Two different kinds of domains are represented by the red and blue colors, respectively. It is important to note that the last stage here is in fact the final one and does not develop any further with time. See also supplementary movie 2.2

typical picture of blocked, almost static domains is shown in Fig. 2.2. The same process in its kinetic form is illustrated in supplementary movie 2.2.

Let us now discuss another example of an interaction which leads to a similar model: the local variation of the order parameter is accompanied by the emittance or absorption of heat (depending on whether the transition is to the low- or high-temperature phase). This results from heat conduction in heating (cooling) of the surrounding regions of space, which, of course, slows the transition process in all cases. This mechanism seems to be universal, and its effectiveness is determined only by the relationship between the rates of the nucleation and heat conduction processes. The local heating (cooling) of a system in a region where a nucleus appears can be taken into account by assuming that the quantity τ in Eq. (2.9) is a function of position and time. The kinetic equation for the order parameter should be supplemented by an equation which describes the evolution of $\tau = \tau(x, t)$. This supplementary equation should be a heat conduction equation with heat removal and with a source $\beta[\rho]$, whose intensity is proportional to the rate of change of the free energy, i.e. $\beta[\rho] \sim \partial\rho/\partial t \cdot \delta\Phi/\delta\rho$.

As a result, we have the following system of connected equations:

$$\partial\rho(x,t)/\partial t = -\gamma\delta\Phi[\rho(x,t)]/\delta\rho(x,t) + \zeta(x,t); \tag{2.13}$$

$$\partial\tau(x,t)/\partial t = \alpha\Delta\tau(x,t) - \partial\rho(x,t)/\partial t \cdot \delta\Phi[\rho(x,t)]/\delta\rho(x,t) + \xi(x,t). \tag{2.14}$$

Before discussing the modeling results, we will show that with some simplification of the model the mechanism under consideration can be described in terms of a single field variable $\rho(x, t)$, which evolves in accordance with an equation similar to Eq. (2.12). The physical arguments which lead to a function like Eq. (2.11) in this case, too, are fairly simple. Each growing domain of the new phase creates a non-uniform temperature field $\tau(x, t)$ around itself. Owing to heat conduction, the temperature at other points in space deviates from the trial temperature, thus altering the conditions for the growth of other domains at those points. This signifies the appearance of an effective long-range field that accompanies the nucleation process in the system. Relating the variation of the temperature field to the order parameter field $\rho(x,t)$, we obtain an energy function $\Phi[\rho]$ like that in Eq. (2.11). In fact, when the fluctuations of $\tau(x,t)$ are "turned on" in a system with a temperature equal to the heat bath temperature τ_0, after one unit of time the mean value of $\tau(x,t)$ will deviate from τ_0 by

$$< \tau > -\tau_0 = \frac{1}{V}\int dx \int_0^1 \partial\tau/\partial t \sim \frac{1}{V}\int \rho^2. \tag{2.15}$$

The reciprocal influence of the domains of the new phase becomes significant when they become so large (and this is evident from the modeling results) that the energy of the domain boundaries between the ordered and the disordered phases can be neglected. As a result, we arrive at a function like Eq. (2.10), which was obtained to describe striction effects in the kinetics of a first-order phase transition. In some cases, this will make it possible, in principle, to disregard the specific mechanism for realizing the long-range effect accompanying the first-order phase transition and to formally analyze models with nonlocalities of the general form.

2.3 Large-Scale Structure of the Fluctuating Field: Universality and Scaling

The tentative structure of the energy function is directly related to the microscopic interactions in the system. In many cases, it can be even analytically derived from the microscopic theory. This means that the coefficients of the expansions used in the kinetic equations have well-defined specific values and in turn should completely determine the density distributions $\rho(x,t)$ and, as a result, the structure of the contact surface.

Most real surfaces, however, have a practically universal (scaling) structure with the power low distribution of the relief. This means that if there is no special reason to produce a different structure, it should appear due to the universal kinetic process which makes the difference in the initial energy functionally negligible.

As we saw in Sect. 2.2, ordered transition is anticipated by nonlinear excitations, which can be interpreted as nucleation centers. The kinetics of the first-order phase transition in different physical systems has been the subject of intensive studies. As a rule, the ordering of a metastable disordered phase is due to the fluctuations produced and, ultimately, to the growth of the nucleus of the stable phase. In a first-order phase transition, there is a change in some order parameter between these two phases, which lowers the free energy as the new phase forms.

The corresponding local energy density $F(\rho)$ must have a metastable minimum and be energetically favorable. However, the free energy is transformed due to the fluctuations. This change is especially essential in the critical region at a second-order transition.

It is well known from the theory of critical phenomena that the fluctuations manifest through the renormalization of critical exponents. But the renormalization group (RG) method allows one not only to perform purely numerical calculations of critical exponents, but also to predict some new effects, which would not be possible with conventional approaches, e.g., the Landau approximation applied above. Among them, there are qualitative effects such as the fluctuation-induced first-order phase transition. This effect takes place in some anisotropic systems where the renormalized free energy $F(\rho)$ undergoes transformations which are typical for the first-order phase transition. The same kind of nucleation centers can be found in a fluctuation-induced first-order transition. However, one can expect that even when the fluctuations are not strong enough to change the transition order, they will somehow manifest. The mean field model is very convenient for analytical studies, but it reduces the fluctuation interaction, and it is impossible to control the correction of the free energy which becomes necessary on account of the neglected fluctuations (Filippov 1994).

A more correct approach was developed almost 50 years ago. It states that the fluctuations renormalize the energy function $\Phi[\rho(x,t)]$ according to the so-called renormalization group (RG) equation, $\partial\Phi/\partial l = \widehat{R}\Phi$. Here we do not need the complex exact form of the RG equation and use its simplified version:

$$\widehat{R}\Phi = -\int d^d r \left\{ \left[(d-2)\frac{\rho\left(\vec{r}\right)}{2} + \vec{r}\nabla_{\vec{r}}\rho\left(\vec{r}\right) \right] \frac{\delta\Phi}{\delta\rho\left(\vec{r}\right)} + \int d^d r' \left[\frac{\delta^2\Phi}{\delta\rho\left(\vec{r}\right)\delta\rho\left(\vec{r}'\right)} - \frac{\delta\Phi}{\delta\rho\left(\vec{r}\right)}\frac{\delta\Phi}{\delta\rho\left(\vec{r}'\right)} \right] \right\} \quad (2.16)$$

where d is the space dimensionality.

The first term in Eq. (2.16) corresponds to a simple scale transformation of the density distribution $\rho(x,t)$ and the second term appears due to the integration over internal fluctuations inside small regions after scale transformations. According to the general RG hypothesis, at the critical point of the phase transition, where the ordering of $\rho(x,t)$ takes place, the functional $\Phi \to \Phi^*$ tends to the fixed point $\partial\Phi^*/\partial l = \widehat{R}\Phi^* = 0$.

Let us now study the time–space evolution of the fluctuating density $\rho(x,t)$ in the critical state. We intend to show that $\rho(x,t)$ on average produces a well-pronounced large-scale structure in spite of its scale invariance. As always, the kinetic equation can be written in the form $\partial\rho/\partial t = -\gamma\delta\Phi^*[\rho]/\delta\rho + \zeta$. At every moment, the average probability $W = \langle w[\rho]\rangle$ of finding the density distribution $\rho(x,t)$ is determined by the function $w[\rho] = \exp(-\Phi[\rho])$. This probability develops with time according to the following equation:

$$\partial W/\partial t = \left\langle \int d^d r \frac{\delta w[\rho]}{\delta\rho} \partial\rho/\partial t \right\rangle . \quad (2.17)$$

By combining Eq. (2.16) with the conditions $\widehat{R}\Phi^* = 0$ and $\partial\rho/\partial t = -\gamma\delta\Phi^*[\rho]/\delta\rho + \zeta$, it becomes clear that at the critical time point the evolution of the probability $W = < w[\rho]>$ is reduced to the following simple scale transformation:

$$\partial W/\partial t = < w[\rho] \int d^d r\left[(d-2)\frac{\rho}{2} + \vec{r}\nabla_{\vec{r}}\rho\right]\frac{\delta\Phi^*}{\delta\rho} > . \qquad (2.18)$$

Physically, this means the following: at the point of the transition energy, the function becomes the universal $\Phi \to \Phi^*$. The kinetic equation based on this function will in time produce new realizations of $\rho(r, t)$ with the same statistical properties but at larger scales. At $t \to \infty$, the structure will become scale-invariant. This structure is generally similar to that found in the intermediate stage of the nucleation process at first-order phase transitions, but it is never completed in static ordering.

To calculate some particular realization of the density at the critical point, one can numerically solve the equation $\partial\rho/\partial t = -\gamma\delta\Phi^*[\rho]/\delta\rho + \zeta$ with the energy density derived from the RG equation in its local approximation:

$$\widetilde{\Phi}[\rho(x)] \approx \int dx\left[\frac{c}{2}(\nabla\rho)^2 + \widetilde{F}(\rho)\right], \qquad (2.19)$$

where $\widetilde{F}(\rho)$ is normalized to the critical temperature $\widetilde{F}(\rho) = F - \tau^*\rho^2 - F(\rho = 0)$ solution of the local version of the RG equation $\widehat{R}\Phi^* = 0$:

$$\widehat{R}F = dF - \frac{d-2}{2}\rho\frac{\partial F}{\partial\rho} + \frac{\partial^2 F}{\partial\rho^2} - \left(\frac{\partial F}{\partial\rho}\right)^2 = 0. \qquad (2.20)$$

Despite its apparent simplicity, Eq. (2.20) is nonlinear and cannot be solved analytically. But in contrast to the functional equation $\widehat{R}\Phi^* = 0$, it is an "ordinary" differential equation and it is easy to find the physical branch F^* of its solution $\widehat{R}F^* = 0$ numerically with very high accuracy. The discrete data array $F^*(\rho_k)$ has a large enough number ($N > > 1$) of points $k = 1, 2, \ldots N$, where each value of $\{\rho_k\}$ defines a unique value of $\{f_k^*\}$.

With good accuracy, this array an be used in the kinetic equation

$$\partial\rho/\partial t = -\gamma[c\Delta\rho + \partial f^*/\partial\rho] + \zeta \qquad (2.21)$$

instead of the analytic formulae for the energy which we had used previously. By repeating the simulation with random initial conditions and with the random time-dependent noise $<\zeta(r,t)> = 0$, $<\zeta(r,t)\zeta(r',t')> = D_\zeta\delta(r-r')\delta(t-t')$, after long-time runs we get an unlimited number of realizations $\rho(r, t)$.

Direct observation of the simulation results shows that every instant density distribution $\rho(r, t)$ contains many nuclei of different sizes. The wider the particular maxima of density that are formed, the longer they survive in the general landscape. This means that at large scales the total process of transformation becomes slower and slower. Besides, stronger correlation between the densities in different spatial positions appears.

For $t \rightarrow \infty$, the density distribution $\rho(r, t)$ tends to the expected scale-invariant structure. One can calculate the correlation function $G(r - r') = <\rho(r, t)\rho(r', t)>$ for a fixed moment t and find this scaling correlation function $G(r - r')$. This means that the power function $G(r - r') = <\rho(r, t)\rho(r', t)> \sim 1/|r - r'|^{\beta}$ depends on the distance $|r - r'|$ between the points. This quite general result confirms the common observation that in many cases, irrespective of the specific features of the system under investigation, the evolution of the surface layer proceeds in a fairly universal manner.

2.4 Chemical Appearance of Fractal Surfaces

One more example of the automatic generation of fractal surfaces arises when a dense layer of one or more products emerges, as a result of a chemical reaction, in the immediate vicinity of the smooth flat interface of the two media in contact. It has been observed that, while growing, this layer becomes more and more porous and rough. Gradually, an essentially inhomogeneous but, as a rule, scale-invariant structure is formed, and the laws governing the growth of this structure are characterized by fractal dimensions and growth exponents (Filippov 1998).

In addition to arousing purely scientific interest, the study of this growth of the corrosion front deserves attention because of its importance from a practical point of view, since the problem is closely linked to that of raising the efficiency of electric batteries. For instance, when a lithium anode is placed into an electrolyte containing $SOCl_2$ as an additive, a porous two-component layer of LiCl and SO_2 is formed at the surface of the anode due to the exceptionally high reactivity of lithium. The presence of such a layer leads to what is known as the lag effect, in which case the element is stored for a long time. Micrographs of the surface layer show that the layer can be considered a combination of a relatively dense initial layer with a subsequent transition to a fractal structure with an ever increasing porosity.

The highly universal properties manifested by different systems of this kind suggest that we can use universal growth models based on a combination of the ideas of continuum field theory and of kinetic equations with a random source. Although fairly common in the theory of phase separation and fluctuation phenomena in phase transitions, kinetic equations with a source of noise should be used cautiously in describing front growth, because, in contrast to phase transitions where the order parameter is generated in the bulk of the system, a random source cannot be considered additive. The generation of a finite density of components forming the front occurs only in the immediate vicinity of an already existing boundary. This means that in the corresponding source in the equation must be multiplicative, i.e., at least contain density as a factor.

However, in recent theoretical studies of phase diagrams and transitions in systems with multiplicative noise, it was noted that the presence of such strong noise can have a dramatic effect on the ordered structure and on the phase diagram, and may lead to the emergence of new nontrivial phases. This means that the model equation should be written in such a way as to exclude additional difficulties associated with this noise.

From an experimental standpoint, the study of fractal corrosion structures is easy since the corrosion front may be directly observed in micrographs and the corresponding 2D distributions of density can be studied explicitly. At the same time, the processes involved are very complex, and despite continuing efforts the theoretical models remain extremely simple, although they presuppose a numerical analysis of the kinetic equations. Usually, only the density of a single distributed quantity that is considered the most important one in each specific case is involved.

In physicochemical processes, however, usually two or more components participate in the reactions. No matter how subtle the description of a system by the single-component approach, the study of the system may be replaced by the analysis of a purely theoretical model. Given the contemporary computer modeling techniques, any attempt to reduce the problem to a single equation is more a tribute to the analytic tradition than a real necessity. Here we will demonstrate the feasibility of moving in this direction by analyzing, as an example, a two-component model formulated for the description of growth and corrosion of a broad class of porous surface layers initiated by chemical reactions.

Below we will examine the simplest two-component case, assuming that we are dealing with chemical reactions that proceed in a system with a contaminated lithium anode. The complete picture of the reactions in such a system is fairly complicated and can be expressed as follows: $Li \rightarrow Li^+ + e^-$, $4Li^+ + 4e^- + 2SOCl_2 \rightarrow 4LiCl + SO_2 + S$.

Actually, we are interested only in the formation of a front consisting of lithium chloride (LiCl) contaminated by the reaction product SO_2 that concentrates near the surface. Bearing all this in mind, we can interpret Eq. 2.9 as an initial equation for the evolution of the density of LiCl which we denote by $\rho_1(r, t)$, with the index 1 indicating the corresponding coefficients and the source of noise. By $\rho_2(r, t)$ we denote the density of SO_2. We model the local repulsion reaction of the products LiCl and SO_2 by a fixed-sign additional term in the effective energy of the system, $V_{12}(\rho_1, \rho_{2,})$, which in the lowest order can be written as $V_{12}(\rho_1, \rho_{2,}) = B\rho_1^2\rho_2^2/2$. Thus, the equation of motion for the first density becomes $\partial\rho_1/\partial t = -\gamma[c_1\Delta\rho_1 + \rho_1(1-\rho_1)][C_1 + \zeta_1(r,t)] - B\rho_1\rho_2^2$. This equation must be augmented with an equation describing the evolution of the second component, $\rho_{2,}$. The second component $\rho_{2,}$, just like the first one, is generated as a result of the same reactions close to the free LiCl surface (which is not contaminated by SO_2). This means that for ρ_2 we must use the same generating term $\rho_1(1 - \rho_1)$ as for ρ_1 : $\partial\rho_1/\partial t = -\gamma[c_2\Delta\rho_2 + \rho_1(1-\rho_1)][C_2 + \zeta_2(r,t)] - B\rho_2\rho_1^2 + f(\rho_2)$.

Here we have allowed for the fact that although both densities, ρ_1 and ρ_2, emerge as a result of the same reaction, the rate of formation of the dense components in ρ_1 and ρ_2 may differ, so that generally $C_1 \neq C_2$. Obviously, the terms linear in ρ_2 cannot ensure that the increase in ρ_2 is stopped and is stabilized at the static limit

($\rho_2 \to 1$). We must also bear in mind that far from the front there is no spontaneous generation of ρ_2, and hence the effective energy is given as $V_2(\rho_2^2) = B\rho_2^2(1-\rho_2^2)$, whose variation yields a combination proportional to the function $f(\rho_2) = \rho_2(0.5 - \rho_2)(1 - \rho_2)$. It contains a barrier that separates the two similar minima at $\rho_2 = 0$ and $\rho_2 = 1$.

It can be shown that in the continuum approximation, with only the lowest harmonics in the energy (and hence with the Laplasian terms $\Delta\rho_{1,\,2}$ in the equations), one has to add generation terms with the step-function cut-off factor $\Theta = \vartheta\left(\int\limits_{|r-r'|<\sigma} dr'\rho(r') - a\right)$. This factor turns on the generation if the density in some neighborhood $|r - r'| < \sigma$ exceeds the critical threshold a. In other words, $\Theta \to 1$ when $\int\limits_{|r-r'|<\sigma} dr'\rho(r') > a$ and $\Theta \to 0$ in the opposite direction.

Ultimately, the kinetic equations take the following form:

$$\begin{aligned}\partial\rho_1/\partial t =& \gamma[c_1\Delta\rho_1 + \rho_1(1-\rho_1)][C_1 + \zeta_1(r,t)]\Theta - B\rho_1\rho_2^2\\ \partial\rho_1/\partial t =& \gamma[c_2\Delta\rho_2 + \rho_1(1-\rho_1)][C_2 + \zeta_2(r,t)]\Theta - B\rho_2\rho_1^2 - \rho_2(0.5-\rho_2)(1-\rho_2).\end{aligned} \tag{2.22}$$

In a purely local approximation, the system of Eq. (2.22) can be reduced to the form of ordinary differential equations:

$$\begin{aligned}\partial\rho_1/\partial t =& \rho_1(1-\rho_1)[C_1 + \zeta_1(r,t)]\Theta - B\rho_1\rho_2^2\\ \partial\rho_1/\partial t =& \rho_1(1-\rho_1)[C_2 + \zeta_2(r,t)]\Theta - B\rho_2\rho_1^2 - \rho_2(0.5-\rho_2)(1-\rho_2).\end{aligned} \tag{2.23}$$

Here the scale of time is normalized to $\gamma = 1$.

The overall structure of the phase portrait is depicted in Fig. 2.3 which presents a physically interesting example. The local system of Eq. (2.23) actually describes the evolution of the densities at each point in space without allowing for an interaction between different points. In this approximation, the interaction is taken into account only via the initial conditions. Specifically, as the front arrives at a certain point in space, both densities, ρ_1 and ρ_2, begin to be generated at that point, so that the physical scenario in the phase portrait in Fig. 2.3 corresponds to the trajectories that emerge in the neighborhood of the trivial point $\rho_1 = \rho_2 = 0$.

The separatrix connecting this point and the saddle point divides the plane $\{\rho_1, \rho_2\}$ into two attracting basins for the two stable, immediately evident fixed points. By studying the behavior of the trajectories that start near the separatrix, we can predict several results of the numerical modeling of the complete equations and, in the final analysis, the properties of real systems. In particular, we can easily predict the role of the source of noise. If the noise is strong, the phase trajectories can pass both above and below the separatrix, irrespective of the scenario according to which the front arrives at the given point.

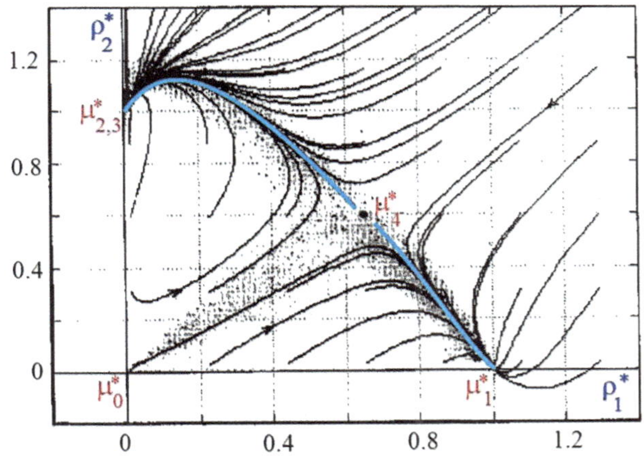

Fig. 2.3 Phase portrait of the local version of the kinetic equations describing the growth of a porous surface. Fixed points corresponding to possible combinations of local densities are marked in red. Two well-pronounced "large rivers" leading to the static final configurations are shown as blue lines. The cloud of small gray points represents specific realizations of the densities numerically found for the exact nonlocal equations

Within a certain time interval after the arrival of the front, both densities, ρ_1 and ρ_2, increase essentially simultaneously and very quickly. This corresponds to the first maximum in the evolution rate

$$W(t) = \left[(\partial\rho_1/\partial t)^2 + (\partial\rho_2/\partial t)^2\right]^2, \tag{2.24}$$

as depicted in Fig. 2.4. Near the saddle point, there can be no further increase in ρ_1 and ρ_2. The rate $W(t)$ rapidly decreases. At the same time, there is a phase separation in the system, with one of the densities, ρ_1 or ρ_2, expelled from the given region in space. This is followed by a sharp increase in $W(t)$, accompanied by a rapid buildup of the remaining component, a process that is slowed down near one of the stable fixed states.

The characteristic double-humped curve representing the evolution rate is indeed observed when the complete system of equations is solved numerically. In accordance with the physics of the problem, the initial condition is selected in the form of a narrow strip of density $\rho_1(r, t)$ near one of the boundaries of the two-dimensional system. Here, the process of generation and separation of the densities $\rho_{1,\,2}(r, t)$ is accompanied by the formation of characteristic dendritic spatial distributions of both densities.

The two densities are generated simultaneously in the vicinity of the front. However, in the absence of noise, the initial density distribution leads to the formation of a completely filled region behind the front. The numerical solution shows that with time the front becomes discontinuous, and the expanding ordered region is transformed into a fractal one. In Fig. 2.5, we use a color scale to show a

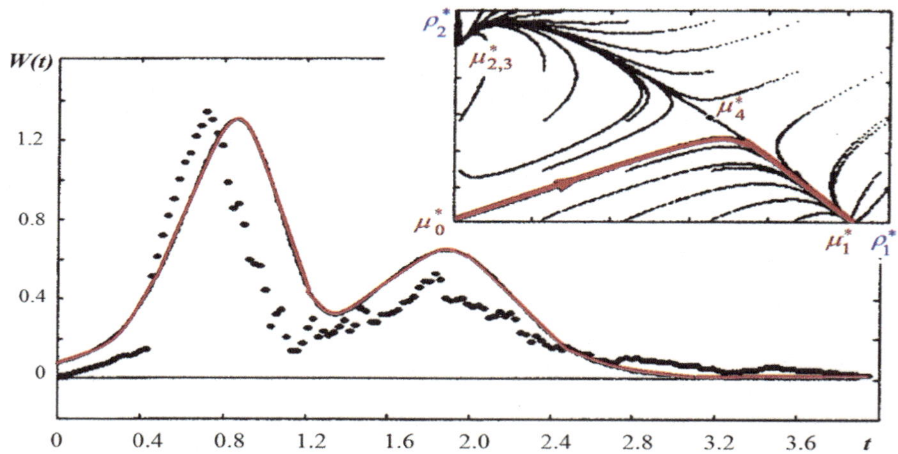

Fig. 2.4 Evolution rate $W(t)$ as a function of time. Good correlation with the phase portrait (shown in the inset) is evident. The particular trajectory for which $W(t)$ is calculated is in red. The slow evolution rate corresponds to the vicinities of the fixed points and to a "large river" leading to a final static configuration. The set of discrete black points represents the mean rate $\widetilde{W}(t)$ numerically found for the exact nonlocal equations

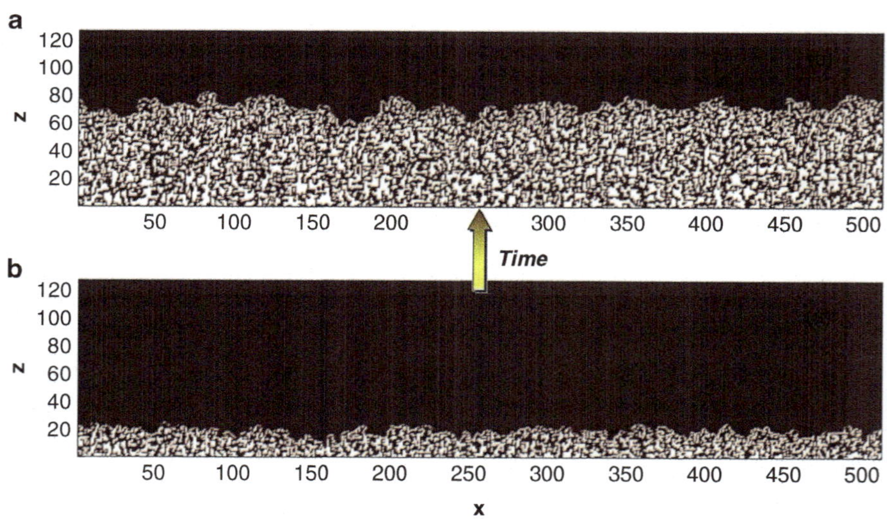

Fig. 2.5 (**a**, **b**) Two intermediate stages of the porous surface layer formation (from panel a to b). With time, such a system can develop a very complex structure featuring not just a fractal surface, but also lacunae with negative curvature, which lead to stronger adhesion of the soft tissue contacting such a surface. See also supplementary movie 2.3.

characteristic fragment of the system with a distribution of the total density that emerges at an intermediate stage in the transition from homogeneous growth to fractal growth. Contaminated regions and regions of active growth are clearly visible. These are characterized by intermediate values of the densities which correspond to the sections in Fig. 2.5 with intermediate shades of the color. The same process in motion is reproduced in supplementary movie 2.3.

Compared to growth processes modeled with only one fluctuating variable, a novel property of this system is the potential emergence of voids behind the front, i.e. regions filled with neither of the two components. When there is only one field, i.e. $\rho_1(r, t)$, such lacunae must be assumed to be filled with the other, "contaminating" component of density, $\rho_2(r, t)$, which in this case is not explicitly present in the equations. Therefore, this model contains additional information about the first field, $\rho_1(r, t)$, which makes it possible to distinguish between the regions occupied by $\rho_2(r, t)$ and the voids.

The mechanism of void formation is clearly visible in Fig. 2.5a, which for a small fragment of the front depicts a typical growth sequence. Three characteristic moments in time are singled out in this figure: the emergence of a dense initial layer, the emergence of the first dendritic protuberances, and the collapse of the first internal pores in a structure with a total density $\rho(r, t) = \rho_1(r, t) + \rho_2(r, t)$. Well-formed voids are clearly visible in Fig. 2.5b.

Void formation is closely related to the ability of the "contaminant" $\rho_2(r, t)$ to block the active sections of the front $\rho_1(r, t)$ and, at least in principle, to terminate the growth process. As the pores collapse, the front usually continues to move in both directions. Naturally, the outer boundary, $\rho_1(r, t)$, is rather insensitive to the presence of a blocked pore somewhere inside the system and continues its forward motion. The inner boundary $\rho_1(r, t)$ surrounding a pore is qualitatively similar to the outer boundary and can move "back" up to the point where it is completely blocked by sections of $\rho_2(r, t)$. The evolution of the system turns out to be exceptionally multifaceted and even a slight variation of the coefficients of the model allows for the reproduction of very realistic configurations of the densities $\rho_1(r, t)$ and $\rho_2(r, t)$.

2.5 Mathematical Creation of Fractal Surfaces

The above section gave an example of creating a fractal surface by growth. Now, let us turn to more formal methods of numerically creating surfaces. Depending on the particular problem, we can use at least two different ways of creating a rough surface that is close to a fractal one.

1. One is to apply Fourier expansion with the harmonics restricted to the area between the maximal and minimal wavelengths. Depending on the number of waves and on both limits, the surface will be more or less close to the real fractal one and include or not include irregularities with small sizes (Filippov and Popov 2007).

2. Another way is to use the random deposition of Gaussian curves. Depending on their number, height dispersion, minimal and maximal widths, the surface will be closer to a fractal one or not.

Many biological surfaces have a structure that resembles a pattern of adjacent objects of some specific scale. These can be quite easily modeled by a set of deposited objects. That is why in our studies we usually choose the second method. In many cases, such surfaces appear as a result of the random deposition of some particles, balls, or blocks of some other objects.

We definitely admit that such a formal model completely ignores the fine structure of adhesion-like weakening at motion by the fracture and restoration of the bonds, and so on. Often only two basic mechanisms are involved in our minimalistic models: attraction to the surface and energy dissipation.

The first mechanism actually restricts the scale of the objects which we include to model the surface since the irregularities have to correspond to the size of the contact points (radius of attraction in the model). In particular, this limitation reflects a situation with adhesive terminal elements (spatulae) of the attachment organs of different animals, which we will discuss in Chap. 4. In this context, surfaces with irregularities much smaller or larger than the spatula size can be treated as almost flat. That is why we normally use surfaces with the irregularities comparable to the spatula. Such a deposition of objects with a fixed size range can be carried out either naturally or artificially. We will not concentrate on any specific cases here and discuss only a couple of more or less abstract examples. Despite the formal mathematical construction such an approach helps in developing simple but realistic numerical models that may be useful, for example, in the study of spatula interaction with various substrate profiles. Formal models can well explain the results of experimental studies and also predict adhesion of animals to real surface profiles, depending on the dimension and stiffness of the spatula.

In the Chaps. 3 and 4, we will study such biological systems using numerically generated surfaces and compare the results obtained with real 3D surface profiles obtained experimentally by the depositing of spheres. In the purely numerical approach, different roughnesses of the surfaces can be achieved by depositing spheres with a defined radius on an originally flat surface, by employing the following procedure.

In two-dimensional space (D = 1 + 1), we take an array of equal spheres or circles of a fixed radius R and place them successively in randomly chosen positions x_n, with $n = 1, 2, \ldots, N$. The number of spheres $N = \|L/R\|$ in the array is an integer value corresponding to the total number of circles of radius R which are necessary to completely cover a system of length L. Each sphere is added virtually to the corresponding segment of the surface:

$$\delta y_n(x; x_n) = \sqrt{\left|R^2 - (x - x_n)^2\right|}. \tag{2.25}$$

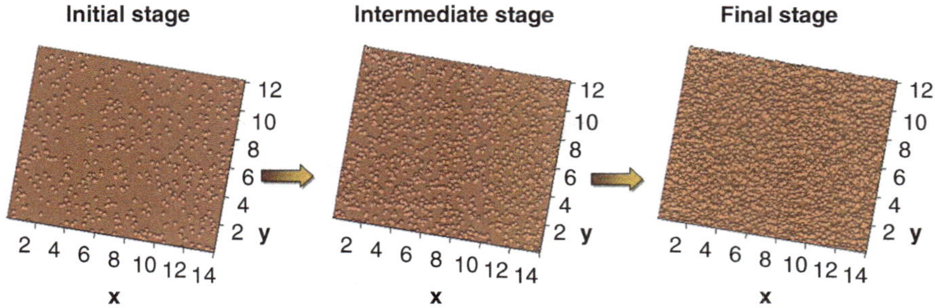

Fig. 2.6 Sequential stages of the formation of a rough surface by the random deposition of spheres with a fixed radius on the originally flat surface. See also supplementary movie 2.4

Due to the random deposition, the spheres can fall either on top of already existing coverage or onto the flat surface. As a result of such a deposition, the total surface gradually accumulates all the spheres $y(x) = \sum_{n=1}^{N} \delta y_n(x; x_n)$, which are generally speaking placed on top of each other. This procedure can also be easily generalized for D = 2 + 1 surfaces. Typical realizations of such surfaces for different radii *R* are presented in Fig. 2.6. The same procedure is presented as a dynamic process in supplementary movie 2.4.

In many cases the specific form of the deposed particles is not very important (or not important at all). In such a case, one can treat the random deposition procedure as an absolutely abstract way to construct the desired surface. One of the simplest ways to do this is to model the rough surface *Z*(*x*, *y*) by a random deposition of Gaussian curves:

$$Z(x, y) = \sum_n G_n(x, y, \{x_n, y_n\}) = \sum_n a_n \exp\left[-\left((x - x_n)^2 + (y - y_n)^2\right)/w_n^2\right]. \tag{2.26}$$

The advantage of this approach is its mathematical transparency since the simple manipulation of the position, amplitude and width of the Gaussian functions will suffice to generate any kind of surface irregularity. When studying the attachment ability of animals, the only prerequisite is that the characteristic scale of the artificial surface irregularities should be smaller or comparable to the scale of the elements interacting with the surface (e.g., the spatulae).

In particular, the typical distance between the hills and valleys of the randomly generated surface $Z(x, y) = \sum_n G_n$ can be regulated by the number of Gaussian curves. An additional convenience of this method is the fact that the amplitude of the asperities cannot be controlled during accumulation of the sum $\sum_n G_n$. Instead, the amplitude of the roughnesses after accumulation is regulated by the normalization $Z(x, y) \rightarrow A(Z(x, y) - \min(Z))/(\max(Z) - \min(Z))$, where the desired amplitude

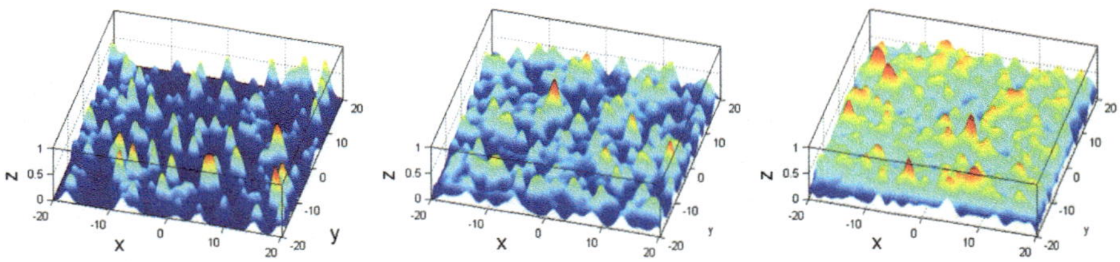

Fig. 2.7 Different variants of rough surfaces generated by the random deposition of Gaussian functions. The number of Gaussian curves used to generate the surface increases from left to right. See also supplementary movie 2.5

A can be chosen between the limit of the flat surface $A = 0$ and values comparable with the characteristic lengths of the system under consideration (Fig. 2.7).

Mathematically created surfaces or potentials *U* are widely used in tribology. In this context, the commonly used Prandtl-Tomlinson model has proven useful for describing shear response in surface forces apparatus (SFA) configurations. This model describes the lateral motion of a plate and has the following form:

$$\partial^2 x/\partial t^2 + \gamma \partial x/\partial t + \partial U(x)/\partial x + K(x - Vt) = 0. \tag{2.27}$$

Here a movable plate of mass M and the center-of-mass coordinate X is pulled by a spring with the spring constant *K*. In the standard Tomlinson equation that is normally applied to study atomically flat surfaces, *U(X)* is the effective periodic potential $U(X) = U_0\cos(2\pi X/b)$ experienced by the plate due to the presence of an embedded system. In some sense, such a simple potential is also a particular realization of the surface pattern appearing on the microscopic level, where its creation is regulated by the process of crystallization, which normally leads to an ideally periodic crystal lattice.

For the general theoretical study below, we use dimensionless units normalized to the characteristic microscopic scales and energies of the particular system under consideration. The parameter γ, as usual, is responsible for dissipation. The spring is connected to a cantilever which moves with a velocity *V*.

The Tomlinson equation is well supported on a microscopical scale, but usually its application is not extended to the mesoscopic scale. However, the surface patterns which we encounter in normal life are dictated by frozen kinetics, which leads to a relief that is quite far from an ideal periodic structure. As a result, the parameters that appear in Eq. (2.27) can be related to the microscopic characteristics of the system, but not correlated with macroscopic friction. This problem is not limited to friction studies, but also appears when dealing with adhesion. In the following chapters, we will see why this is important for the study of attachment devices of animals, which have to adapt to real surfaces on an intermediate mesoscopic scale, not just at the microscopic level, where patterns can be treated as either flat or periodic.

The key problem here is that the mesoscopic structure of frictional surfaces is of a fractal character and cannot be characterized by a certain wave vector (or even a few vectors), as is normally done in applications of the Tomlinson model. On the other hand, it would be inconvenient to reproduce each time one of the physical or chemical processes described in this chapter. Yet it is important to know how such surfaces appear, or rather it becomes important if the studied surface really has a specific substructure with a periodicity important for the modeling process.

If we can forget about the underlying process, however, and accept the naive hypothesis that the majority of surfaces are simply fractal, it is natural to extend the model to patterns generated according to a very formal mathematical definition of the fractal.

So, let us consider a fractal potential of the form:

$$U(x) = U_0 \int_{q_1}^{q_2} dqc(q) \cos(qx + \zeta), \quad c(q) = q^{-\beta}. \tag{2.28}$$

Here, q_1 and q_2 are characteristic cut-off wave vectors and $\zeta\ (x)$ is the random phase that we assume to be δ-correlated: $<\zeta\ (q)\zeta(q')> = 2\pi\delta(q-q')$. For the majority of physically interesting systems, $\beta \approx 0.9$. Below we will keep this value for definiteness.

For further study, it is convenient to proceed to a discrete representation of the integral in Eq. (2.2), $\int dqc(q) \to \sum$, with a discrete step between the wave vectors Δq, which is determined by the smallest vector q_1 corresponding to an inverse maximal length l_{max} of the system, which normally equals L, $l_{max} = L$. The total number N_{tot} of the terms in the sum is given by $N_{tot} = q_2/q_1 \equiv q_2/\Delta q$. The discrete approach is best suited for our further numerical studies since it allows us to adjust the potential to different scales by including the number of the modes into the summation: $N_{modes} < N_{tot}$.

The Tomlinson model which is based on an analytical definition of the potential is much simpler for computer implementation. The force is calculated by using an analytical formula at each time step, and the procedure can be formally extended to infinite time/space runs. The modified fractal model operates with the data array $U_{fractal}(x) = U\{x_j\}$, where $j = 1, 2, \ldots N_{modes}$. This means that its numerical generation has to be extended to an infinite run, too. For this reason, instead of Eq. (2.2), we use the following differential definition of the force caused by the fractal potential: $\partial U_{fractal}(x)/\partial x = U_0 \Delta x \sum_j q_j c(q_j) \sin(q_j x + \xi)$. This allows us to extend $U_{fractal}(x)$ infinitely each time the x-coordinate runs out of the array bonds. For the numerical procedure, this means that the modified Tomlinson equation,

$$\partial^2 x/\partial t^2 + \gamma \partial x/\partial t + \partial U_{fractal}(x)/\partial x + K(x - Vt) = 0, \tag{2.29}$$

is actually extended by an additional differential equation defining $\partial U_{fractal}(x)/\partial x$, which is solved in parallel.

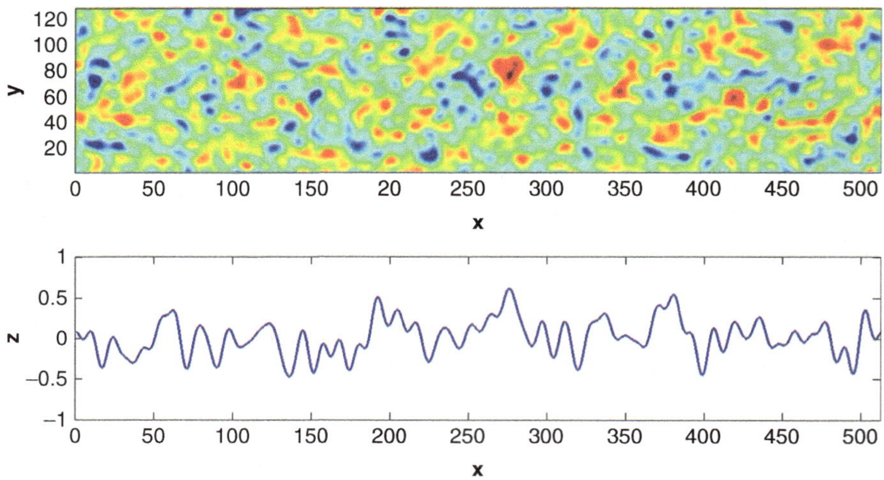

Fig. 2.8 Fractal 3D surface generated by the formal application of Fourier transformation as shown by the color map. The red color corresponds to the higher vertical coordinate. The second graphic presents a 2D cross-section of the same surface along one horizontal line (the central line in this particular case). Both representations of the surface can be used in numerical experiments, depending on the particular problem

To study the scale dependence of the friction force, one can generate a set of fractal potentials for different numbers of modes ($N_{modes} < N_{tot}$). This number defines a cut-off wave vector $q_{cutoff} = \Delta q N_{modes}$ and the corresponding cut-off wave length $\lambda_{cutoff} = 2\pi / q_{cutoff}$ of the potential. Potentials with different N_{modes} can be treated just like the same potential "measured" with different spatial resolution. All space scales larger than the cut-off wave length are included in the potential $U_{fractal}(x)$ and should be treated explicitly in the frame of the dynamic model, like Eq. (2.29).

An analogous procedure based on Fourier transformation can be extended to generate a realistic 3D surface covering the 2D plane. A typical fractal surface generated by the formal application of the Fourier transformation is shown by the color map in Fig. 2.8. The red and blue colors color correspond to the higher and lower vertical coordinates respectively. The second graphic in Fig. 2.8 presents a 2D cross-section of the same surface along one of the horizontal lines. In some numerical experiments each one of these cross-sections can be used as a particular realization to accumulate statistics.

The generation of structured surfaces is not limited to friction or adhesion problems. In principle, surface modification and functionalization through nanostructures is a well-known way of producing desirable properties of various materials. However, it is not an innovation since it has been employed by nature for a long time.

2.6 The Combination of Discrete and Continuous Techniques

The main idea behind using a combination of discrete and continuous techniques is to solve an originally continuous problem by combining a discrete approach which is used to achieve some final, or at least some almost stationary, configuration with further continuous density calculations.

A combination of discrete and continuous approaches may be applied to many different chemical and physical systems. For example, it can be applied to study the magnetic ordering in quasi 2D systems or the growth of surface structures. In all these cases, the real process involves an interaction of many spatially distributed densities, and its direct simulation would require extremely time-consuming calculations.

A famous example of such an approach is the study of the topological phase transition in quasi 2D superconducting systems, where the extremely complex evolution of the superconducting vortices was replaced to some extent by the motion of "charged particles" (Kosterlitz and Thouless 1973). This approach was found to be extremely fruitful and was even awarded the Nobel Prize in 2016.

An important biological example of this approach is the study of the surface nanostructure of snake skin which is supposed to reduce both abrasion and friction, as has been previously shown for similar artificial nanostructures (Kovalev et al. 2016). On the ventral surface of snakes, this nanostructure typically consists of arrays of nano-dimples. One of the most widespread structure types in biological systems is the hexagonal arrangement, which provides the highest density. Visually, the arrays of nano-dimples in snake skin resemble such a structure (Fig. 2.9).

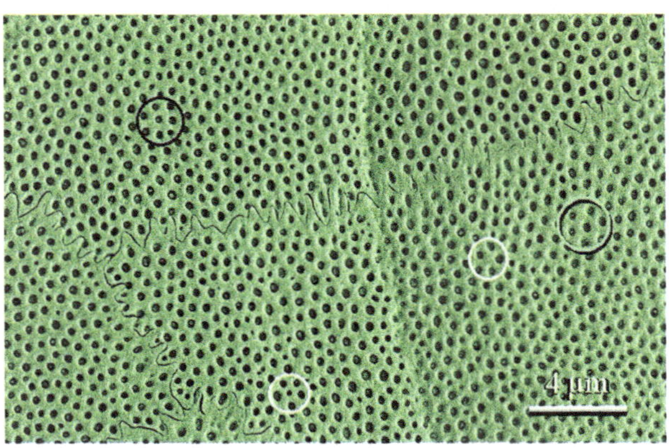

Fig. 2.9 SEM image of a ventral scale in the tail of the snake *Morelia viridis*. The green circle marks a typical hexagonal arrangement of dimples around a central one, whereas the white and black circles mark five- and sevenfold symmetrical arrangements of dimples, respectively. From Kovalev et al. (2016)

However, it is well known that biological systems do not exhibit a perfectly ideal arrangement. In Chap. 8 we will quantitatively analyze the imperfections (disorder) of various hexagonal nanostructure arrangements, e.g., in snake skin, using correlation analysis. In the following, we will suggest a simple mathematical model that helps to explain the differences in these correlation functions and to obtain some additional information with respect to the mechanisms of pattern formation in biological systems. In this model, a structural arrangement similar to that found in snake skin appears due to the "freezing" of an overdamped relaxation. It is caused by weak repulsion between structures that initially follow a random distribution. Subsequently, correlation analysis is applied to this model and its results are compared with the results of the correlation analysis of the real nanostructure arrangement in snake skin.

The power spectrum and the distribution of distances between the nearest neighbors can be calculated for snake skin in the same manner as for any crystal. This calculation revealed that, although the structure of snake skin seems hexagonally ordered, the power spectrum does not demonstrate pronounced peaks, even in the first correlation ring. This is due to the absence of highly oriented domains with a prominent arrangement. At the same time, there is a characteristic distance between the nearest neighbors of around 0.4 μm. This matches the fact that the first ring in the correlation function, which reflects the maximum probability of the distances, is well pronounced.

It may be assumed that isotropy, i.e. no specific preferential directions in the dimple arrangement in snake skin, is important for maintaining certain tribological properties. The presence of a preferential direction would imply stress enhancement in some direction, and, therefore, facilitation of material failure, as well as stronger abrasion.

The standard approach to weakly disordered hexagonal systems normally specifies the preparation of the predefined hexagonal arrangement, which is then slightly perturbed from the originally perfect positions by random noise. In our present approach, the particular amplitude of Gaussian random shifts may be appropriately adjusted to fit the numerically found distribution of the distances close to those experimentally observed in real snake skin.

Despite this, the power spectrum of the snake nano-dimples obviously has a different symmetry compared to the simulated one. This result means that the real spatial distribution cannot be simulated by such a naive, slightly randomized hexagonal configuration. Rather, the real arrangement appears to be the result of some self-organization process caused by an interaction between the nanostructures.

In line with the approach presented here, one can generate, as an initial condition, an array of randomly distributed dimples and cause short-range repulsion between them by applying some potential. In this case, the system of dimples will evolve according to the standard equation of motion for their centers: $\partial \vec{r}_i / \partial t = -\gamma \sum_j \partial U / \partial \vec{r}_{ij}$.

But here we are actually not interested in an exact pattern, but rather in the shape of the distribution of distances and in the symmetry of the correlation function. These can be easily calculated at every single step of the simulation.

The total number of dimples in the numerically generated array is completely predefined by the number of dimples in a given area of the real system. As a result, in the course of the kinetic process, the mean distance between the dimples will remain fixed by the restricted size of the observed area, and the only parameters that will change during the relaxation are the symmetry and the shape of the distribution of distances. From the mathematical point of view, this means that only one fitting parameter, i.e. the width of the final distribution of distances, remains in the simulation. The relaxation process proceeds up to the moment when the width of the distribution of distances coincides with the one found experimentally. This automatically defines the moment when the simulation has to be terminated by immediately freezing the final configuration that is statistically close to the one observed in the real system.

Another well-known example of biologically produced surface structures are super-hydrophobic surfaces that gain this functional feature due to nanostructures consisting of crystals or wax particles. In materials science, surface coatings with regularly arranged globular particles have been produced by colloidal lithography, a technique utilizing the self-assembly of nano-particles on a substrate surface. However, the range of producible patterns and properties of these coatings as well as their durability are still quite limited. Therefore, studying thes close problems in biology by numerical modeling seems to be promising for generating new solutions.

A very interesting and non-trivial problem was raised by the study of water repellence and the ultrastructure of covering granules. For a deeper study of the related questions, see Chap. 8. Here, we will only deal with the question of complex patterns which can self-organize on the spherical surfaces of colloidal particles. Direct observation shows that the distribution of asperities on the surface of spheres is species-specific and probably depends on the interactions of the substances forming these structures.

In order to gain a better understanding of the process of self-assembly and self-arrangement of nano-particles on spherical microstructures, we will now apply a theoretical approach. In the spirit of the previously described kinetic approach, we will introduce a numerical model that allows us to test the effect of different interactions between the particles on the morphology of the final structure. Our results show a good correspondence with the numerically found structures.

As in the previous example of snake skin, the model of the systems under consideration here is organized as a combination of both discrete and continuous approaches. From a general (physical, chemical and biological) point of view, one can expect hat the observed structures can be expected to appear in some kinetic process during which an initially more or less uniformly distributed substance continuously redistributes and solidifies in 3D space. As above, our goal is to minimize the calculation time and to simultaneously gain a good understanding of the process in the frame of a relatively simple model.

In the particular case of structuralized colloidal particles, the problem is complicated by the specific topology of the surface, because the redistribution of the densities and their further solidification take place on spheres (Filippov et al. 2017). This means we are dealing with a kind of phase transition (or phase separation) inside a spherical layer. As far as we know, such a problem with geometrically frustrated assemblies has not been widely studied in conventional physics and chemistry. However, in biological systems such geometries are widespread and therefore must be properly studied.

The topological complexity of the problem forces us to formulate the numerical model as simply as possible, i.e. to replace an originally continuous problem with a combination of a discrete model (to find an almost stationary configuration) and further continuous density calculations.

Such an approach is also applicable to Tamme's problem where the study of kinetic ordering on a sphere can be replaced by finding an equilibrium ordering of "*N* equal charges" on the sphere. In all these cases, it is important that the approach used will lead to a discrete distribution of the maxima or minima of the densities, which are then "dressed", i.e. combined with the continuous field with an expected density distribution.

However, it is important to note that in its classical mathematical form Tamme's problem is treated as an optimization problem which can be formulated as follows: for a given natural number n, place n points on the surface of a sphere in a way that maximizes the shortest distance between any two points. This question normally dictates maximal simplification of the problem (e.g., by the limited number of points which allows for an analytical solution, or by restricting their interaction to simple Coulomb interaction).

Our goal here, however, is almost the opposite. We are trying to find out which particular structures will appear via different interactions and how to reproduce the observed structures. These structures are certainly not optimal, and in many cases the distribution of the final densities, being generated by frozen kinetics, will be extremely inhomogeneous. It is similar to studying the structural transitions between different phases in solid state physics. However, it differs from the standard physical studies due to the atypical confinement of the geometrically frustrated assemblies of particles on the spherical surface.

In light of these considerations, we will organize the procedure as follows. First, we randomly distribute some number of particles (nucleation centers for the densities) on the surface of the sphere. At this stage, the most important parameters are the radius R_S of the sphere, the number of particles N, and the characteristic distance R_0 of the repulsion between them.

The model and the numerical procedure will be described in more detail in Chap. 8. Here we will only mention that the total interaction can be formulated alternatively as a pure short-range repulsion term $U_1\left(\left|\vec{r}_j - \vec{r}_k\right|\right)$, or as a combination of both a repulsion and an attraction term $U_2\left(\left|\vec{r}_j - \vec{r}_k\right|\right)$. The second variant leads to a typical effective potential with minimum

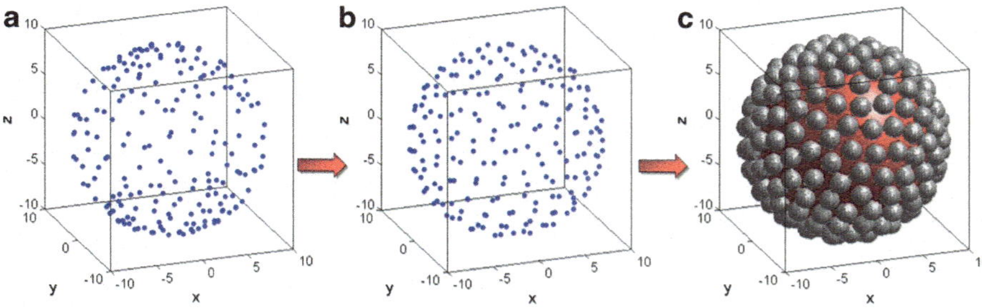

Fig. 2.10 Combination of the discrete and continuous approaches in a single model. A discrete array of interacting points randomly placed on a sphere (**a**) dynamically rearranges into a frozen, almost static configuration (**b**), which is "dressed" by continuous density distributions (**c**)

$U_{Interaction}\left(\left|\vec{r}_j-\vec{r}_k\right|\right)=U_1\left(\left|\vec{r}_j-\vec{r}_k\right|\right)+U_2\left(\left|\vec{r}_j-\vec{r}_k\right|\right)$, which determines an equilibrium distance between the particles which may be reached in long-time asymptotics.

The nontrivial aspect of the situation here is that the movement of the particles is confined to the sphere. From the mathematical point of view, this means that we have to apply a strong potential, $U_{Sphere}\left(\left|\vec{r}_j-\vec{R}_S\right|\right)$, which attracts them to the surface of the given radius R_S. If this potential is strong enough, they practically cannot leave the spherical surface, despite the (relatively weak) repulsion $U_{Sphere}\gg U_1$ between the particles.

The equation of motion for the "particles" thus takes the form: $\partial\vec{r}_j/\partial t=-\gamma\sum_k\partial\left[U_{Interaction}\left|\vec{r}_j-\vec{r}_k\right|+U_{Sphere}\left|\vec{r}_j-\vec{r}_k\right|\right]/\partial\vec{r}_k$.

During this routine, the particles dynamically rearrange with a rate which gradually decreases with time. This is another example of the "large river effect" described in the previous chapter. A particular realization of this procedure is shown in Fig. 2.10.

As usual, this slowing-down process may be controlled by calculating the mean velocity of the particles and stopping the procedure when the velocity becomes negligible. As discussed above, the particles will never reach the real "ground state". This is especially true for a spherical surface. But after sufficient time they reach some stationary distribution which reflects what we expect for the "large river" (or frozen kinetics) in a real system. The final stage of the simulation consists of "dressing" the spherical surface $\rho_j\left(\vec{r}\right)$ and the nucleation centers (particles) with continuous density distributions.

The total density around the nucleation centers may be determined by summing up all the particles $\rho_{Total}\left(\vec{r}\right)=\sum_j\rho_j\left(\vec{r}\right)+\rho_S\left(\vec{r}\right)$. It can be proven that each numerically found pattern is determined by the relationship between the area of the sphere and the size of the particles (or their density on the surface) on the one side, and the interaction between the particles on the other side.

According to the general discussion in the previous chapter, we cannot directly visualize the obtained continuous density distributions as a projection on the 2D surface of the plots. But, in standard manner, we can plot the surfaces corresponding to any constant density and compare them with the data of real structures. Some of the theoretical and naturally observed configurations will be shown in Chap. 8.

References

Filippov AE (1994) Large scale structure of fluctuating order parameter field. J Stat Phys 75:241–252

Filippov AE (1998) Two-component model for the growth of porous surface layers. J Exp Theor Phys 87:814–822

Filippov AE, Gorb SN (2019) Methods of the pattern formation in numerical modeling of biological problems. Facta Univ Ser Mech Eng 17(2):217–242

Filippov AE, Popov VL (2007) Fractal Tomlinson model for mesoscopic friction: from microscopic velocity-dependent damping to macroscopic Coulomb friction. Phys Rev E 75:27103

Filippov AE, Wolff JO, Seiter M, Gorb SN (2017) Numerical simulation of colloidal self-assembly of super-hydrophobic arachnid cerotegument structures. J Theor Biol 430:1–8

Kosterlitz JM, Thouless DJ (1973) Ordering, metastability and phase transitions in two-dimensional systems. J Phys C Solid State Phys 6:1181–1203

Kovalev A, Filippov AE, Gorb SN (2016) Correlation analysis of symmetry breaking in the surface nanostructure ordering: case study of the ventral scale of the snake *Morelia viridis*. Appl Phys A Mater Sci Process 122(253):3–6

Kuzovlev YE, Soboleva TK, Filippov AE (1993a) Structure and evolution of new phase nuclea at first order phase transitions. J Exp Theor Phys 76:858–867

Kuzovlev YE, Soboleva TK, Filippov AE (1993b) Formation of the filli structure at nucleation processes. J Exp Theor Phys Lett 58:357–362

Landau LD (1937) Theory of phase transitions. Zh Eksp Teor Fiz 7:19–32

Landau LD, Khalatnikov IM (1965) On the anomalous absorption of sound near a second order phase transition point. In: ter Haar D (ed) Collected papers of L.D. Landau. Elsevier Inc, pp 626–629

Patashinskii AZ, Pokrovskii VL (1979) Fluctuarion theory of phase transitions. Pergamon Press, Oxford

Popov VL, Heß M (2015) Method of dimensionality reduction in contact mechanics and friction. Springer, New York

Chapter 3
Clusterization of Biological Structures with High Aspect Ratio

Abstract In this chapter, we will concentrate ourselves on the effect of so-called clusterization, which appears in biological and biologically-inspired fibrillar adhesion systems. If the arrays of fiber-like structures are flexible enough to allow efficient contact with natural rough surfaces, after few attachment-detachment cycles, the fibers tend to adhere one to another and form clusters which are much larger than original fibers. Due to this effect, they are less flexible and form much worse contacts with rough surfaces. Prevention of the clusterization is very important for the effectiveness of both natural and artificial adhesives. It is known that the arrays of setae of real gecko or beetle are much less susceptible to this problem than artificial ones. Here we provide numerical modeling of two different solutions of this problem that were previously experimentally studied in biological systems: gradients of the fiber stiffness and spatial structure of the fibers distributed in 3D space.

In this chapter, we will concentrate on the so-called clusterization effect, which appears in biological and biologically inspired fibrillar adhesion systems. This effect (or rather its absence) is very important for the effectiveness of both natural and artificial adhesives. If artificially produced arrays of fiber-like structures are flexible enough to allow efficient contact with natural rough surfaces, the fibers often tend to adhere to each other and to form clusters after a few attachment–detachment cycles. Normally, such clusters are much larger than the original fibers, and, since they are less flexible than single fibers, they will form less efficient adhesive contacts, especially on rough surfaces. The main problem here is that the forces responsible for clusterization are the same intermolecular or capillary forces which attract the fibers to fractal surfaces of natural substrates. However, the setae of geckos or beetles are much less susceptible to this problem because in biological systems it is solved in two different ways:

Electronic supplementary material The online version of this chapter (https://doi.org/10.1007/978-3-030-41528-0_3) contains supplementary material, which is available to authorized users.

A. E. Filippov, S. N. Gorb, *Combined Discrete and Continual Approaches in Biological Modeling*, Biologically-Inspired Systems 16,
https://doi.org/10.1007/978-3-030-41528-0_3

1. The adhesive tarsal setae of beetles feature pronounced gradients in material composition and properties along their length (Peisker et al. 2013). It has been previously hypothesized that these gradients represent an evolutionary optimization which increases the performance of the adhesive system by enabling effective adaptation to rough surfaces while simultaneously preventing a lateral collapse of the setae. Modeling experiments have clearly demonstrated that fibers with short soft tips and stiff bases are better at maximizing adhesion and minimizing clusterization in multiple attachment–detachment cycles than fibers with longer soft tips on stiff bases and fibers with stiff tips on soft bases (Gorb and Filippov 2014). This not only shows the crucial role of material gradients along the setae in beetle fibrillar adhesive system, but leads to the prediction that similar gradients will have convergently evolved in various lineages of other arthropods.
2. One of the possible reasons for the effectiveness of the gecko foot-hair is that the ends of the setae have a more sophisticated non-uniformly distributed 3D structure than existing artificial biomimetic systems.

On the other hand, biomimetic systems based on the use of vertically aligned carbon nanotubes (VACNTs) show a clusterization effect which stabilizes the system by decreasing the friction coefficient (Schaber et al. 2015b). Previous research demonstrated that VACNTs exhibit strong frictional properties (Schaber et al. 2015a). Experiments indicated a strong decrease of the friction coefficient from the first to the second sliding cycle in repetitive measurements on the same VACNT spot, but stable values in consecutive cycles. VACNTs form clusters under shear applied during friction tests, and self-organization stabilizes the mechanical properties of the arrays.

In the following, we will examine and describe these three mechanisms in detail, using a combination of experimental research and numerical modeling. An important problem when combining experimental and theoretical research is that in some cases we can easily reproduce a system in the form of a theoretical model, but not as an experimental system and in other cases, we can set up some experimental system but will need some nontrivial numerical tricks to reproduce it as a theoretical model. That is why, for the sake of clarification, we will sometimes jump from some real system to a toy experimental or theoretical model in order to explain our approach.

3.1 Adhesion without Clusterization Due to a Material Gradient

3.1.1 Fibrillar Adhesive Systems of Insect Feet

The contact formation of insect adhesive pads on various substrates depends on the pad's ability to adapt to different surface topographies. The quality of contact may be increased by various micro- and nanostructural features of the biological structures (Gorb 2001; Gorb and Beutel 2001; Gorb et al. 2002; Creton and Gorb 2007; Voigt et al. 2008). For example, in hairy adhesive pads of insects, their hierarchical organization enables the formation of multiple contacts that increase the overall length of the total peeling line (Varenberg et al. 2010). Crack trapping mechanisms of adhesive systems with multiple contacts provide advantages on rough substrates (Hui et al. 2004). Thin tape-like contact tips of hairs (setae) in combination with applied shear forces lead to the formation of a maximal real contact area without slip within the contact (Filippov et al. 2011).

Another important feature for contact formation is material flexibility, because flexible material can increase the real contact area between the pad and the substrate at relatively low normal load. However, increases in material flexibility may lead to a decrease of the mechanical stability of structures made of such materials (Borodich et al. 2010). That is why adhesive setae consisting of very soft material would buckle and cluster together (Jagota and Bennison 2002; Spolenak et al. 2005a, b). Such a clusterization/condensation of setae would undo the functional advantages of multiple contacts. Thus, the material properties of such setae represent an optimization problem which has been solved in the course of biological evolution by the emergence of gradients of thickness and mechanical properties. Thickness gradients have been discovered in various adhesive insect setae by numerous scanning electron microscopy studies (Gorb 2001). Gradients of material structure and local mechanical properties in tarsal setae of the beetle *Coccinella septempunctata* and the presence of a material gradient at the level of each single seta were shown by Peisker et al. (2013). The Young's modulus of the beetle seta ranges from 1.2 MPa at the tip (Peisker et al. 2013), where high proportions of the elastic protein resilin are incorporated (Weis-Fogh 1960, 1961), to 6.8 GPa at the setal base, where sclerotized cuticle material is dominating. This gradient likely represents an evolutionary optimization which increases the performance of the adhesive system by enabling effective adaptation to rough surfaces while simultaneously preventing lateral collapse of the setae. However, this hypothesis is difficult to prove experimentally, using native biological specimens. That is why we decided to test it by a numerical simulation, asking the following questions: (1) Does the presence of a material gradient along the setae contribute to proper contact formation if compared to setae with different bulk material properties without such a gradient? (2) Does the gradient reduce the clusterization of setae?

In the following, we will first introduce the known properties of insect setae and then describe our modeling procedure.

3.1.2 Structure and Material Properties of Insect Setae

Confocal laser scanning microscopy (CLSM) analyses revealed the presence of resilin in adhesive tarsal insect setae, as indicated by blue autofluorescence (Michels and Gorb 2012; Peisker et al. 2013). At the tips, the material composition of the setae is strongly dominated by resilin, whereas the central and proximal parts mainly consist of other materials (other proteins and very likely chitin), indicated by the dominance of green and red autofluorescence in these structures (Fig. 3.1).

The transition from the soft resilin-dominated distal parts to setal sections which mainly consist of stiff materials is characterized by a rather pronounced longitudinal gradient orientated in the dorso-ventral direction. The Young's modulus of the setal material in the distalmost section of the tip is low (1.2 ± 0.3 MPa), whereas at the

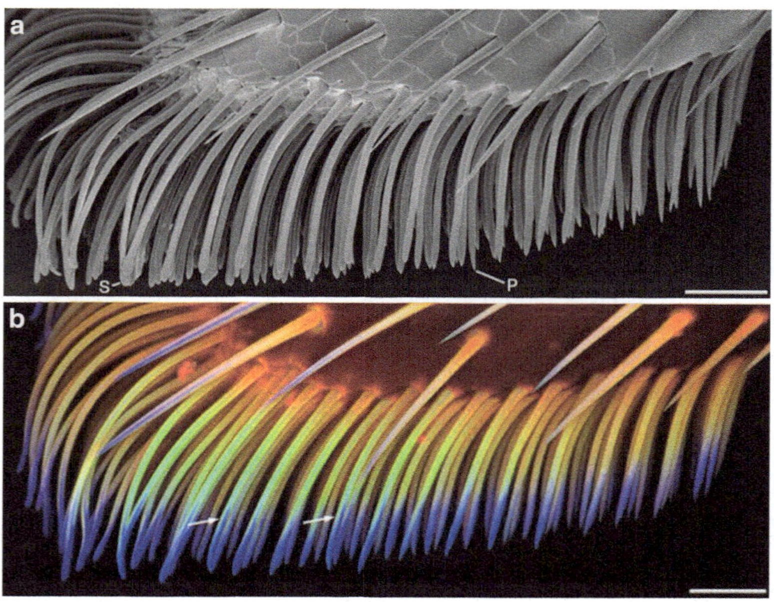

Fig. 3.1 Morphology and material composition of adhesive tarsal setae in the foreleg of a female *Coccinella septempunctata*, lateral view. (**a**) Scanning electron micrograph (SEM). (**b**) Confocal laser scanning microscopy (CLSM) maximum intensity projection showing an overlay of four different autofluorescences: red/yellow/green = stiffer cuticle, blue = resilin. The arrows indicate the dorso-ventral material gradient in exemplary setae. S, exemplary spatula-like seta; P, exemplary seta with a pointed tip. Scale bars = 25 mm. From Peisker et al. (2013) (by permission of the Nature Publishing Group)

base of the seta it is considerably higher (2.43 $\pm$ 1.9 GPa) (Peisker et al. 2013). This information was used in the numerical model presented below.

3.1.3 Mathematical Model of Insect Setae with Gradients of Mechanical Properties

To model the material properties of the setae, we applied one of the discrete minimalistic models described in Chap. 1. It includes the following elements:

- An array of initially parallel fibers attached to a hard planar base.
- Elasticity of the fibers $F_{elastic}$ continuously varies along their length and can change from very soft to much stiffer or almost rigid (but still with some degree of flexibility). Longitudinal $\vec{F}_{jk}^{\parallel}$ and transversal $\vec{F}_{j}^{\perp}$ stiffness of the fibers are simulated by the following interaction between the segments: $\vec{F}_{jk}^{\parallel} = K^{\parallel}\left(\vec{r}_j - \vec{r}_k\right) \times \left[1 - \left(\vec{r}_j - \vec{r}_k\right)^2/dr^2\right]$, and $\vec{F}_{j}^{\perp} = K^{\perp}\left(2\vec{r}_j - \vec{r}_{j+1} - \vec{r}_{j-1}\right)$.

In this particular case, it is enough to restrict ourselves to a two-dimensional model, where the values $\vec{r}_j = \{x_j, y_j\}$ are the coordinates at the beginning of the segment j; $k = j \pm 1$. The longitudinal force, $\vec{F}_{jk}^{\parallel}$, is described by a two-minima potential which tends to keep a distance between the points $\vec{r}_j$ and $\vec{r}_{j\pm1}$, close to the equilibrium length of the segment dr. The transversal force $\vec{F}_{j}^{\perp}$ keeps $\vec{r}_j$ close to the mean value between its nearest neighbors, $\left(\vec{r}_{j+1} + \vec{r}_{j-1}\right)/2$, and tends to keep the angle between the neighboring segments close to 180°.

The adhesive surfaces at the fiber ends attract the ends of neighboring fibers by molecular and capillary forces. For definiteness and for the sake of simplicity, we reduce this particular simulation by the gradient of Morse potential $U_{VdW}(r) = U_o(1 - \exp(-r/r_0))^2$ with physically reasonable amplitude $U_0 = 10\ nN \cdot nm$ and the minimum located at the distance $r_0 = 0.01\ \mu m$ from the surface.

The rough surface of the substrate, to which the fibers attach, can be modeled by different approaches as discussed in Chap. 2. Its particular structure is not important for this particular study where we are mainly interested in the interaction between the fibers responsible for clusterization. So we can simply assume that the substrate surface has a fractal structure with a given Fourier spectrum and amplitude of roughness. Thus, it can simply be generated numerically and defined by the real part of $Y(x) = A \iint dq_x C(q) \exp(iq_x x + \zeta)$ with scaling spectral density, where A is the amplitude of the surface roughness, i is an imaginary unit, q_x are Fourier components along the x direction and ζ is a random phase.

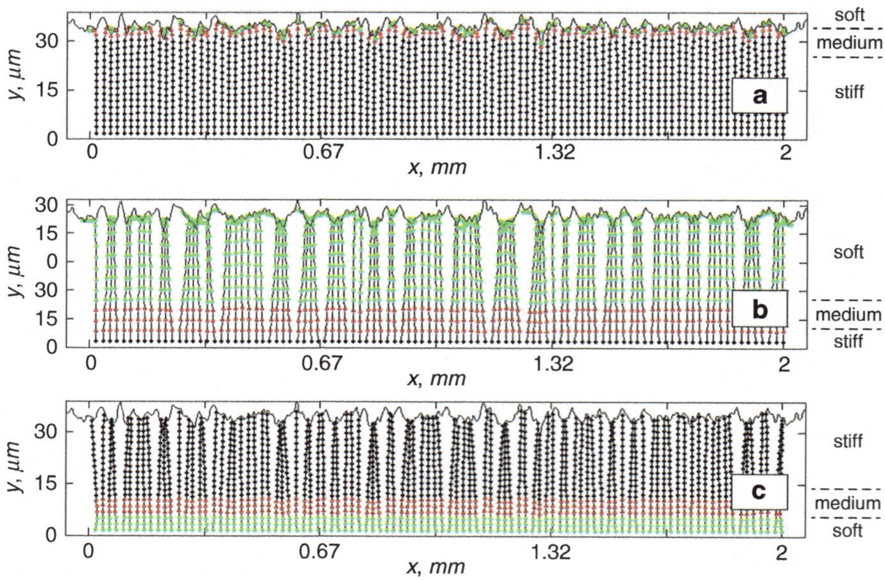

Fig. 3.2 Three typical configurations of the fibrillar structure (setal array) attached to a stiff support (below the plotted setal length) and in adhesive contact with a random fractal surface (continuous line above the setae). This numerical model was used to mimic biological setal arrays as shown in Fig. 3.1. The plots show three different types of fibers: (**a**) stiff fibers with short elastic ends, (**b**) long elastic fibers connected to the base by short stiff roots, and (**c**) stiff fibers with soft elastic segments near the base. Stiff, medium and soft segments are represented by black, red and green circles, respectively. (From Gorb and Filippov 2014)

Details of how to proceed when modeling the profile $Y(x)$ have been described in Chap. 2 as well as in previous papers (Filippov and Popov 2007a, b; Popov et al. 2007). In the current literature (Persson and Gorb 2003) it is generally accepted that the majority of physical surfaces have a scale-invariant spectrum $C(q) = 1/q^{\beta}$ with the exponent $\beta \approx 0.9$. The amplitude of the numerical "surface" is taken to be comparable with the radius of van der Waals interactions, $A = r_0$.

The soft parts of every fiber, which normally are physically thin and flexible, interact with the corresponding regions of other fibers of the array. This interaction force has the same origin as the attraction to the hard surface $U_{interact}(r) = U_o(1 - \exp(-r_{jk}/r_0))^2$ and comparable characteristic parameters U_0, r_0. For the simplicity of the model, we reduce the mutual interaction of the fibers to the interaction between the nearest neighbors: $r_{jk} = |r_j - r_{j\pm1}|$. For this particular problem, one can neglect the effects of inertia and treat the system as an over-damped one. In this approximation, the equation of motion does not contain second-time derivatives and can be formally written in the form $\partial r/\partial t = F$, where the force accumulates all above interactions $F = F_{elastic} + F_{VdW} + F_{interact}$.

The conceptual structure of the model is illustrated in Fig. 3.2. The rough surface of the substrate is shown by the continuous upper curves in panels a–c.

In order to understand the potential functional role of the material gradients found in beetles (Peisker et al. 2013), we examine three different kinds of fiber arrays: (a) long and stiff fibers with short elastic ends; (b) long elastic fibers connected to the basal plate by short and hard roots; (c) relatively stiff fibers with short and soft elastic filaments connected to the base. These variants are shown in panels (a), (b), and (c) of Fig. 3.2, respectively.

To illustrate the varying stiffness of the fiber segments in Fig. 3.2, we formally divided the stiffness into three categories: (1) close to the maximal stiffness (black circles), (2) less than half of the maximal stiffness (a region around y_0 with width Δ) (red circles), and (3) less than 0.1 of the maximal stiffness (green circles).

In all these cases the stiffness of the fibers continuously varies along the vertical coordinate. To simulate it, we apply the smooth step function $\Theta(y) = 1/[1 + \exp(-(y - y_0)/\Delta)]$ with regulated positions of bend y_0 and width Δ. This function tends to 1, when $y \ll y_0$, and gradually goes to zero at the opposite limit. This allows us to model all the above-mentioned cases in a common approach.

Our numerical procedure is organized as follows. Originally unperturbed arrays of parallel fibers attached to a hard horizontal base are brought into contact with a numerically generated fractal surface. The fibers distort due to their interaction with the surface as well as with their neighbors. Many fibers will be attracted to the same individual asperities of the surface. This attraction increases their mutual interaction, in contrast to the original unperturbed state.

The time-dependent distortions of the fibers, as well as the variation of the interaction forces, can be recorded to control the process of contact formation and to stop it when the system reaches a certain stationary configuration. Afterwards, we can remove the substrate surface and allow the system to relax spontaneously to some new stationary state.

Many of the fibers which had been attracted to the same asperities of the surface will still interact strongly with each other and remain close together, assembling into local *clusters*. Mutual attraction *between* the fibers competes with the elastic forces *inside* the fibers which try to return them to the straight position, and the whole array to its original structure with parallel organization. Any further scenario of the setal arrangement certainly depends on the relationship between these forces and their spatial distribution. In some cases, the structure can return back to the original state, but in some cases it cannot. If so, the fibers remain assembled into strongly confined bunches (the so-called *clusterization/condensation* phenomenon). This phenomenon is very important from a practical point of view, because a clustered system will not attach efficiently to a new surface during the next contact event.

Qualitative results of this clustering process in the systems shown in Fig. 3.2 are summarized in Fig. 3.3, where they are presented after their detachment from the surface and relaxation to the static state. The kinetic behavior of the model array of setae/fibers during an attachment–detachment cycle for all three systems is presented in supplementary Movies 3.1, 3.2 and 3.3.

Figures 3.2 and 3.3 and the movies show clearly that, in contrast to the strongly clustered systems with long flexible (b) or long hard filaments (c), the system in

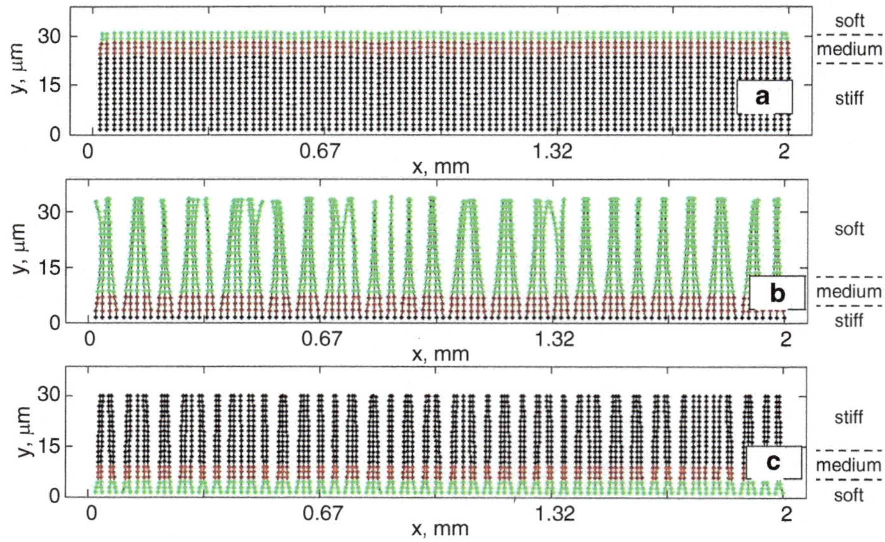

Fig. 3.3 The same system as presented in Fig. 3.2 after detachment from the fractal surface and relaxation to the static state. Plots (**a**) to (**c**) clearly show the difference between systems with either long elastic (**b**) or long stiff filaments (**c**) and the system with stiff filaments and short soft ends (**a**) which practically returns back to its original configuration. Stiff, medium and soft segments are represented by black, red and green circles, respectively. (From Gorb and Filippov 2014)

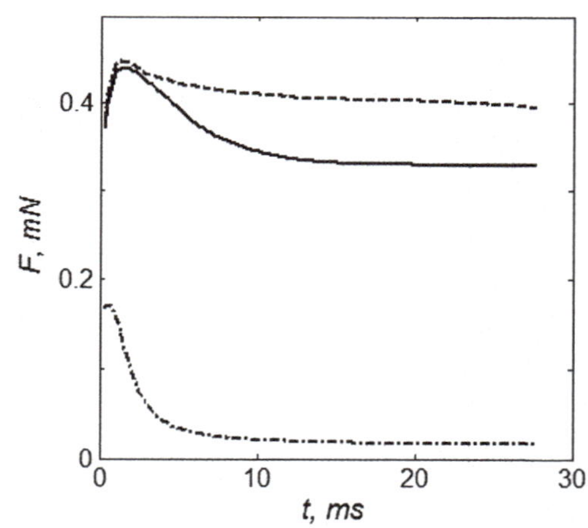

Fig. 3.4 Time-dependent vertical forces during the attachment of initially unperturbed systems to a fractal surface. The solid, dashed and dash-dotted lines correspond to cases (**a**), (**b**) and (**c**) of the previous figures, respectively. (From Gorb and Filippov 2014)

(a) with long hard filaments with short flexible ends practically returns back to its original configuration. This observation leads to a very important question: for a complete return to the original state after relaxation, it seems important to have short flexible ends of the fibers (in comparison to their complete length), but will the deformation of these ends in the attached state be enough to produce sufficiently strong attachment forces?

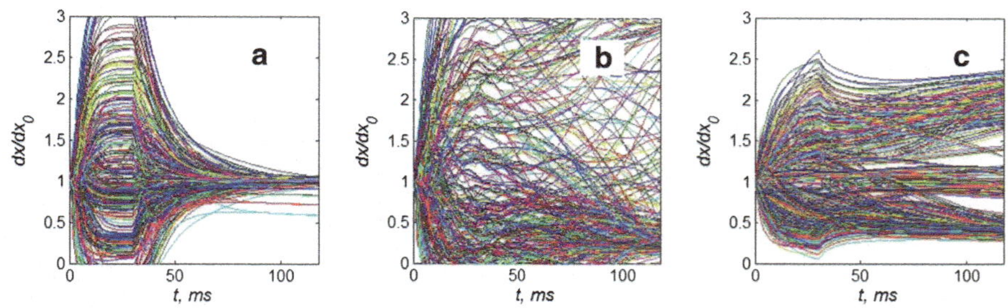

Fig. 3.5 Temporal development of arrays $\{dx_j\}$ of distances $j = 1, 2, \ldots N_x$ between the ends of neighboring fibers, $dx_j = x_{j+1} - x_j$, during a single attachment–detachment cycle for the same cases (**a–c**) as above. All the distances are normalized to the distances of the original unperturbed system: $dx_j = dx_0$ at $t = 0$. Each line in the plots corresponds to a time-dependent distance between one pair of neighbors, $dx_j = x_{j+1} - x_j$. In the attached state, all the filaments tend to a configuration which represents a certain compromise between fiber stiffness, adhesion to the surface and mutual interaction between the fibers. After detachment, the system relaxes into an asymptotic configuration corresponding to a compromise between the stiffness and mutual interaction of the fibers only. (From Gorb and Filippov 2014)

To compare the vertical attraction forces in all three cases (a–c), we modeled them numerically over the complete system during the entire time interval of attachment (Fig. 3.4). Maximal forces in the cases (a) (solid line in Fig. 3.4) and (b) (dashed line) are comparable. The potential barrier (difference between the maximum force at the beginning and its minimum after adaptation to the surface) is even higher in case (a). Qualitatively, this effect appears to be due to the fact that the flexible filaments are too long in case (b). And long hard filaments rotating around flexible roots, as in case (c), cannot perfectly adapt to the surface. As a result, the maximum attachment force here is much lower than in cases (a) and (b).

Now, to obtain time-dependent information about the deformation of the fibers, we calculate the array $\{dx_j\}$, $j = 1, 2, \ldots N_x$ of the distances between the contact ends of the nearest neighbors $dx_j = x_{j+1} - x_j$. Figure 3.5 shows the temporal development of every such array (cases a–c) during a complete attachment–detachment cycle. Each line in the plots corresponds to one particular time-dependent distance between a pair of closest neighbors, $dx_j = x_{j+1} - x_j$. All these distances are normalized to the distances of the original unperturbed system, so $dx_j = 1$ at $t = 0$.

The history of the process is clearly visible in the plots. When some fibers are attracted to the same asperities of the surface and form clusters, the distance between their ends tends to zero: $dx_j = x_{j+1} - x_j \to 0$. At the same time, the distance between fibers belonging to different clusters increases. This distance will correlate with characteristic distances between the asperities, but it will be random for a random fractal surface. Finally, the configuration of the fibers represents a complex compromise between (1) the stiffness of the fibers, (2) the fractal structure of the surface, and (3) the strength of the mutual interactions between the fibers.

When the external surface is removed, the system of fiber arrays relaxes into a new final configuration which is driven by a compromise between the stiffness and

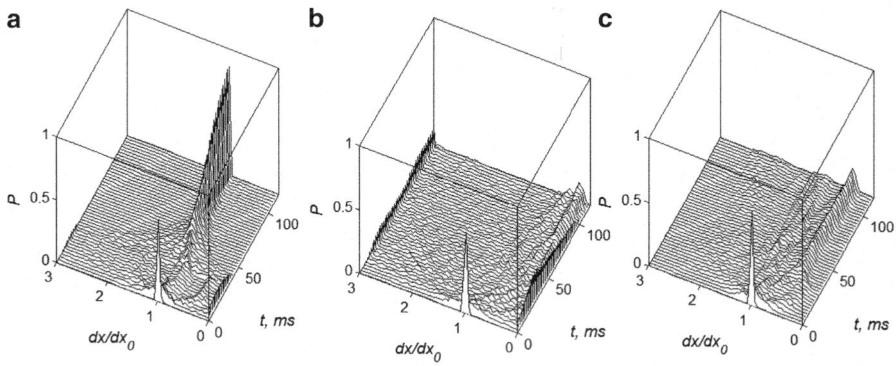

Fig. 3.6 Statistical analysis of the plots presented in Fig. 3.5. The sequences of the histograms show the temporal development of the probability $P = P(dx)$ of finding a particular value of the distance $dx_j = x_{j+1} - x_j$ between two neighboring fibers. The cases (**a**)-(**c**) are the same as above. Starting from an unperturbed configuration (initial peak of probability around $dx_j = dx_0$), all the systems evolve to smooth distributions $P(dx)$. In the clustered attached state (cases **a** and **b**) the probabilities have well pronounced maxima at $dx \approx 0$. After detachment from the external surface, all the systems tend to distributions $P(dx)$ which perfectly agree with the observed final states shown in Figs. 3.3 and 3.5. (From Gorb and Filippov 2014)

the mutual interaction between the fibers only. If stiffness dominates, the system will return to the original unperturbed state. The time-dependent history of this process is clearly presented in Fig. 3.5a. It is interesting to note that in case (c) the stiff fibers, despite their strong elastic energy, cannot completely return back to their initial state; they remain glued together at their top ends.

To analyze these results statistically, one can calculate the probability $P = P(dx)$ of finding a particular value of the distance $dx_j = x_{j+1} - x_j$ between closest neighboring fibers. This was done for a sequence of discrete time steps and is summarized in Fig. 3.6. The cases (a)–(c) in this figure are the same as above.

These statistical data help to clarify the information presented in Fig. 3.5. The initial probability peak around $dx = 1$ corresponds to an almost unperturbed configuration at the time of first contact with the surface. Over time, in all three types of systems the filaments are deformed into configurations with a smooth distribution $P(dx)$. This means that different distances $dx_j = x_{j+1} - x_j$ appear with similar probabilities. In time, in systems with soft ends (a and b) many fibers are attracted to the same asperities of the surface. As a result, the probabilities show well pronounced maxima close to $dx \approx 0$. After detachment, all the systems tend to asymptotic probability distributions, which perfectly agree with the observed final configurations shown in Figs. 3.3 and 3.5.

3.1.4 Functional Significance of Gradients of Material Properties

Generally, gradients of material properties are well known in biological materials (Vincent 2002). Pure bulk materials are almost absent in biology, where proper combinations of different constituents often lead to mechanical gradients with important functional significance. This has been previously shown for insect cuticle (Barbakadze et al. 2006), snake skin (Klein et al. 2010), human teeth (Wang and Weiner 1997; Fong et al. 2000), and other materials.

Gradients have also recently been found in the smooth attachment devices of insects (Perez Goodwyn et al. 2006). Interestingly, the gradients in the smooth pads of locusts and bushcrickets are different from the gradients reported in the pads of ladybird beetles (Peisker et al. 2013). Smooth adhesive pads consist of a softer core covered by a stiffer layer, whereas hairy pads have the opposite arrangement: stiffer bases are combined with softer distal parts. Both types of gradients combine conformability to the surface roughness of the substrate and resistance to the environment.

The opposite directionality of these material gradients can be well explained by the differences in pad architecture. Smooth pads consist of branching rods or cellular foams, which, in combination with fluid-filled spaces between the solid structures, hold the shape of the pad. This principle is combined with the presence of a relatively stiff superficial layer terminating the fibers. This layer keeps the distances between the fiber tips at some constant value (and, in species living in arid environments, protects the pad from desiccation) (Perez Goodwyn et al. 2006; Gorb 2008). In hairy pads, the adhesive setae are not terminated by a continuous layer and can potentially buckle and cluster together (Jagota and Bennison 2002; Spolenak et al. 2005a, b). As strong clusterization leads to a decrease of the functional advantages provided by multiple contacts (Varenberg et al. 2010), it is reduced by the presence of gradients of thickness (Gorb 2001) and mechanical properties (Peisker et al. 2013).

While the disadvantages of purely stiff and purely soft fiber arrays are intuitively clear, it is difficult to assess the advantages of various gradients from the fiber base to the fiber tip (soft to stiff = downstream gradient and stiff to soft = upstream gradient). Our modeling experiment clearly demonstrated that fibers with short, soft tips and stiff fiber lengths (i.e., a *short upstream gradient*) show advantages with regard to maximizing adhesion and minimizing clusterization in multiple attachment–detachment cycles if compared to fibers with longer soft tips on stiff bases (*long upstream gradient*) and fibers with stiff tips on soft bases (*downstream gradient*). As mentioned above, such *short upstream gradients* were recently described in beetles (Peisker et al. 2013); however, we can predict that similar gradients must have convergently evolved in various lineages of arthropods.

3.2 Adhesion without Clusterization Due to a Non-uniformly Distributed 3D Structure

According to the general ideas presented in Chaps. 1 and 2 and, we studied the clusterization problem in Sect. 3.3.1 in a simplified manner, using a 1 + 1-dimensional model. However, one of the possible reasons for the effectiveness of gecko foot-hairs is that the ends of the setae have a more sophisticated, non-uniformly distributed 3D structure which differs from both the over-simplified theoretical models and the existing artificial systems.

Below, we will therefore numerically simulate a more realistic 3D spatial geometry of non-uniformly distributed branches of nanofibers of the setal tip of the gecko foot-hair, study its attachment–detachment dynamics, and discuss its advantages in comparison with a uniformly distributed geometry.

3.2.1 Hierarchical Structure of the Gecko Adhesive Setae

The ventral side of gecko toes bears so-called lamellae with arrays of 3–5 μm thick setae, which are further subdivided at their tips into 100–1000 single nanofibers ending with flattened tips (spatulae) of about 200 nm in width and length (Hiller 1968; Autumn et al. 2000, 2002; Huber et al. 2005; Rizzo et al. 2006) and about 15 nm in thickness (Persson and Gorb 2003) (Fig. 3.7). Such a subdivision of a large adhesive contact into many single separate contacts leads to an enhancement of the adhesive force of this fibrillar system due to a variety of reasons (Arzt et al. 2003; Persson 2003; Varenberg et al. 2010). This effect is also enhanced by the specific spatula-like shape of the single contacts (Gao and Yao 2004; Spolenak et al. 2005a; Filippov et al. 2011).

As above, one of the important problems of the synthesis of artificial fibrillar adhesives inspired by the gecko foot-hair is clusterization (Jagota and Bennison 2002; Northen and Turner 2005; Yurdumakan et al. 2005). Normally, such clusters are much larger than the original fibers and, due to their lower flexibility, form less efficient contacts with external surfaces, especially rough ones (Persson 2003). However, the optimization of such variables as fiber radius, fiber aspect ratio, material properties, and contact shape can lead to an optimal solution, which has been previously described (Spolenak et al. 2005b).

However, arrays of real gecko setae are much less susceptible to this problem, despite their high aspect ratio. This is due to the fact that every single seta has a hierarchical structure in itself (Fig. 3.7b, c), and that each level of this multilevel architecture is not prone to clusterization (Bhushan et al. 2006; Kim and Bhushan 2007). Another possible reason for the low clusterization tendency of real setae is that their ends have a much more sophisticated, non-uniformly distributed 3D structure (Fig. 3.7) than any existing artificial system (Northen and Turner 2005; Yurdumakan et al. 2005) or any system considered in existing models (Kim and Bhushan 2007).

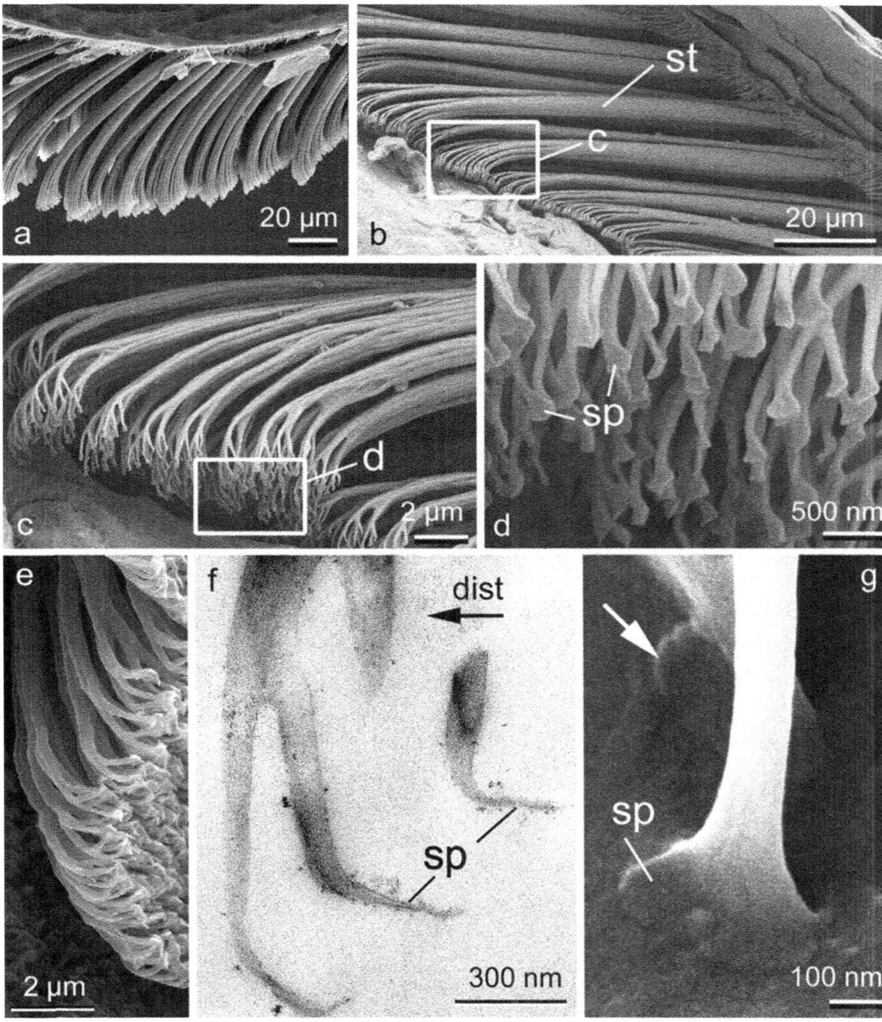

Fig. 3.7 Hierarchical organization of the gecko attachment system: (**a**) longitudinal section of the gecko toe with the lamella (thin horizontal keratinous film) covered with setae in a non-contact state; (**b**) setae (st) in contact with an external substrate; (**c**, **d**) setae branching into single nanofibers terminating in spatulae (sp); (**e**–**g**) spatula orientation in a state without (**e**, **f**) and with contact (**g**) to an external surface. White rectangles indicate areas which are zoomed wider in the following images. Black arrow in (**f**) indicates the distal direction of the toe (valid for all images in this plate). White arrow in (**g**) points to a spatula flipping from the non-contact to the contact orientation. (From Filippov and Gorb 2014)

Here we report the results of a study in which we numerically simulated the 3D spatial geometry of non-uniformly distributed branches of nanofibers of the gecko setal tip and studied its attachment–detachment dynamics. We will then discuss its advantages as compared to a uniformly distributed geometry.

3.2.2 Mathematical Model of Contact Formation by Gecko Setae

According to the nature of this particular problem we here choose a discrete numerical model which describes the dynamics of a set of elastic fibers attached to one, slightly wider, rigid root fiber, which is initially situated at some angle $\varphi_{t=0} = \varphi_0$ and at some distance from the contact surface. The conceptual structure of the model is shown in Fig. 3.8. For definiteness, we assume the number of the elastic fibers, N_x, to be 10 ($N_x = 10$) and each fiber to be composed of fifty elastic segments ($N_y = 50$), each with a length of $dR = 0.04$ μm, so as to achieve a total fiber length that is in accordance with the measurements on SEM images (cf. Figure 3.7).

The fibers are modeled with longitudinal ($K^{\|}$) and transverse ($K^{\perp}$) stiffness, with $K^{\|} = K^{\perp}$. Transverse stiffness tends to keep the angle between neighboring segments close to 180°, and longitudinal stiffness is responsible for its reaction on the extensions along the fibers. Any deformation of fibers gives rise to elastic forces proportional to their stiffness. The longitudinal force, $\vec{F}_{jk}^{\|}$, is here described by a double-well potential, which is constructed so as to keep the distance between the nodes $\vec{R}_j$ and $\vec{R}_{j\pm1}$ close to the equilibrium length of the segment dR : $\vec{F}_{jk}^{\|} =$

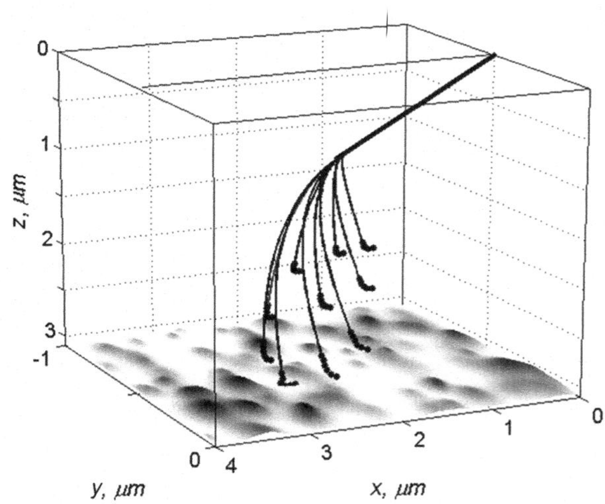

Fig. 3.8 Conceptual 3D structure of the model system. All the segments are numerically defined to mimic the real terminal structure of the gecko foot hair. The bold straight line corresponds to the rigid root segment (seta). The thin lines represent flexible filamentous nanofibers terminating in spatulae. The bold points at the ends of the curved lines represent extremely flexible spatula regions. The fractal roughness of the contact substrate was generated according to the standard procedure described in the text and is represented by the gray-scale pattern in the XY plane. The Z coordinate of the whole plot is flipped upside-down, so the filaments are not obscured by the surface and the spatial 3D structure is clearly visible. (From Filippov and Gorb 2014)

$K^{\|}\left(\vec{R}_j - \vec{R}_k\right)\left[1 - \left(\frac{\vec{R}_j - \vec{R}_k}{dR}\right)^2\right]$, where $\vec{R}_j$ is a position vector of the middle of the segment (the node) j; $k = j \pm 1$. This form of the equation for the longitudinal force was chosen since it is linear in the case of small displacements and increases non-linearly for large displacements. The combination of these two effects causes a minimum of effective potential at the equilibrium length dR. The second force, $\vec{F}_j^{\perp}$, is directly proportional to the lateral deflection and tends to keep $\vec{R}_j$ close to the mean value between its nearest neighbors: $\left(\vec{R}_{j+1} + \vec{R}_{j-1}\right)/2$ $\vec{F}_j^{\perp} = K^{\perp}\left(2\vec{R}_j - \vec{R}_{j+1} - \vec{R}_{j-1}\right)$. It must be noted that the "transverse" force $\vec{F}_j^{\perp}$ was chosen in its present form because it is easy to realize it numerically in order to prevent any bending of the fibers, but it is not restricted purely to bending force, since it may include a longitudinal component as well.

We modeled the initial structure of the fibers in such a way as to mimic as far as possible the real form of the setae as shown in Fig. 3.7. Each of the fibers is elastically attached to the rigid root fiber (bold straight line in Fig. 3.8). The fibers have different lengths and orientations. As a result, their ends vary in three directions: parallel to the line of the root and in two orthogonal directions. To better mimic the natural structure, we numerically generated individual initial positions for every single segment. The resulting configuration is shown in Fig. 3.8. It also reproduces the correct orientation of the terminal parts (spatulae) of the fibers, which are slightly curved back in the direction of the root.

According to our main hypothesis, the different positions of the fiber ends of the realistic fascicle-like spatial structure of the seta and their orientation to the surface are important for the functional efficiency of the structure. So, our aim was to study how the attachment scenario depends on the 3D configuration and, as a result, how important parameters (fraction of attached spatulae and tendency to clusterization at an equilibrium) correlate with the 3D spatial structure of the seta.

In the real system, an additional degree of freedom is derived from the rotation of the relatively rigid root fiber. To reproduce this feature in the model, the rotational stiffness (B) is introduced, which dynamically tends to keep the root angle φ close to φ_0. In the first approximation, the rotational force acting on the root angle is linearly proportional to the difference $\varphi_0 - \varphi : f^{\varphi} = B(\varphi_0 - \varphi)$.

At $B > 0$, the force $f^{\varphi} = B(\varphi_0 - \varphi)$ in this equation tends to attract the angle φ to its equilibrium value φ_0. As a result, the whole system rotates towards the substrate when the fibers are attracted to it. If there is no force present between the surface and the fibers, the system gradually returns to a shape close to the initial one depicted in Fig. 3.8.

Real surfaces of the substrates the fibers attach to have a semi-fractal structure with a well-known Fourier spectrum and amplitude of roughness. As discussed above (Sect. 3.2), this structure can be simulated by the self-affine fractal given by the real part of the function $Z(x, y) = A \iint dq_x dq_y C(q) \exp\left(iq_x x + iq_y y + \zeta(x, y)\right)$ with scaling spectral density $C(q)$. Here, A is the amplitude of surface roughness, i is an imaginary unit, $q_{x,\,y}$

are Fourier components along the x and y directions, $q = \sqrt{q_x^2 + q_y^2}$ and $\zeta(x, y)$ is a δ-correlated random phase. Details of the procedure for generating the profile $Z(x, y)$ have been described in several previous papers (e.g., Filippov and Popov 2007a, b; Popov et al. 2007). It is currently accepted (cf. Persson and Gorb 2003) that the majority of physical surfaces have a scale-invariant spectrum $C(q) = 1/q^{\beta}$ with the exponent $\beta \approx 0.9$.

The attraction of the setal tip to the external surface is caused by intermolecular (van der Waals) forces. For the sake of simplicity, we also simulate it by the gradient of Morse potential, $U_{VdW}(r) = U_o(1 - \exp(-r/r_0))^2$, where $r_j = \left|\vec{r}_j - Z(x, y)\right|$, with the physically reasonable amplitude $U_0 = 10$ nN × nm and the minimum located at the distance $r_0 = 0.01$ μm from the surface. The amplitude of the numerically generated surface is taken to be in the range of the radius of the van der Waals interaction, $A = r_0$.

The thin and flexible parts of every fiber interact with corresponding regions of other fibers of the array. The interaction force has the same origin as the fibers' attraction to the stiff substrate $U_{interact}(r) = U_o(1 - \exp(-r_{jk}/r_0))^2$ and thus comparable characteristic parameters U_0, r_0. For the sake of simplicity, we reduce the mutual interaction of the fibers to the interaction of the closest neighbors: $r_{jk} = \left|\vec{r}_j - \vec{r}_{j\pm1}\right|$. This force is responsible for the clusterization of the fiber structure, and our goal here is to study whether a specific spatial distribution of the fibers can prevent it.

Our numerical experiment is organized as follows. We place the fascicle-like seta in a position where the distance between the surface and the last segment (closest to the surface) is comparable with r_0, allowing it to move according to standard equations of motion $\partial r_j/\partial t = F^j$, where the total force accumulates above the interactions $F^j = F^j_{elastic} + F^j_{VdW} + F^j_{interact}$. When the process starts, the segment closest to the surface is normally attracted first. Due to the elastic and molecular interactions with all the other segments, they will follow. At the other end, the segments are attached to the root thus transferring the interaction to it. The last segment, being fixed to the root at point zero, cannot completely move with all the other ones, but it can rotate to minimize the distance of its free end to the surface.

3.2.3 Functional Significance of a Non-uniform Geometry

To simplify the visualization and plotting of the results, we will first discuss the complete process of the setal alignment for a 2D system in the {x, z} plane. In this case, the momentary positions of the segments at some discrete points in time as well as trajectories of all the joint points connecting the segments can be simultaneously depicted in a simple planar plot. A typical scenario of the motion is presented in Fig. 3.9, where the momentary positions and the trajectories are represented by black and gray lines, respectively.

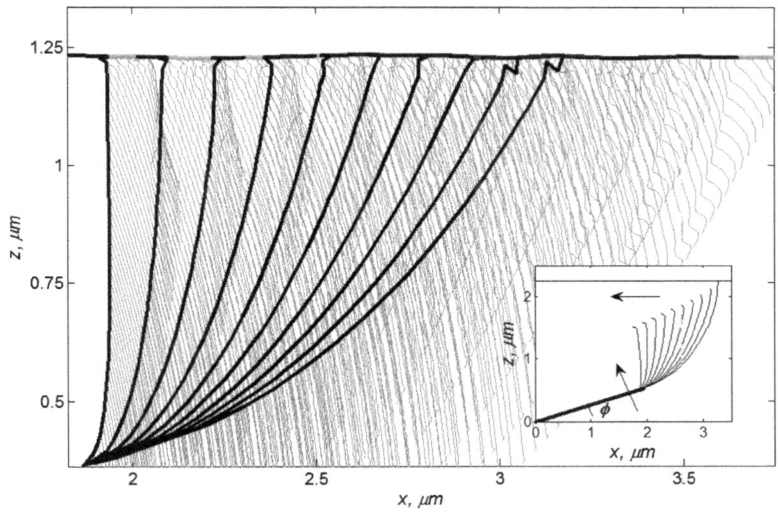

Fig. 3.9 Conceptual picture of the model in the 2D plane {x, z}. The fibers and their trajectories are shown by bold black and thin gray lines, respectively. The initial configuration is presented in the inset. The arrows show two additional directions of motion which appear due to the attraction of the fibers to the external surface: rotation of the rigid root (seta) and shear along the x-coordinate. (From Filippov and Gorb 2014)

Rotation of the root segment is the key difference between this scenario and the ordinary attraction of a regular grid of vertical fibers. As is clearly visible in Fig. 3.9, this rotation leads to a directed proximal shift of the attachment point along the surface and results in an effective healing effect. In other words, additionally to the vertical attraction to the surface, all the fibers will simultaneously slightly move along the surface. Previously, it was experimentally and theoretically demonstrated that such shear movements enhance the adhesion of thin tapes (as represented by the spatulae) (Autumn et al. 2000; Filippov et al. 2011).

Such a motion will have an important effect in the real biological system, because it automatically leads to a contact process which involves more and more fractions of the attached spatulae, and, most importantly, without any active control by the animal. Also, it is important to note that due to this process all terminal spatulae will turn in a direction opposite to the direction of motion. When they attach to the surface by their very tips, shear motion presses them against the already established contact points. As a result, the spatulae will flip around themselves. This rotation produces typical configurations that are clearly visible in SEM images of real gecko hairs attached to the surface (Fig. 3.7g), where the spatulae are usually turned in a direction opposite to their original orientation (Fig. 3.7e, f).

As mentioned above, a shift of the spatulae in the vertical direction should prevent their clusterization after their detachment from the surface. The main cause of clusterization in this system is the fact that the spatulae tend to be attracted to the same asperities protruding from the substrate surface. So, when the fascicles of

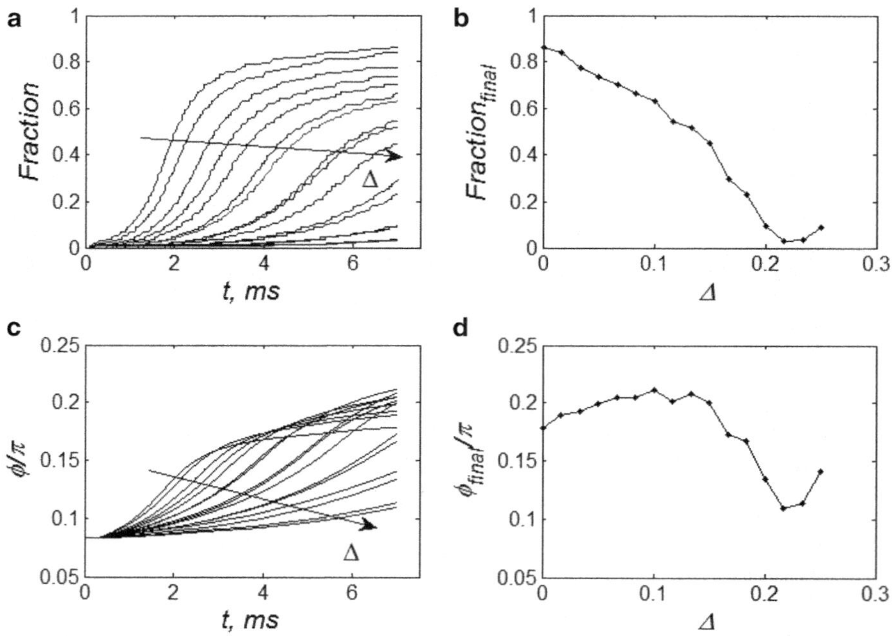

Fig. 3.10 Temporal development of (**a**) the fraction of attached fibers and (**b**) of the angle φ in the case of non-interacting fibers. Final values for the fraction of attached fibers and the angle φ at different values of Δ are shown in subplots (**c**) and (**d**), respectively. The tendencies with variation of Δ in the time-dependent plots are represented by the arrows. (From Filippov and Gorb 2014)

the nanofibers start to detach from the surface, the fibers are located close to each other and will remain attracted to each other even after complete detachment. As a result, the system will remain modified and will not operate properly in the next attachment cycle.

The 3D spatial configuration of the setae with their branches of nanofibers and terminal spatulae has been optimized by natural selection. Our numerical experiment confirms the intuitive expectation that if the flexible ends of the nanofibers are shifted away from the substrate in the vertical direction, the stored elastic energy tends to return them to their original position, easily separating them. On the other hand, at a given angle φ_0, a very strong shift in the vertical direction is necessary for some of the fibers to prevent their reattachment to the surface. It leads to a formal optimization problem.

To solve this problem, we compared the numerical solution of the equations of motion $\partial r_j/\partial t = F^j$ with and without mutual interaction between the fibers. The results are summarized in Figs. 3.10 and 3.11, respectively, where the time-dependent fraction of the attached nanofibers and the rotation angle φ are presented along with their final values at different Δ, which corresponds to the difference between the maximal and minimal length of the fibers normalized to the maximal length.

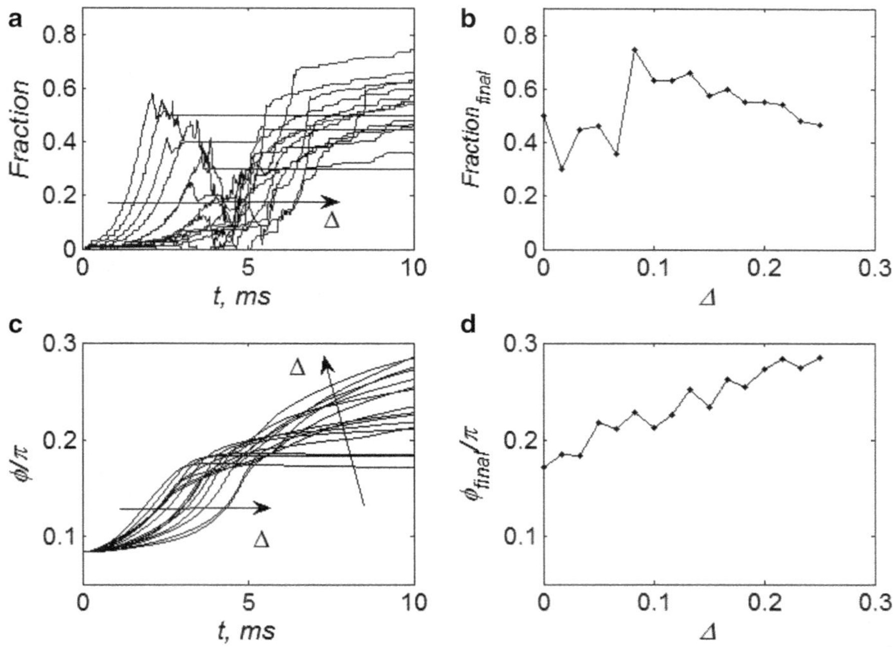

Fig. 3.11 Temporal development of (**a**) the fraction of attached fibers and (**b**) the angle φ in the case of interacting fibers. Final values for the fraction of attached fibers and the angle φ at different values of Δ are shown in subplots (**c**) and (**d**), respectively. The tendencies with variation of Δ in the time-dependent plots are represented by the arrows. (From Filippov and Gorb 2014)

In the case of non-interacting fibers, the results are quite trivial. The fraction of attached fibers gradually decreases with time. The time-dependent behavior of the angle φ is also quite predictable here. When the very first fibers attach to the surface, the angle φ starts to grow. Elasticity works against this rotating force and, in principle, stops the angle φ at some final value at $t \to \infty$. If the (upper) fibers relatively distant from the surface are strongly shifted in the vertical direction, they do not participate in the attraction, and it prevents the rotation as well. The final angle φ at $t \to \infty$ is defined by a balance of the forces caused by the very few attached fibers and falls down at some shift Δ.

One important disadvantage of the 2D model is that it does not reproduce the distance between the spatulae in y-direction, orthogonal to the rotation plane {x, z}. However, this coordinate becomes extremely important when all the fibers contact the surface. As mentioned above, rotation of the structure leads to a shearing effect which aligns all the spatulae in one direction. As a result, in the 2D model the spatulae seem to overlap on the surface in many places and their mutual attraction and the tendency to clusterization are overestimated.

Quite a different model picture appears when the fibers are interacting. In this case, the system has a clear optimum of Δ, which is visible in Fig. 3.11. The main reason for this is that for very small values of Δ ($\Delta \to 0$) the fibers are attracted to

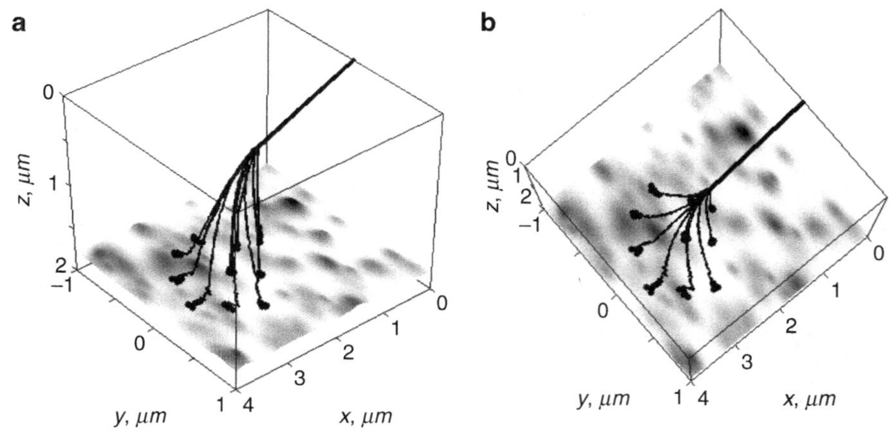

Fig. 3.12 Two different views of the same attached configuration of the modeled seta, shown in (**a**) isometric and (**b**) almost vertical projection. 3D structure of the model system in Fig. 3.8 with all spatulae attached. For explanations see Fig. 3.8. (From Filippov and Gorb 2014)

each other in the space immediately above the substrate surface and tend to form clusters. Such clusterization reduces their ability to attach to the substrate in the next attachment cycle. At the opposite limit, when $\Delta \rightarrow \infty$, the interaction of the fibers becomes weak and the fraction of attached fibers decreases. However, as found in the numerical experiment, the attraction of the remaining fibers to the attached ones still favors the rotation, and the final angle φ grows.

In reality, the fibers move apart from each other, as shown in the conceptual 3D picture (Fig. 3.8) in the {x, y} projection, and they remain separate even when coming into contact with the substrate. Moreover, numerical experiments show that the specific configuration of forces which occurs in the contact process turns the spatula {x, y} plane to the opposite radial orientation around central point, where all of them remain attached to the root segment. Two different views of the typical final attached configuration are shown in Fig. 3.12a, b. Since they remain separate on the surface substrate, the spatulae can easily return to their initial positions shown in Fig. 3.8. Thus, their specific configuration prevents their clusterization and allows a practically infinite number of attachment–detachment cycles. The dynamics of the attachment–detachment cycle in 2D and 3D space is presented in the supplementary Movies 3.4 and 3.5, respectively.

Overall, this model shows that in addition to well known hierarchical organization of the seta (Bhushan et al. 2006; Kim and Bhushan 2007), the non-symmetric and non-uniform distribution of its nanofibers help to prevent clusterization. Such a sophisticated, non-uniformly distributed 3D structure additionally provides a rotation-induced proximal shear of the spatulae, which is crucial for their contact formation and, generally, for the formation of adhesive contacts in thin, spatula-like tapes (Filippov et al. 2011). Since these effects are entirely based on the particular geometry of the seta, this geometry would certainly be of value in practical applications, since it is a simple, purely mechanical effect that would enhance the performance of artificially made fibrillar adhesives.

3.3 Adhesion with Clustering Behavior

Clustering of fibers of biomimetic attachment systems is a crucial factor limiting their effective adhesive forces (Geim et al. 2003; Jagota and Bennison 2002). Clustering occurs when the adhesive forces between the fiber tips are stronger than the forces required to bend the fibers (Spolenak et al. 2005a, b). In biological fibrillar attachment devices, as exemplified above, different strategies have evolved to avoid clustering of the attachment hairs. In geckos, one such strategy is the sophisticated hierarchic three-dimensional arrangement of the contact elements of the single foot-hairs (Filippov and Gorb 2014). In insects, another strategy is a gradient of material properties in the attachment hairs from stiff at the bottom to soft at the tip (Gorb and Filippov 2014). However, clustering of fine fibers may represent another, very interesting and unexpected way of stabilizing fiber arrays. A certain degree of clustering of the fibers of attachment devices may contribute to their adaptation to different macro- and micro-roughnesses of substrates as previously suggested for the subdigital anti-slip setae of chameleon feet (Spinner et al. 2014).

In the following Section, we will examine the clustering behavior of 1 mm long arrays of vertically aligned multi-walled carbon nanotubes (VACNTs) firmly bound to a substrate. VACNTs are good candidates for mimicking gecko foot-hairs (Yurdumakan et al. 2005) since the thickness of the single carbon nanotubes (between 5 and 20 nm) is well in the range of the contact elements of gecko setae (Persson and Gorb 2003).

3.3.1 Carbon Nanotube Arrays as an Approach to Bioinspired Adhesives

Vertically aligned carbon nanotube arrays (VACNTAs) are microstructures that consist of carbon nanotubes perpendicularly oriented to a substrate surface. VACNTAs are widely used in a range of applications. When shear forces are applied, VACNTAs form clusters, and this self-organization stabilizes the mechanical properties of these clustered arrays. With increasing load – in the range between 300 μN and 4 mN as normally applied to the array surface during friction tests – the size of the clusters increases, while the coefficient of friction decreases (Schaber et al. 2015a).

To better understand the experimentally obtained results (Schaber et al. 2015a), we formulated a minimalistic model which reproduces the main features of the system with a minimum of adjustable parameters. We calculated the van der Waals forces between the spherical friction probe and the bunches of arrays using the well-known Morse potential function to predict the number of clusters, their sizes, the instantaneous and mean friction forces, and the behavior of the VACNTs during consecutive sliding cycles and at different normal loads (Schaber et al. 2015b).

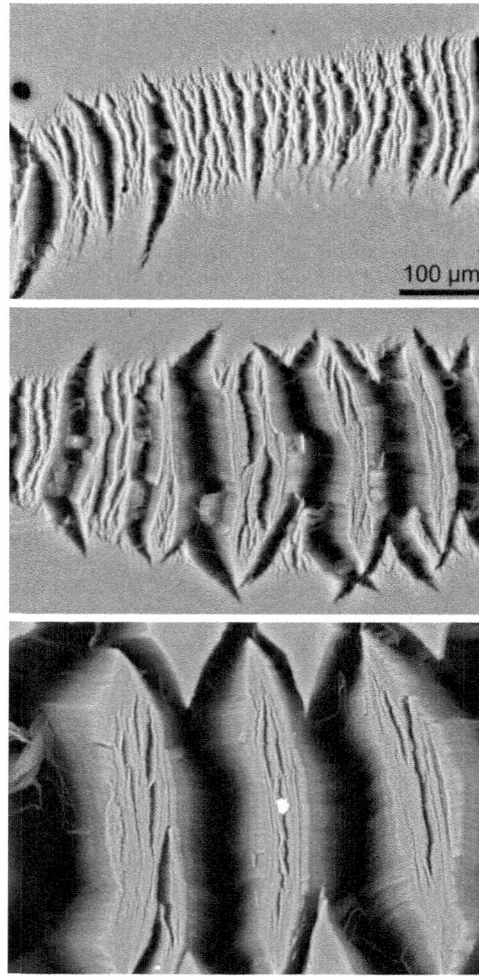

Fig. 3.13 Clusters of VACNT after friction tests with normal loads of 289 μN, 673 μN, and 3860 μN, respectively (from top to bottom), seen on the surface of the arrays. (From Schaber et al. 2015b)

The experimentally determined coefficient of friction on pristine surfaces of VACNT arrays is very high (much above the value 1). It decreases with consecutive cycles of sliding of the probe along the array at the same location to remarkably stable values. This decrease of friction is accompanied by clustering of the VACNT, which stabilizes the mechanical properties of the arrays (Schaber et al. 2015a, b). The size of the clusters depends on the normal force applied on the sample during friction tests (Fig. 3.13).

Previous simulations dealing with friction on vertically aligned fibrous materials mainly relied on finite element models (FEM) and molecular dynamics approaches. One such model analyzed the static friction on a vertically aligned micro-fibril array made of poly-dimethylsiloxane covered with a flat 4 μm film. The model predicted strong enhancement of friction compared to a control without the underlying fibrils (Liu et al. 2009) Using molecular dynamics and FEM simulations, Hu et al. (2010) showed an increase of shear forces in VACNT blocks with increasing length of the entangled tips of the single CNT within the array. Lou et al. (2008) simulated the interface between the probe tip used in atomic force microscopy (AFM) (diameter 4 nm) and very short (1.5 nm) VACNTs arranged in a super lattice. The study revealed that the stick-slip behavior during friction tests is largely caused by the penetration of the tip into the valleys between the single CNTs, which leads to a strong interaction of the sides of the tubes with the sides of the probe. Using another atomic scale model, Landolsi et al. (2010) calculated the stick-slip behavior of a micro-spherical AFM probe on VACNT arrays with 30 nm protruding length of the CNTs.

To predict the results gained in these experiments and to expand our knowledge about the experimentally not accessible interface between the probe and the surface of the VACNT arrays, we here present a numerical model using the Morse potential function (Morse 1929). The model calculates van der Waals interactions between the spherical probe and the bundles of VACNTs of the arrays to simulate their clustering and its effect on the coefficient of friction during repeated friction tests. Experimental data were obtained, as described by Schaber et al. (2015a, b), by friction tests comprising five consecutive sliding cycles on the same location of the pristine surface of VACNT samples using a sapphire sphere as the probe (diameter 1.5 mm).

In the pristine condition, the single nanotubes are assumed to be bundled together more or less uniformly to parallel bunches, each of which contains quite a large number of tubes (Fig. 3.14a). The base end of every tube, and of a bunch, respectively, is stationary fixed to the substrate. The only degree of freedom remaining for a bunch is to rotate with respect to the fixation point. When an external force deflects the bunch, it tends to return to its original vertical orientation.

To avoid time-consuming calculations but still to simulate realistic collective behavior of CNT arrays, we limit the model of each single fiber to the one-dimensional chain of bunches in the line of interaction with the external force. Each bunch is represented by its central point on top of the surface. For briefness, these points are hereafter called "bunches". In the model, the effective elastic force that returns the bunches to their equilibrium vertical positions corresponds to their rotation under load.

3.3.2 Mathematical Model of the Clustering of Nanotube Arrays

For simplicity's sake, we will reduce the problem to a one-dimensional model where the deformation of the cluster only depends on its position in the direction of motion of the spherical indenter labeled by the coordinate x. We will denote

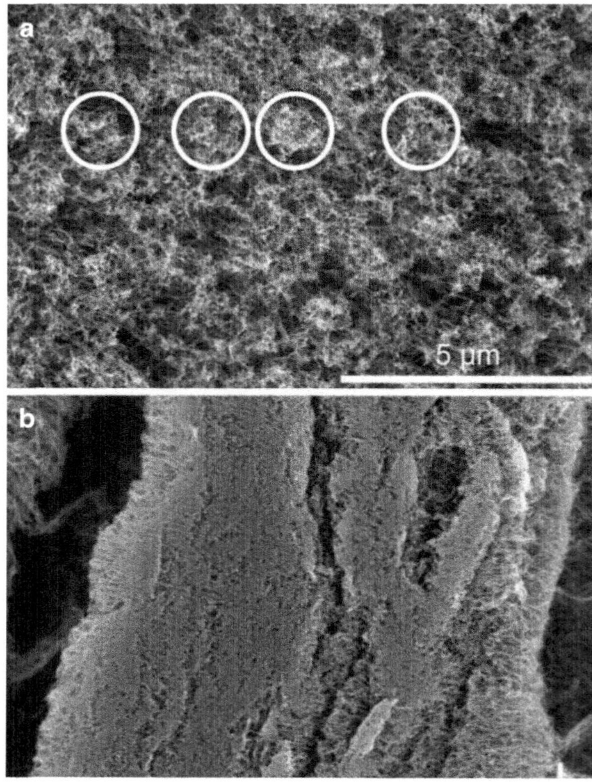

Fig. 3.14 (**a**) Pristine surface of a VACNT array, with white circles indicating some bunches of CNT. (**b**) Clustered surface of the same sample after the friction experiment (same magnification as in a). (From Schaber et al. 2015a)

the alignment (represented by an array in the numerical model) with the equilibrium positions $x_0 = \{x_0^j\}$ of the tubes. In the basic variant of the model, one can assume all the bunches to be equivalent and placed equidistantly. In this case, $x_0 = \{x_0^j\} = dx_0[1, \ldots, N]$, where N is the total number of bunches, $x_0 = L/N$, and L is the length of the system. As described above, when the instant positions of the tubes $x = \{x^j\}$ are shifted from their original positions $x_0 = \{x_0^j\}$, a set of elastic forces appears: $f^j_{elastic} = k(x_0^j - x^j)$. The phenomenological elastic constant k effectively simulates the rotational force appearing at the point where the rigid bunch attaches to the solid substrate. It is important to note that we reduced the description for the one-dimensional model, so all the values in the equation are scalar and the indices describe only the numbers in the array, but not the vector coordinates.

With regard to the real experimental setup, the assumption of a regular lattice of equivalent bunches is not correct. Direct observation shows that there are quite a number of differently multi-walled CNTs with different diameters and different distances between their centers (Fig. 3.14a), forming bunches of different proportions. To integrate this feature into the model, the distance $\{dx_0^j\}$ between the closest neighbors is randomly varied from one pair of tubes to the next, $x_0^{j+1} - x_0^j = dx_0^j$, with the total length of the system equal to the sum $L = \sum_{j=1}^{N} dx_0^j$. The numerical realization of the array $\{dx_0^j\}$ must contain all possible kinds of distances $\{dx_0^j\} = \{dx_0^j(1), dx_0^j(2), \ldots, dx_0^j(n)\}$, where n is the total number of possible distances randomly placed with statistical weights P_k with $k = 1, 2, \ldots, n$ corresponding to the empirical probability of finding this realization in the real array $\sum_{k=1}^{n} P_k = 1$.

Therefore, our model now accounts for different sizes and elasticities of the tubes. Direct numerical simulation showed that this modification changes some particular quantitative results only, but does not influence the general qualitative behavior. Therefore, in the further description we will neglect these differences and refer to the regular system only.

The single CNTs interact with the moving spherical probe and with each other by van der Waals forces. In turn, the spherical probe is moved by another elastic force produced by the external cantilever and it interacts with the integral collective force of the bunches. These mesoscopic interaction forces preserve the general properties of the Lennard-Jones potential of the original intermolecular forces, but in the present numerical model it is more convenient, and in some sense even more realistic, to simulate them as simply as possible by using some effective potential.

One common representation for intermolecular potentials well suited for numerical simulations is the Morse potential (Morse 1929). In standard notations it can be written as follows: $U_{VdW}^{jk} = D\left[1 - \exp\left(-a\left(r^{jk} - r_0\right)\right)\right]^2$. Here, D is the depth of the potential minimum, defined in relation to the infinite intermolecular distance $r^{jk} = |x^j - x^k|$, r_0 is the position of the potential minimum, and the parameter $a = (f_{VdW}/2D)^{1/2}$ controls the width of the potential at a given value of the force constant. The force is defined as a derivation, $f_{VdW} \equiv -\partial U_{VdW}/\partial r|_{r=r_0}$, which is calculated for the point of the potential minimum. Let us stress also that "in relation to the infinite intermolecular distance $r^{jk} = |x^j - x^k|$" here refers to a standard physical definition of potential energy, where it is formally calculated from its asymptotic value at distance going to infinity. It does not refer to an actual infinite distance.

The interaction of the bunches with the moving sphere is comparable to their mutual interaction. The corresponding potential can be written in the analogous form $U_{VdW}^{ball} = D^{ball}[1 - exp\,(-A(R^j - R_0))]^2$, with another set of parameters A, D^{ball}, R_0 and the distances $R^j = |X - x^j|$ between the bunches and the instant positions of the rigid sphere with the coordinate of center of mass X. A schematic diagram of the model defined by Eq. 3.1 is shown in Fig. 3.15.

All other forces, which remain out of our control, must be included by an interaction of the system with an external "thermostat", simulated by a set of δ

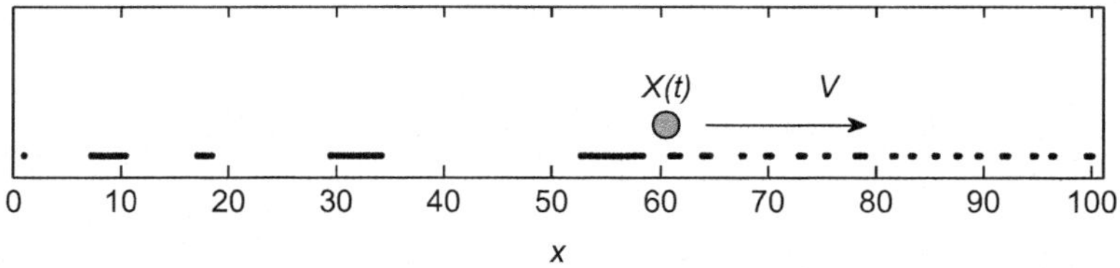

Fig. 3.15 Schematic diagram of the model defined by Eq. (3.1) with a snapshot of the process of clustering at an intermediate stage of the simulation. The tops of bunches are represented by black dots. The sphere (gray circle) is shown in its current position with $X(t)$ in the center of mass $X(t)$. V is the current velocity of the (external) cantilever scanning the system in both directions with positive and negative velocities $V(t) = \pm V$. The bunches at their current coordinates $x = \{x^j\}$ are shifted from their original positions $x_0 = \{x_0^j\}$ and attracted to each other by the elastic forces $f^j_{elastic} = k(x_0^j - x^j)$. The nearest bunches interact with each other by intermolecular Morse potentials: $U^{jk}_{VdW} = D[1 - \exp(-a(r^{jk} - r_0))]^2$. The interaction of the bunches with the moving sphere is comparable to their mutual interaction and described by qualitatively the same potential $U^{ball}_{VdW} = D^{ball}[1 - exp(-A(R^j - R_0))]^2$ with another set of parameters defined in the text. (From Schaber et al. 2015b)

correlated in time and space random sources: $\langle\zeta^j\rangle = 0$, with intensity σ: $\langle\zeta^j\zeta^k\rangle = \sigma\delta_{jk}$, and an energy dissipation described by the phenomenological damping constant γ.

The equations of motion can now be written in the final form:

$$\begin{aligned} \frac{\partial^2 x^j}{\partial t^2} &= -\,\partial U^j_{ball}/\partial r - \sum_{k=j\pm1} \partial U^{jk}_{VdW}/\partial r - \gamma\frac{\partial x^j}{\partial t} + \zeta^j; \\ \frac{\partial^2 X}{\partial t^2} &= K(X_{ext}(t) - X) - \sum_k \partial U^k_{ball}/\partial R^k - \gamma_{ball}\frac{\partial X}{\partial t}, \end{aligned} \qquad (3.1)$$

where K is the elastic constant and $X_{ext}(t)$ the instant position of the external cantilever. This system of equations can be numerically solved using standard MatLab software.

3.3.3 Functional Significance of CNT Clusterization in Multiple Attachment–Detachment Cycles

We can forecast the following qualitative behavior of the system before any numerical simulation: at one limit, when the bunches are strongly attached, which results in almost no rotation from their original positions and weak interaction between the bunches as well as with the sphere, we expect them to remain close to their initial positions $x_0 = \{x_0^j\}$. At the opposite limit, strong interaction with the sphere will especially disrupt the positions of the bunches in its close vicinity. If the mutual interaction of the bunches in relation to the elastic force $f^j_{elastic} = k(x_0^j - x^j)$ is strong enough, and they are disturbed, the bunches will tend to a new equilibrium distance r_0, which is normally considerably smaller than the initial $dx_0 = L/N$ in the array $x_0 = \{x_0^j\}$. Let us note that the total number of the tubes is fixed, and the distance between them averaged over the whole system remains equal to dx_0. The only compromise expected in this case is that the system must split into a number of relatively large clusters separated by almost vacant gaps.

Thus, the moving sphere pushes or attracts bunches, depending on their position in front or behind the sphere, and gradually causes them to bundle into clusters. Assembled bunches follow the sphere for a while, and new bunches are added along its way. When the size of the cluster surrounding the sphere becomes too large for a balance between the van der Waals and the elastic forces, the sphere will not be able to direct the movement of the complete cluster anymore. Some of the bundles will "opt out" and return as close as possible to their original positions.

Besides the interaction with the sphere, mutual interaction between direct neighbors can also produce a spontaneous assembling of small groups containing very few bunches. Our observation shows that this kind of instability of the perfect trial lattice appears for the soft system with an elasticity of $k \cong 0.01$. Such small values of elasticity are necessary to model the experimentally observed strong effect of clustering. Figure 3.15 shows such a system with partially formed groups of bunches in front of the sphere and extended clusters behind it.

The time-dependent scenario of the clustering can be recorded and quantitatively described by accumulating and plotting the complete set of time-dependent arrays $x^j = x^j(t)$. Figure 3.16a shows a typical result of this procedure. It is directly evident how the initially equidistant trajectories $x^j(t)$ evolve to the dense bundles corresponding to the clusters. Figure 3.16b shows the time-dependent friction force.

The ends of the clusters can formally be defined as those points where the instant distances $dx_{jj\,\pm\,1}(t)$ between the bunches are much longer than dx_0; for example, $dx_{jj\,+\,1}t \gg dx_0$ and $dx_{jj\,-\,1} \cong r_0 < dx_0$ for the right end of the cluster.

The model allows us to define a procedure for calculating the total number of the clusters, the mean number of bunches per cluster (= mean size of the cluster) and the mean length (averaged distance between the left and right ends) of the clusters. These time-dependent values and the corresponding number of clusters are presented in Fig. 3.16c and d, respectively. The dynamics of the system leading to the plots presented in Figs. 3.16 and 3.17 are shown in the supplementary Movie 3.6.

The "size" and the "length" of the cluster relate to the density of the tubes inside it, which is in turn directly related to two spatial scales of the problem dx_0 and r_0 as well as to the relationship between elastic and interaction forces. The rigidity of the original nanotubes influences the interaction between the bunches; higher rigidity shifts the equilibrium distance between them from r_0 to the trial value dx_0, and vice versa. Nanotubes freely rotating on their basal points favor the formation of larger clusters. This observation provides a good criterion for validating our simulation results.

Important information is also provided by the temporal development of the friction force during each sliding cycle of the same region, by its standard deviation

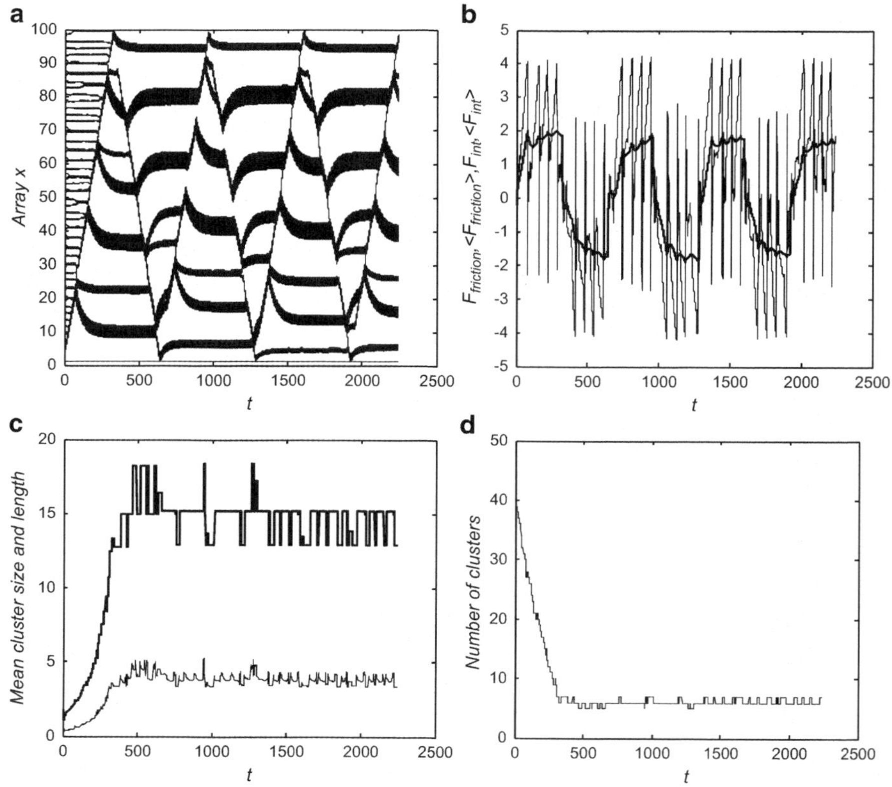

Fig. 3.16 Typical evolution of the time-dependent values in the presence of clustering during five sliding cycles. (**a**) Transformation of the array $x^j = x^j(t)$. (**b**) Friction force (the force value averaged over each current scan is shown by the bold line). (**c**) Mean size (bold line) and length of the cluster. (**d**) Number of clusters produced by five sliding cycles. (From Schaber et al. 2015b)

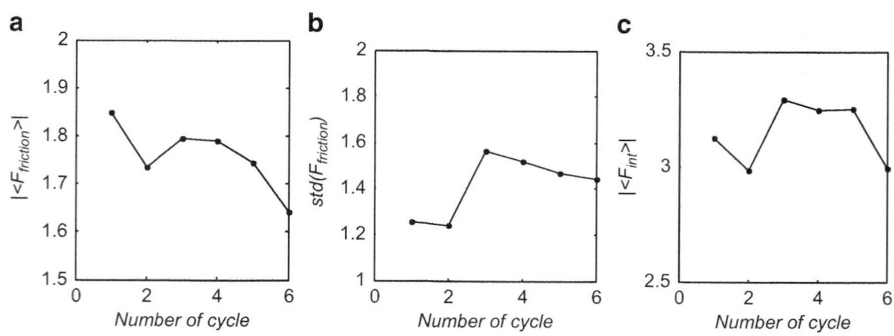

Fig. 3.17 (**a**) Absolute friction forces of VACNT arrays, (**b**) their standard deviations, (**c**) interaction strengths between the sphere and the bunches (characterizing the effective mean contact area) integrated over each sliding cycle. (From Schaber et al. 2015b)

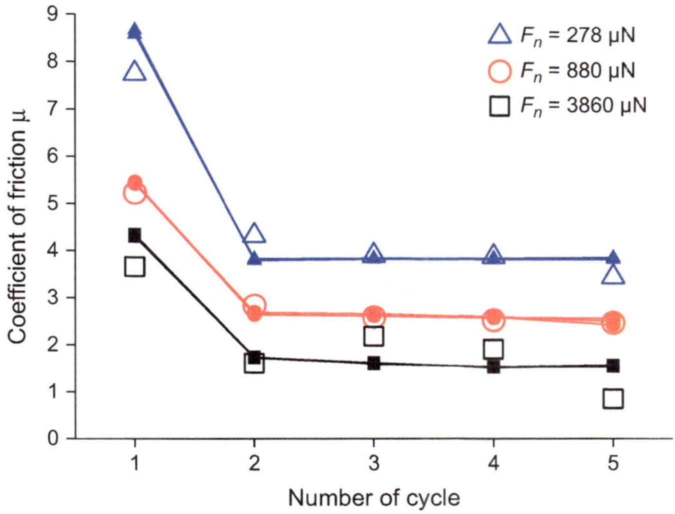

Fig. 3.18 Modeled friction coefficients for five consecutive sliding cycles on the same array of bunches (small solid symbols) and experimentally determined values (open symbols) at three different mean normal loads F_n. The coefficient μ was 0.61 to best match the experimental data for 278 μN normal load, 0.54 for 880 μN, and 0.46 for 3860 μN. The model data represent the results of three executions of the computer code with the same parameters. The differences between the runs are hardly visible. (From Schaber et al. 2015b)

describing how pronounced the stick-slip effect is, and by the variation of the size of the effective contact area, which characterizes a number of bunches situated close to the sphere and interacting strongly with it (as estimated on the basis of the absolute value of the corresponding interaction force). From a physical point of view, it is interesting to determine and compare the mean values of all these variables derived from the modeling of consecutive scans. These values are summarized in Fig. 3.17 and are in good correlation with experimental observations.

The factor $U_{VdW}^{ball} = D^{ball}[1 - exp\,(-A(R^j - R_0))]^2$ in the model reflects the strength of interaction between the spherical probe and the bunches after the first sliding cycle at different experimental normal loads, thus matching the model with the empirical results. The model well predicts the experimentally obtained friction coefficients and their decrease with consecutive sliding cycles (Fig. 3.18). The friction coefficient decreases monotonously with increasing mean normal load F_n, from 0.61 at the smallest F_n of 278 μN down to 0.46 at the largest F_n of 3860 μN.

The model also predicts the experimentally observed decrease of the number of clusters (corresponding to an increase of their sizes) with increasing normal loads (Fig. 3.19a). The same is true for the decrease of the friction coefficient with increasing normal force (Fig. 3.19b). This agreement of the simulated and the experimental data makes it possible to use the model for predictions beyond the existing experimental results. Thus, by using the model we can calculate the

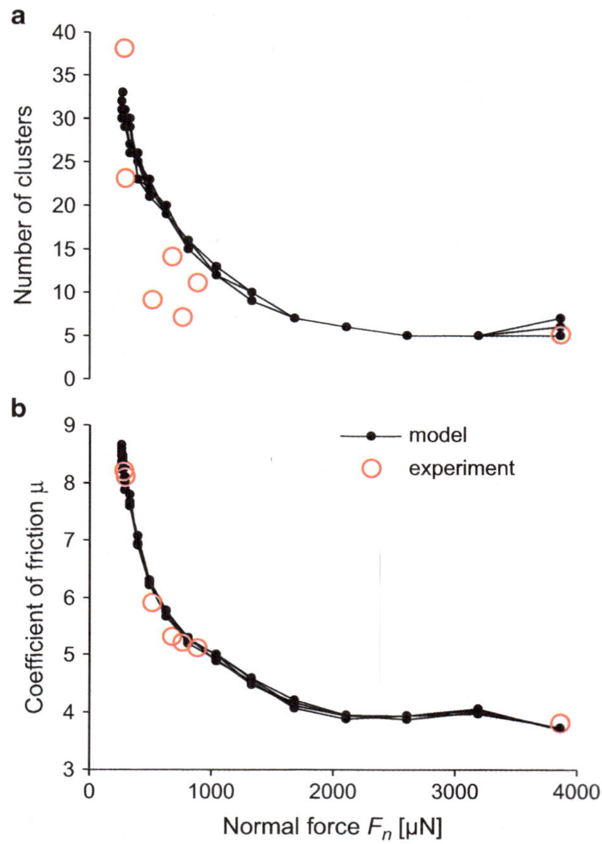

Fig. 3.19 Predicted (small solid dots) and experimentally determined (open circles) values of (**a**) the number of clusters after five consecutive sliding cycles and (**b**) the friction coefficient during the first sliding cycle at different normal loads F_n on the surface of the arrays. The model data are derived from six runs of the model with the same parameters. (From Schaber et al. 2015b)

coefficients of friction and the numbers of clusters of VACNT arrays for any normal load requested.

The model clearly shows that the clusters already form during the first sliding cycle of the friction tests, which corresponds well with the strong decrease of the friction coefficient from the first to the second cycle (Figs. 3.17 and 3.19). Therefore, there is strong evidence that the cluster structure is responsible for the stability of the high friction coefficient in consecutive cycles. In our friction testing setup, the coefficient of friction practically did not change with normal loads exceeding approximately 2 mN, nor did the number of clusters along the given sliding length (Fig. 3.18).

Regarding possible applications of VACNT arrays as biologically inspired anti-slip and attachment devices based on fibrillar adhesion, the simulation shows that the stable friction values are predominantly achieved by the conditioning during the first cycle. Furthermore, the model implies that VACNT arrays can handle a wide range of loads for applications where high friction is needed.

References

Arzt E, Gorb S, Spolenak R (2003) From micro to nano contacts in biological attachment devices. Proc Natl Acad Sci U S A 100:10603–10606

Autumn K, Liang YA, Hsieh ST, Zesch W, Chan WP, Kenny TW, Fearing R, Full RJ (2000) Adhesive force of a single gecko foot-hair. Nature 405:681–684

Autumn K, Sitti M, Liang YA, Peattie AM, Hansen WR, Sponberg S, Kenny TW, Fearing R, Israelachvili JN, Full RJ (2002) Evidence for van der Waals adhesion in gecko setae. Proc Natl Acad Sci U S A 99:12252–12256

Barbakadze N, Enders S, Gorb S, Arzt E (2006) Local mechanical properties of the head articulation cuticle in the beetle *Pachnoda marginata* (Coleoptera, Scarabaeidae). J Exp Biol 209:722–730

Bhushan B, Peressadko AG, Kim TW (2006) Adhesion analysis of two-level hierarchical morphology in natural attachment systems for 'smart adhesion'. J Adhes Sci Technol 20:1475–1491

Borodich FM, Gorb EV, Gorb SN (2010) Fracture behaviour of plant epicuticular wax crystals and its role in preventing insect attachment: a theoretical approach. Appl Phys A Mater Sci Process 100:63–71

Creton C, Gorb SN (2007) Sticky feet: from animals to materials. MRS Bull 32:466–468

Filippov AE, Gorb SN (2014) Spatial model of the gecko foot hair: functional significance of highly specialized non-uniform geometry. Interface Focus 5:20140065

Filippov AE, Popov VL (2007a) Fractal Tomlinson model for mesoscopic friction: from microscopic velocity-dependent damping to macroscopic Coulomb friction. Phys Rev E75:027103

Filippov AE, Popov VL (2007b) Flexible tissue with fibres interacting with an adhesive surface. J Phys Condens Matter 19:096012 (10pp)

Filippov AE, Popov VL, Gorb SN (2011) Shear induced adhesion: contact mechanics of biological spatula-like attachment devices. J Theor Biol 276:126–131

Fong H, Sarikaya M, White SN, Snead ML (2000) Nano-mechanical properties profiles across dentin-enamel junction of human incisor teeth. Mater Sci Eng C 7:119–128

Gao H, Yao H (2004) Shape insensitive optimal adhesion of nanoscale fibrillar structures. Proc Natl Acad Sci U S A 101:7851–7856

Geim AK, Dubonos SV, Grigorieva IV, Novoselov KS, Zhukov AA, Shapoval SY (2003) Microfabricated adhesive mimicking gecko foot-hair. Nat Mater 2:461–463

Gorb SN (2001) Attachment devices of insect cuticle. Kluwer Academic Publishers, Dordrecht

Gorb SN (2008) Smooth attachment devices in insects. In: Casas J, Simpson SJ (eds) *Advances in insect physiology: insect mechanics and control*, vol 34. Elsevier Ltd./Academic, London, pp 81–116

Gorb SN, Beutel RG (2001) Evolution of locomotory attachment pads of hexapods. Naturwissenschaften 88:530–534

Gorb SN, Filippov AE (2014) Fibrillar adhesion with no clusterisation: functional significance of material gradient along adhesive setae of insects. Beilstein J Nanotechnol 5:837–845

Gorb SN, Beutel RG, Gorb EV, Jiao Y, Kastner V, Niederegger S, Popov VL, Scherge M, Schwarz U, Votsch W (2002) Structural design and biomechanics of friction-based releasable attachment devices in insects. Integr Comp Biol 42:1127–1139

Hiller U (1968) Untersuchungen zum Feinbau und zur Funktion der Haftborsten von Reptilien. Z Morphol Tiere 62:307–362

Hu S, Jiang H, Xia Z, Gao X (2010) Friction and adhesion of hierarchical carbon nanotube structures for biomimetic dry adhesives: multiscale modeling. ACS Appl Mater Interfaces 2:2570–2578

Huber G, Gorb SN, Spolenak R, Arzt E (2005) Resolving the nanoscale adhesion of individual gecko spatulae by atomic force microscopy. Biol Lett 1:2–4

Hui CY, Glassmaker NJ, Tang T, Jagota A (2004) Design of biomimetic fibrillar interfaces: 2. Mechanics of enhanced adhesion. J R Soc Interface 1:35–48

Jagota A, Bennison SJ (2002) Mechanics of adhesion through a fibrillar microstructure. Integr Comp Biol 42:1140–1145

Kim TW, Bhushan B (2007) Effect of stiffness of multilevel hierarchical attachment system on adhesion enhancement. Ultramicroscopy 107:902–912

Klein M-C, Deuschle J, Gorb S (2010) Material properties of the skin of the Kenyan sand boa *Gongylophis colubrinus* (Squamata, Boidae). J Comp Physiol A 196:659–668

Landolsi F, Sun Y, Lu H, Ghorbel FH, Lou J (2010) Regular and reverse nanoscale stick-slip behavior: modeling and experiments. Appl Surf Sci 256:2577–2582

Liu J, Huy CY, Jagota A, Shen L (2009) A model for static friction in a film-terminated microfibril array. J Appl Phys 106:053520

Lou J, Ding F, Lu H, Goldman J, Sun Y, Yakobson BI (2008) Mesoscale reverse stick-slip nanofriction behavior of vertically aligned multiwalled carbon nanotube superlattices. Appl Phys Lett 92:203115

Michels J, Gorb SN (2012) Detailed three-dimensional visualization of resilin in the exoskeleton of arthropods using confocal laser scanning microscopy. J Microsc 245:1–16

Morse PM (1929) Diatomic molecules according to the wave mechanics II. Vibrational levels. Phys Rev 34:57–64

Northen MT, Turner KL (2005) A batch of fabricated dry adhesive. Nanotechnology 16:1159–1166

Peisker H, Michels J, Gorb SN (2013) Evidence for a material gradient in the adhesive tarsal setae of the ladybird beetle *Coccinella septempunctata*. Nat Commun 4:1661

Perez Goodwyn P, Peressadko A, Schwarz H, Kastner V, Gorb S (2006) Material structure, stiffness, and adhesion: why attachment pads of the grasshopper (*Tettigonia viridissima*) adhere more strongly than those of the locust (*Locusta migratoria*) (Insecta: Orthoptera). J Comp Physiol A 192:1233–1243

Persson BNJ (2003) On the mechanism of adhesion in biological systems. J Chem Phys 118:7614–7621

Persson BNJ, Gorb S (2003) The effect of surface roughness on the adhesion of elastic plates with application to biological systems. J Chem Phys 119:11437–11444

Popov VL, Starcevic J, Filippov AE (2007) Reconstruction of potential from dynamic experiments. Phys Rev E 75(6):066104

Rizzo NW, Gardner KH, Walls DJ, Keiper-Hrynko NM, Ganzke TS, Hallahan DL (2006) Characterization of the structure and composition of gecko adhesive setae. J R Soc Interface 3:441–451

Schaber CF, Heinlein T, Keeley G, Schneider JJ, Gorb SN (2015a) Tribological properties of vertically aligned carbon nanotube arrays. Carbon 94:396–404

Schaber CF, Filippov AE, Heinlein T, Schneider JJ, Gorb SN (2015b) Modelling clustering of vertically aligned carbon nanotube arrays. Interface Focus 5:20150026

Spinner M, Westhoff G, Gorb SN (2014) Subdigital setae of chameleon feet: friction-enhancing microstructures for a wide range of substrate roughness. Sci Rep 4:5481

Spolenak R, Gorb SN, Gao H, Arzt E (2005a) Effects of contact shape on the scaling of biological attachments. Proc R Soc Lond A 461:305–319

Spolenak R, Gorb SN, Arzt E (2005b) Adhesion design maps for bio-inspired attachment systems. Acta Biomater 1:5–13

Varenberg M, Pugno NM, Gorb SN (2010) Spatulate structures in biological fibrillar adhesion. Soft Matter 6:3269–3272

Vincent JFV (2002) Arthropod cuticle: a natural composite shell system. Compos Part A Appl Sci Manuf 33:1311–1315

Voigt D, Schuppert JM, Dattinger S, Gorb SN (2008) Sexual dimorphism in the attachment ability of the Colorado potato beetle *Leptinotarsa decemlineata* (Coleoptera: Chrysomelidae) to rough substrates. J Insect Physiol 54:765–776

Wang RZ, Weiner S (1997) Strain-structure relations in human teeth using Moire fringes. J Biomech 31:135–141

Weis-Fogh T (1960) A rubber-like protein in insect cuticle. J Exp Biol 37:889–907

Weis-Fogh T (1961) Molecular interpretation of the elasticity of resilin, a rubber-like protein. J Mol Biol 3:648–667

Yurdumakan B, Raravikar NR, Ajayan PM, Dhinojwala A (2005) Synthetic gecko foot-hairs from multiwalled carbon nanotubes. Chem Commun 2005:3799–3801

Chapter 4
Contact Between Biological Attachment Devices and Rough Surfaces

Abstract Spatula is an individual adhesive element situated at the tip of an adhesive hair in geckos, spiders and some insects. Usually, the contact area and work of adhesion depends on the type of the substrate. It is interesting to note that the interaction of a spatula-like hair tip with the substrate is practically independent on the numerical approach by which the substrate is modeled. However, in the experiments on real animals, as well as in numerically-modeled adhesion of this type of contact systems, pull-off force drops at some particular substrate roughness. In this chapter, we present a numerical model, which is capable to explain experimentally found effect of the adhesion drop at the scale of spatula. Besides we apply our numerical approach to study dynamics of spatular tips during contact formation and show that the contact area of the tips increases under applied shear force, especially, when spatulae are misaligned prior to the contact formation. The shear force has an optimum, when maximal contact is formed, but no slip occurs. In such a state, maximal adhesion can be generated. Another factor influencing attachment is the pad secretion, which flow on rough substrates is studied numerically here. The obtained results demonstrate that an increase in the density of the substrate microstructures leads to an increase in fluid loss from the pad. Additional numerical study, discussed in this chapter, deals with adhesive properties of plant fruits used for dispersal. These fruits can readily stick using secretion provided by the set of glands arranged in a smart manner radially at the distal end of the cut-cone-shaped fruit.

Electronic supplementary material The online version of this chapter (https://doi.org/10.1007/978-3-030-41528-0_4) contains supplementary material, which is available to authorized users.

A. E. Filippov, S. N. Gorb, *Combined Discrete and Continual Approaches in Biological Modeling*, Biologically-Inspired Systems 16,
https://doi.org/10.1007/978-3-030-41528-0_4

In this chapter, we will discuss different aspects of adhesive contact between biological attachments devices and various substrates. (1) Dimensional aspects of the spatulate tips of hairy attachment pads. (2) Shear-induced adhesion of spatulate attachment devices. (3) Wet adhesion of adhesive pads with fluid secretion. (4) Adhesive properties of plant fruits used for dispersal.

A general understanding of adhesion failure of setal attachment pads in contact with some particular substrate roughness is important for the construction of biologically inspired adhesive and anti-adhesive surfaces. Below, we present a numerical model which is capable of explaining the experimentally discovered effect of a decrease in adhesion at the level of an individual adhesive element (so-called *spatula*) situated at the tip of an adhesive hair (seta). The model involves the interaction of an elastic thin film/plate (spatula) with a rough substrate in reduced 2D space.

According to the approach presented in the previous chapters, the contact area may be defined in terms of two different types of model substrates. The surface topography may either be composed of fused particles of different sizes or obtained from a real rough substrate by Gaussian convolution (in other words, by integration with Gaussians of varied width).

It is also interesting to compare the results for these two model substrates with those obtained for surface profiles of real substrates of different known roughnesses (measured via white light interferometry), which can be utilized as a reference. Interestingly, the interaction of a spatula with the substrate was found to be practically independent of the substrate model type. However, similar to the observations in experiments with real animals, numerically modelled adhesion suddenly decreases at some particular substrate roughness. It could be shown that this decrease in adhesion is related to incomplete contact between the spatula and the substrate whenever the characteristic roughness wavelength is in the same range as the size of the spatula (Gorb 2001; Huber et al. 2007; Voigt et al. 2008; Wolff and Gorb 2012).

An interesting feature of the problem under consideration here is the fact that most biological hairy adhesive systems of insects, arachnids and reptiles that are involved in locomotion do not rely on flat punches on their tips, but rather on spatulate structures (Scherge and Gorb 2001; Arzt et al. 2003; Spolenak et al. 2005; Varenberg et al. 2010). Several hypotheses have previously been proposed to explain the functional importance of this particular contact geometry. Among them are the following:

1. enhanced adaptability to rough substrates;
2. contact formation by shear force rather than by normal load;
3. increase of the total peeling line due to using an array of multiple spatulae;
4. contact breakage by peeling off.

In the present chapter we will apply our numerical approach to examining the dynamics of spatulate tips during contact formation on rough substrates. We will show that the contact area of the tips increases with the applied shear force, especially when spatulae are misaligned prior to contact formation. As expected, the applied shear force has an optimum where maximal contact is formed but no slip occurs. In such an equilibrium state, maximal adhesion can be generated. This demonstrates the crucial role of spatulate terminal elements in biological fibrillar adhesion.

Another interesting aspect is the experimentally observed adhesion failure of setal attachment pads in contact with some particular substrate roughness. We will present a numerical model which is capable of explaining this effect involving the interaction of an elastic thin film/plate (spatula) with a rough substrate in reduced 2D space. According to the approach presented in the previous chapters, the surface topography may either be modeled by fused particles of different sizes or obtained from a real rough substrate by Gaussian convolution. Interestingly, the interaction of a spatula with the substrate was found to be practically independent of the substrate model type. However, similar to the observations in experiments with real animals, numerically modelled adhesion decreases due to incomplete contact between the spatula and the substrate at the point when the characteristic roughness wavelength is similar to the size of the spatula (Gorb 2001; Huber et al. 2007; Voigt et al. 2008; Wolff and Gorb 2012).

Another factor influencing the attachment ability of insect adhesive pads is proposed to be pad secretion. It is quite understandable that surface roughness strongly reduces the adhesion forces of insect pads. Partially, this effect has been explained by decreased contact area and rapid fluid absorption from the pad surface by rough surfaces. However, so far it has been unclear how the fluid flows on rough substrates with different roughness parameters and surface energy. In the present chapter, we will numerically study the fluid flow on rough substrates during contact formation.

The adhesive properties of plant fruits used for dispersal are another biological phenomenon that will be discussed in the context of this chapter. The fruits of *Commicarpus helenas,* for example, can readily stick to various surfaces including the skin, fur, and feathers of potential dispersal vectors using secretion provided by the set of glands arranged radially at the distal end of the truncated-cone-shaped fruit. Field observations show that this particular geometry promotes self-alignment of the fruit to various surfaces after initial contact has been established by just one gland. Such self-alignment in turn leads to an increase in the number of contact points and in the adhesive contact area.

Below, we will examine this particular geometry from a theoretical point of view by probing the adhesion ability of geometries with 2–7 radially distributed attachment points. The results show that the radial arrangement provides rapid alignment to the surface. A robust adhesion can be achieved with just five adhesive points; any further increase does not substantially improve the performance. These results are important not only for understanding the functional morphology of biological adhesive systems, but also for the development of technical self-aligning adhesive devices.

4.1 The Role of Dimension in the Adhesive Properties of Spatula-Like Biological Attachment Devices

4.1.1 The Significance of Roughness with Regard to Attachment Capabilities

Adhesive setae, which have evolved independently several times in different lineages of animals, provide effective attachment to different surfaces during locomotion (Gorb and Beutel 2001; Varenberg et al. 2010). Their adhesive performance is partially due to the terminal contact elements called spatulae, which are known to be responsible for adaptability to various surface roughnesses (Persson and Gorb 2003; Peressadko and Gorb 2004; Eimüller et al. 2008; Filippov et al. 2011) (Fig. 4.1a–d). In their natural environment, geckos (Huber et al. 2007), spiders (Wolff and Gorb 2012), and insects (Gorb 2001; Gorb and Gorb 2002; Voigt et al. 2008; Bullock and Federle 2011) are able to attach to unpredictable surfaces of very different roughness which should normally reduce the adhesion ability of their hairy attachment devices. In several previous papers, it has been shown that there is a range of roughness (henceforth called *critical roughness*) that leads to a strong decrease in the animal's attachment ability depending on the size of the end plates (spatulae) that contact the substrate (Persson and Gorb 2003; Peresadko and Gorb 2004; Wolff and Gorb 2012). In experiments with the fly *Musca domestica* (Peressadko and Gorb 2004), the beetles *Leptinotarsa decemlineata* (Voigt et al. 2008) and *Gastrophysa viridula* (Gorb 2001; Voigt et al. 2008), the spider *Philodromus dispar* (Wolff and Gorb 2012), and the lizard *Gekko gecko* (Huber et al. 2007), the critical roughness was characterized by asperities ranging from 0.3 to 1.0 μm (Fig. 4.1e–g).

Adhesion between the spatula and the substrate depends on the real contact area. Even if the spatulae consist of relatively stiff materials, they are still quite flexible due to their minimal thickness (15–20 nm in geckos, 30–40 nm in spiders, and 100–200 nm in insects). Such a tape-like geometry has a very low bending stiffness and therefore can adapt well to the vast majority of natural surface profiles (Persson and Gorb 2003; Varenberg et al. 2010) and generate strong attractive forces (friction, adhesion). At a higher degree of roughness (asperity size ranging from 9.0 to 12.0 μm), the substrate profile effectively comes close to a smooth substrate when compared with the spatula length and width (Fig. 4.1e–g). Of course, in this case the real contact area and the resulting attractive forces are usually somewhat lower than on substrates with a smooth surface, because the spatula can adapt only to some extent.

On surfaces with asperity sizes between 0.3 μm and 1.0 μm (critical roughness), the spatula is larger than the individual asperities, and an intimate contact between spatula and substrate cannot be achieved, which in turn results in a reduction of the real contact area and, thus, adhesion (Peressadko and Gorb 2004; Huber et al. 2007; Voigt et al. 2008; Bullock and Federle 2011; Hosoda and Gorb 2011; Wolff and Gorb 2012). The degree of force reduction tends to depend on the dimensions of the

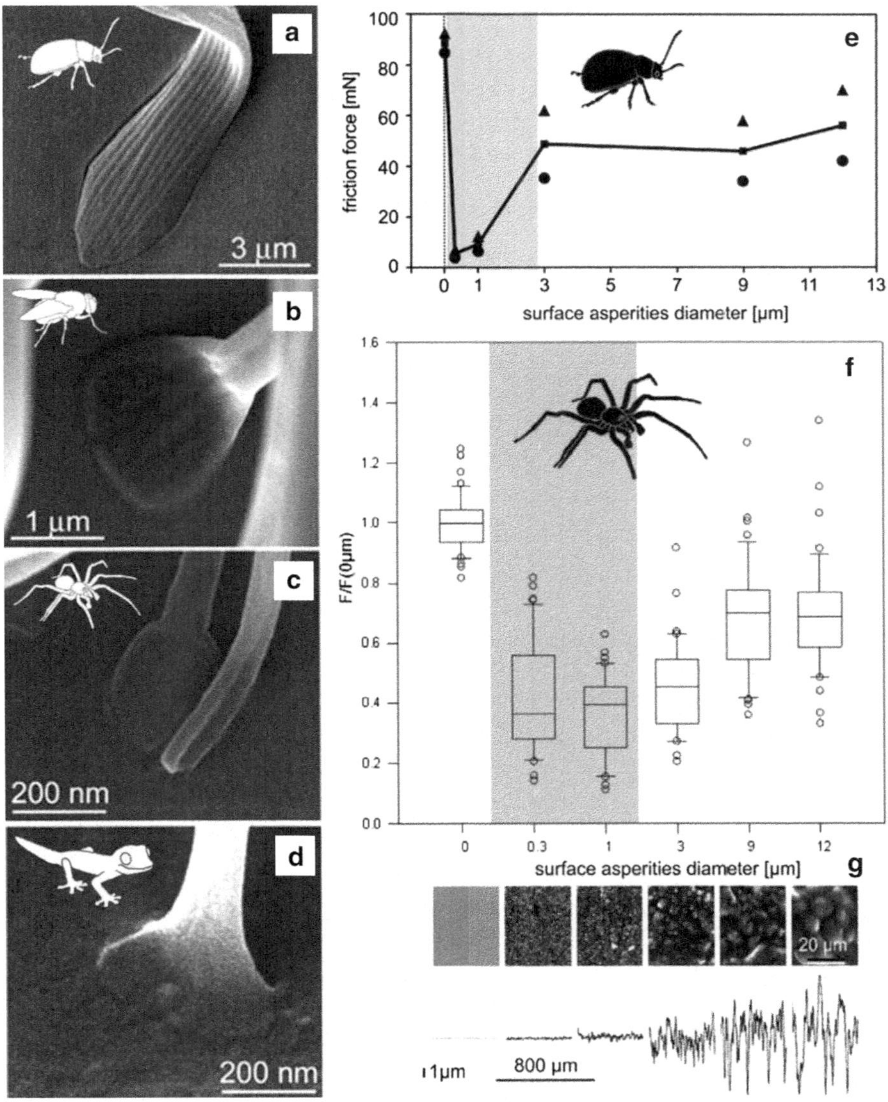

Fig. 4.1 (**a–d**) Spatula structures in different animals: the beetle *Gastrophysa viridula* (**a**), the fly *Calliphora vicina* (**b**), the spider *Cupiennius salei* (**c**), and the gecko *Gekko gecko* (**d**) (Varenberg et al. 2010). (**e** and **f**) Attachment abilities of beetles and spiders: (**e**) Friction forces obtained in the centrifugal force experiment with male (circles) and female (triangles) specimens of the beetle *Leptinotarsa decemlineata* (from Voigt et al. 2008). The line connects the average values for both sexes. (**f**) Maximal traction force achieved by the spider *Philodromus dispar* on epoxy resin surfaces with different asperity sizes. Data are normalized to the individual average of traction force maxima obtained on the smooth surface. Box endings correspond to the 25th and 75th percentiles; the line within shows the median; error bars define the 10th and 90th percentiles; outliers are marked by circles (from Wolff and Gorb 2012). (**g**) SEM images and profiles of tested substrates (from Wolff and Gorb 2012). The grey boxes within the graphs **e–g** indicate the *critical roughness*.

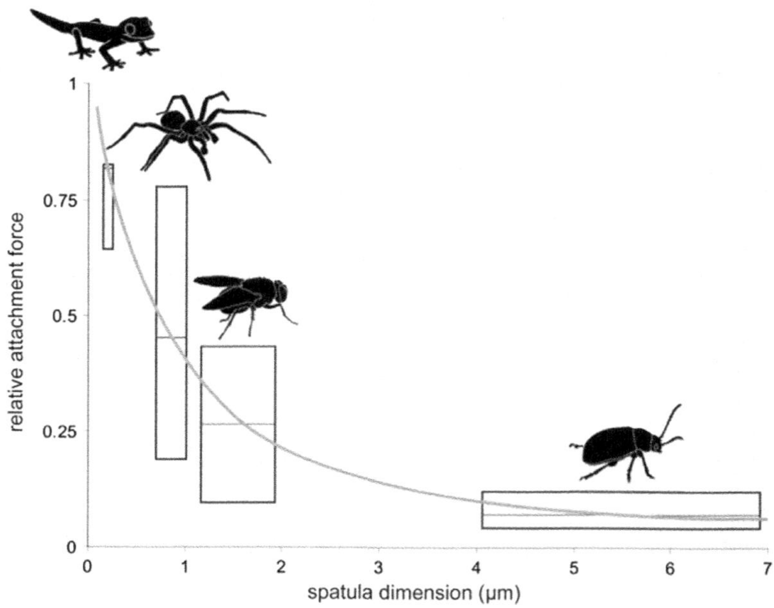

Fig. 4.2 Comparison of the attachment forces exhibited by the hairy attachment devices of lizards (*Gekko gecko*) (Huber et al. 2007), spiders (*Philodromus dispar*) (Wolff and Gorb 2012), flies (*Musca domestica*) (Peressadko and Gorb 2004) and beetles (*Leptinotarsa decemlineata*) (Voigt et al. 2008) on substrates with 0.3 µm asperity size. The relative attachment force decreases with increasing size of the contact-forming element. (From Wolff and Gorb 2012)

terminal elements: the larger the spatulae, the stronger the force reduction (Fig. 4.2). Thickness is another aspect which is crucial for contact formation with rough substrata. Insect spatulae are not just larger than those of geckos and spiders, but also thicker (beetle: 400 nm – Eimüller et al. 2008; fly: 180 nm – Bauchhenss 1979, Gorb 1998a; spider: 30–40 nm – Wolff and Gorb 2012; gecko: 20 nm – Persson and Gorb 2003; Huber et al. 2005a, b).

Here, we aim at developing a simple, but realistic 2D numerical model of animal spatula interaction with various substrate profiles. This model can well explain the results of previous experimental studies and predict the adhesion of different animals to actual surface profiles depending on the dimension and stiffness of their spatulae. We studied the model on numerically generated surfaces and on real 3D surface profiles obtained by using white light interferometer. The obtained results are interesting for studies of biological adhesive and anti-adhesive surfaces, but also for the development of adhesive and anti-adhesive biomimetic surfaces.

4.1.2 Contact Formation with Numerically Generated Rough Surfaces

To simulate the process of attachment, we applied two different modifications to the model, one of them based on numerically generated surfaces with different characteristic scales and the other one on a 3D white-light-interferometric scan of the real surface.

In the first approach, different degrees of roughness of the surfaces were achieved by depositing spheres with a defined radius on the originally flat surface. To simulate this process numerically, we modified a procedure of random deposition as outlined in Chap. 2. For a given radius *R*, we used an array of equally sized circles in 2D space (instead of the spheres in 3D space) and placed them sequentially into randomly chosen positions x_n, where $n = 1, 2, \ldots, N$. The number of the circles was defined by the total number of circles of radius *R* necessary to uniformly cover the length $L : N = \|L/R\|$. Each sphere was virtually added to the following segment of the surface:

$$\delta y_n(x; x_n) = \sqrt{\left|R^2 - (x - x_n)^2\right|}. \tag{4.1}$$

Due to the random deposition, the circles can fall either on top of already existing coverage or onto empty spaces. As a result of such a deposition, the total surface is gradually covered by circles, $y(x) = \sum_{n=1}^{N} \delta y_n(x; x_n)$, which are, generally speaking, placed on top of each other. Typical realizations of such surfaces for different radii *R* are presented in Fig. 4.3 by solid (blue) lines.

The spatula of a hairy attachment pad was simulated in the same manner as in previous studies (see the review by Popov et al. (2016) and references therein). It was represented as a set of elastically connected segments with strong longitudinal stiffness $k^{\|}$, preventing extension and compression of the practically rigid segments, and two components of elastic forces, $\vec{f}^{elastic} = \{f^{\|}, f^{\perp}\}$, acting between the neighboring nodes (joints) $\vec{r}_j$.

The longitudinal force, $\vec{f}^{\|}_{jj\pm1}$

$$\vec{f}^{\|}_{jj\pm1} = k^{\|}\left(\vec{r}_j - \vec{r}_{j\pm1}\right)\left[1 - \left|\left(\vec{r}_j - \vec{r}_{j\pm1}\right)/r_{0,jj\pm1}\right|^2\right] \tag{4.2}$$

tends to conserve the distance between the nodes $\vec{r}_j$ and $\vec{r}_{j\pm1}$ close to the equilibrium $r_{0jj\pm1}$. The equilibrium array $r_{0jj\pm1}$ is calculated from the trial study of an initial distribution of joints producing a realistic configuration. This array has to be taken as an initial condition. Due to the attraction factor $1 - \left|\left(\vec{r}_j - \vec{r}_{j\pm1}\right)/r_{0,jj\pm1}\right|^2$ in the longitudinal force (cf. Eq. 4.2), the spatula returns to its original form when there is no more interaction with the surface.

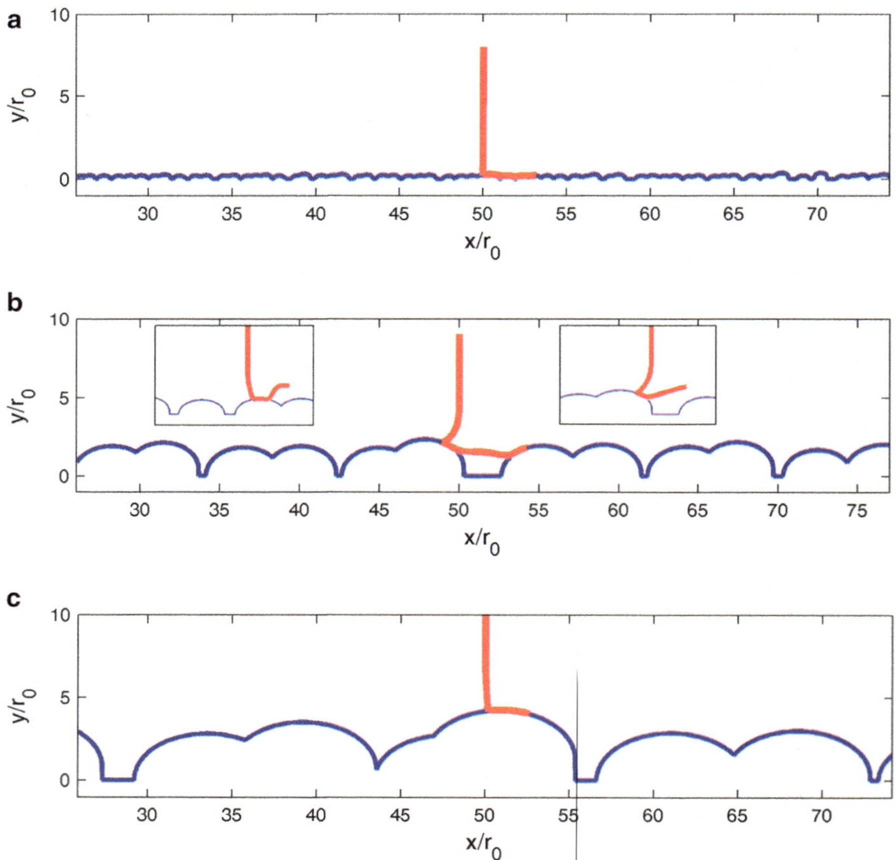

Fig. 4.3 Typical contact events of the spatula with (**a**) small, (**b**) intermediate, and (**c**) large surface asperities. The two inserts in (**b**) show different versions of a partial contact with a surface with intermediate size asperities. (From Kovalev et al. 2018)

As in all the previous models, we assume that there is a transverse stiffness in the joints (nodes in 2D) between the segments:

$$\vec{f}_j^{\perp} = k^{\perp}\left(2\vec{r}_j - \vec{r}_{j+1} - \vec{r}_{j-1}\right), \tag{4.3}$$

which represents the system's resistance to bending. Normally, transverse stiffness is much weaker than longitudinal stiffness, $k^{\|} > > k^{\perp}$.

Also, it should be noted that in a real system the spatula seems to have varying stiffness along its length. This property was also taken into account in the present analysis by using values of $k^{\perp}$ that vary along the system from segment to segment. From preliminary experiments we expect that $k^{\perp}$ should be represented by a nonlinear function which is almost constant at the basis of a seta and quickly declines near the tip (Peisker et al. 2013). For simplicity and definiteness in this particular analysis, the value of $k^{\perp}$ that we used for the vertical (basal) part of the model spatula was 10 times higher than the value used for the horizontal part (contact plate).

At small distances, all setae are attracted to the substrate due to either van der Waals interaction or surface tension in liquid bridges (Gilett and Wigglesworth 1932; Stork 1980a). In both these cases, the interaction strength rapidly decreases with increasing distance to the surface. For simplicity and definiteness in this particular study, we assume that the purely attractive adhesive force exponentially decreases with distance:

$$\vec{F}_s = F_0 \exp\left(-\left|\vec{r}_j - \vec{r}_k\right|/r_0\right). \tag{4.4}$$

Here, the vectors denote the nodes between the spatula segments, $\vec{r}_j$, and the nodes of the surface, $\vec{r}_k$, respectively. The symbol r_0 describes a characteristic distance of the interaction. Normally, r_0 varies between very few and a few dozens of nanometers. In the numerical study, this distance represents the natural scale of the problem under consideration. So, it will be further used to normalize all the distances in the study ($r_0 = 1$ μm).

Figure 4.3 illustrates the general idea of the study. For every particular radius R, we generated a large number of realizations of a rough surface, and we brought at least one spatula segment (with the fixed particular length S) into contact with the generated surfaces. Anyway, the introduction of an attracting surface is equivalent to the determination of the equilibrium spatula state. Due to the attraction force F_s, the number of spatula segments in contact with the surface increases with time. Taking into account that at small length scales the motion of such systems is normally overdamped (Diens and Klemm 1946), the temporal development of the system can be described by simple relaxation equations:

$$\partial \vec{r_j}/\partial t = (\gamma\tau)^{-1}\sum_j \left[\vec{F}_s^j + \vec{F}_{elastics}^j\right], \tag{4.5}$$

where the summation $\sum_j \left[\vec{F}_s^j + \vec{F}_{elastic}^j\right]$ accumulates all the elastic and surface forces acting on each spatula segment $j = 1, 2, \ldots, j_{\max}$, γ is a constant with the dimension N·m^{-1}, and τ is a characteristic relaxation time.

Both the instant interaction force and the fraction of attached spatula segments were monitored during our numerical experiment. The attraction force competes with the elastic force which tries to return the system to its original unperturbed configuration. After a certain time interval, the system will reach some final level of attached segments. All the spatula segments will come into contact with the substrate if interaction forces prevail. If elastic deformation prevails, only some fraction of segments will be in contact when the system reaches equilibrium.

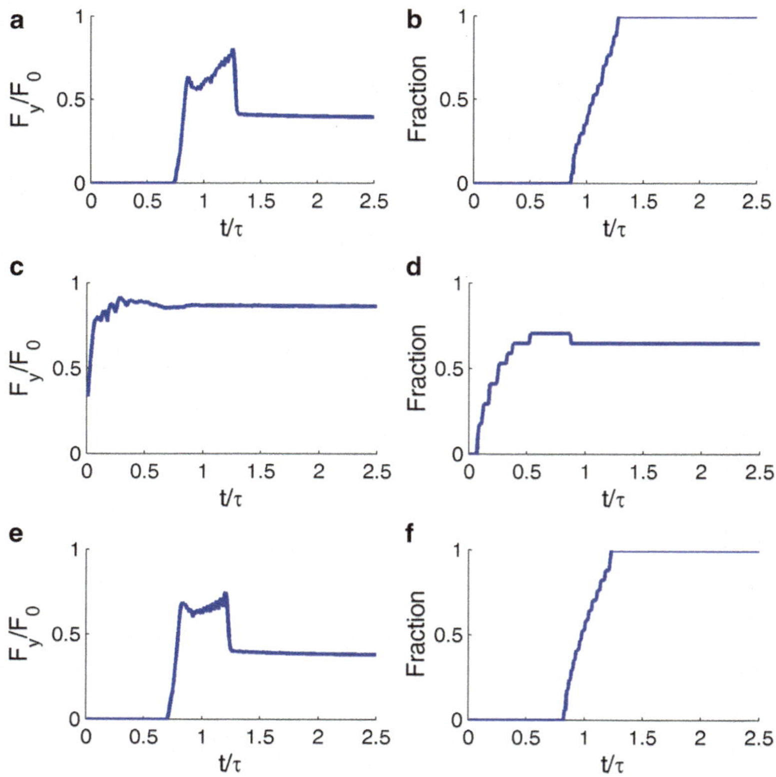

Fig. 4.4 Temporal development of the interaction force (**a**, **c**, **e**) and the fraction of attached segments (**b**, **d**, **f**) corresponding to the contact events in Fig. 4.3. τ is a characteristic relaxation time for Eq. 4.5. (From Kovalev et al. 2018)

The time-dependent curves of the interaction force and of the fraction of attached segments demonstrate that at sufficiently long time intervals, the curves will approach an equilibrium state. This process is clearly visible in Figs. 4.3 and 4.4. When the curves reach a plateau, the simulation may be stopped and the equilibrium values can be determined. Since the individual rough surfaces are randomly generated, the equilibrium values will vary from one numerical experiment to the next. Thus, repeated simulations should be performed to properly estimate the average values.

As Fig. 4.4 shows, there is a clear difference between the temporal development of spatula adhesion on surfaces composed of particles of intermediate size – in the range of the size of the spatula itself – and on surfaces composed of particles with either larger or smaller sizes. Whenever the irregularities of the substrate are extremely small, spatula deformation is practically independent of the substrate features since all the spatula segments are close to the surface and almost perfectly

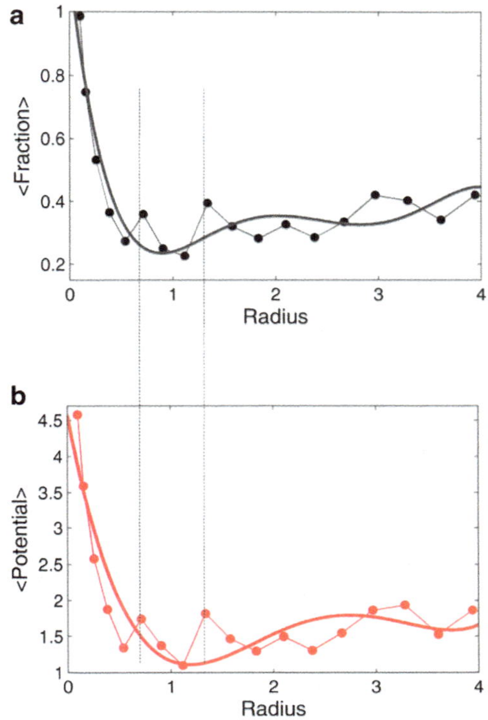

Fig. 4.5 Scale dependence of (**a**) the fraction of attached spatula segments and (**b**) the work (potential) of adhesion for substrates generated by randomly deposited spheres with characteristic radius R at a given spatula length S. The vertical lines mark region of weak adhesion. (From Kovalev et al. 2018)

attracted to it (Fig. 4.3a). In the case of large particles, the spatula is almost always positioned on the smooth region of one particle, which it can match by just some slight deformation (Fig. 4.3c). This matching between the substrate surface and the spatula does not always work, when the spatula length S and the characteristic particle size R are comparable ($S \sim R$). Some examples of such "unlucky" configurations are illustrated in Fig. 4.3b. Decrease in the equilibrium fraction of the attached segments coincides with the drop in the average adhesion work (potential) of adhesion. This region is marked by the vertical lines in Fig. 4.5.

4.1.3 *Contact Formation on Rough Surfaces Created by Gaussian Convolution*

The surface modeling approach for studying the spatula contact with a real 3D substrate presented in Sect. 4.1.2 represents a qualitatively consistent procedure since it produces fixed aperity sizes. In contrast, the application of Gaussian convolution allows to smoothen the surface, and in this way to simulate different

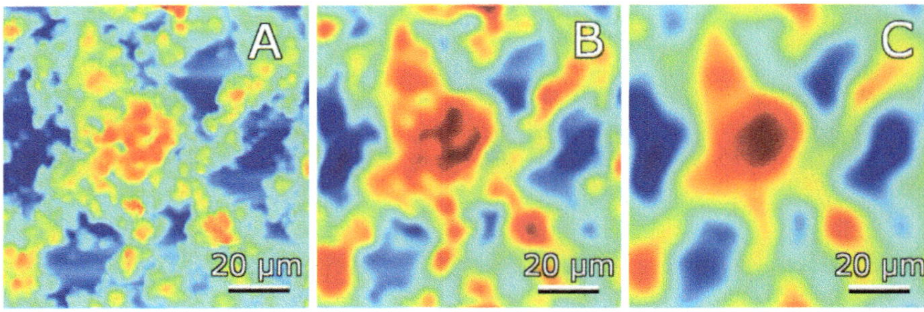

Fig. 4.6 (**a**) Original 3D surface and (**b** and **c**) two different stages of its smoothening using Gaussian convolution . Gaussian half width on the half height equals 5 μm (**b**) and 20 μm (**c**). Blue and red colors correspond to the minimum and maximum of the surface height, respectively. (From Kovalev et al. 2018)

Fig. 4.7 Typical contact of the spatula with a 2D profile from the surface presented in Fig. 4.6b. The insert shows an alternative variant of incomplete attachment of the spatula to the surface. (From Kovalev et al. 2018)

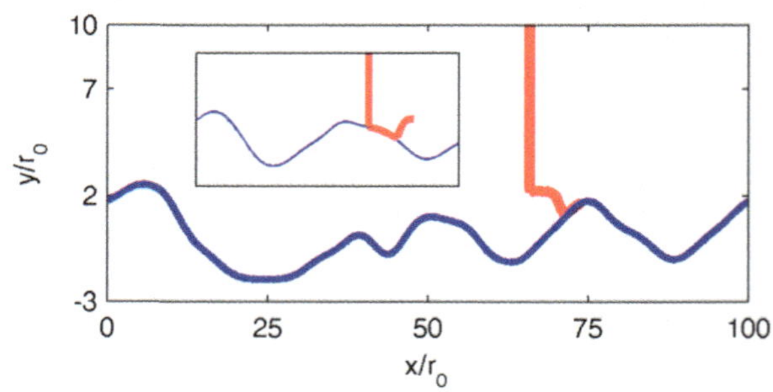

characteristic asperity sizes. For example, panels (b) and (c) of Fig. 4.6 present two different stages of the smoothening of an original surface (Fig. 4.6a) by Gaussian convolution.

A series of 2D profiles was extracted from the surface produced in such a way. A particular example of spatula interaction with a substrate with asperities of the most interesting size (i.e. in the range of the spatula size) is presented in Fig. 4.7. It shows a typical 2D profile of the 3D surface presented in Fig. 4.6b. The insert shows one of the alternative variants with incomplete attachment of the spatula to the surface. Qualitatively, the interaction between the spatula and the substrate looks very similar to those we have seen for surfaces created with particles of a fixed size.

With a substrate modeled in this way the spatula configuration strongly varies from one surface profile to the next, and therefore the spatula–surface interaction has to be calculated for a relatively large number of different surface profiles. We modeled 26 realizations, and each of the surfaces was independently smoothened using Gaussians with varied width. The average equilibrium fraction of the attached spatula segments and the average work of adhesion on differently smoothened surfaces are presented in Fig. 4.8. Similar to the results for the model surfaces presented above (Fig. 4.5), we found pronounced minima in the work of adhesion and the fraction of attached spatula segments.

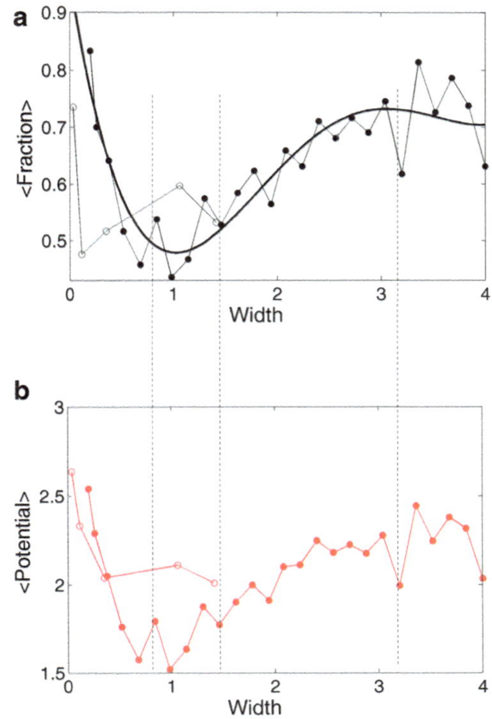

Fig. 4.8 Scale dependence of (**a**) the fraction of attached spatula segments and (**b**) the work (potential) of adhesion for scans of surfaces gradually smoothened by Gaussian convolution with varied width σ (●, closed circles) and of real surfaces (o, open circles) at a given spatula length S. The vertical lines mark region of varied adhesion. (From Kovalev et al. 2018)

4.1.4 Contact Formation with Real Substrates of Different Roughness

To verify the results obtained in Sect. 4.1.3 for surfaces created with Gaussian convolution, further spatula contact simulations were performed with 3D scans of five different polishing papers consisting of resin-bonded particles with sizes of 0.3, 1, 3, 9, and 12 μm, respectively (Buehler, Lake Bluff, IL, USA). For each substrate, 150 individual line scans, and five surface profile scans were used for the simulation. Color maps of the typical profiles of the five different polishing papers are depicted in Fig. 4.9. For a comparison with the results obtained by Gaussian convolution, the scale dependence of the fraction of attached spatula segments and the work of adhesion on these real surfaces are also provided in Fig. 4.8.

The numerical simulations all demonstrated a decrease in adhesion for a certain width of substrate roughnesses. This correlates well with the effects observed in adhesion experiments with real animal adhesive pads. The decrease in the model is found to be related to an incomplete contact formation between the spatula and the substrate if the latter is composed of particles with a size comparable to the size of the spatula. The spatula contact area is larger and the adhesion stronger on substrates with either larger or smaller surface roughnesses.

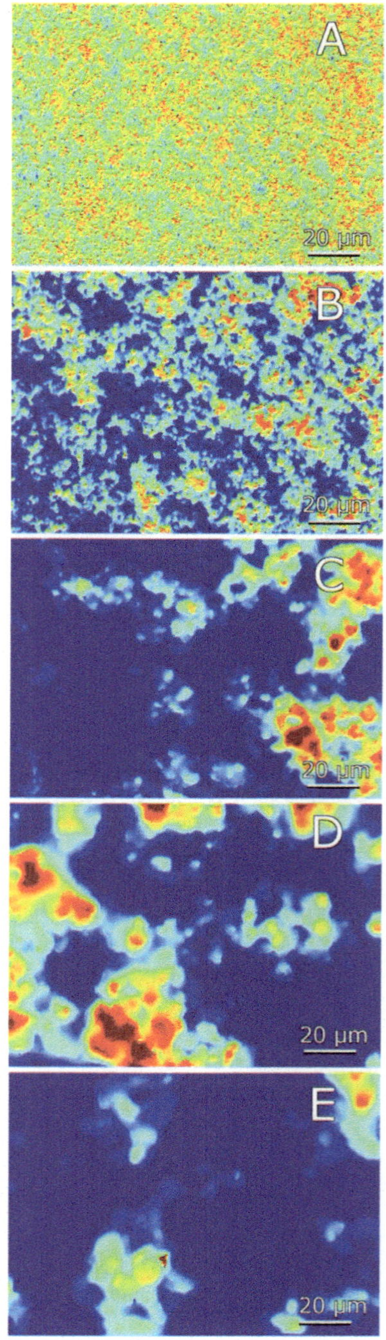

Fig. 4.9 Color maps of typical profiles of five different polishing papers used in the numerical simulations of rough surfaces as described in the text. (From Kovalev et al. 2018)

4.1.5 Biological Consequences of Roughness-Dependent Attachment Capabilities

The advantages of hairy adhesive pads of animals on rough substrates have been demonstrated in many experimental studies (Arzt et al. 2003; Persson and Gorb 2003; Gorb 2005; Kovalev et al. 2012; Heepe and Gorb 2014). It is also well known that their adhesion capability decreases at some specific substrate roughness (Voigt et al. 2008). By applying a numerical simulation it was possible to identify the relationship between the spatula size and the substrate roughness as the reason for this phenomenon. According to the model, this decrease in adhesion is associated with only a partial contact of the spatula due to the increased probability of "unlucky" configurations of the spatula in contact with substrates of some particular range of roughness. This conclusion can be drawn from the specificity of the spatula contact geometry derived from studying the model (Figs. 4.3 and 4.7).

A more precise characterization of the spatula geometry in contact with the substrate may be implemented in the model by determining the fraction of the attached spatula segments. We could see that the temporal development of the spatula fraction in contact with the substrate and of the attraction force is closely related to the model specific characteristic time. However, this temporal development is different for different substrate roughnesses. While the attraction force is more relevant for the contact building process, the spatula potential is more relevant for the contact breaking process, which is closely related to the work of adhesion and presents another way of characterizing adhesion.

Interestingly, on the substrates on which the spatula exhibits the lowest work of adhesion, only half of the spatula segments are in contact with the substrate according to the model. On substrates which are relatively smooth at the length scale of the spatula and composed of particles that are approximately three times larger/smaller than those on the substrate with the lowest work of adhesion, the fraction of the attached spatula segments is greater by around one and a half, while the work of adhesion is higher by just one third (Fig. 4.5). The same results were obtained when using scans of real rough surfaces and scans with surfaces smoothened by Gaussian convolution (Fig. 4.8). This approach thus validates Gaussian convolution as a method suitable for producing model rough surfaces for the numerical modeling of adhesion.

Our simple model correctly predicted the decrease in adhesion on substrates with some specific roughness; however, it underestimated the adhesion reduction in comparison with real experiments on animals (Gorb 2001; Peressadko and Gorb 2004; Voigt et al. 2008; Wolff and Gorb 2012). Thus, the attachment measured in experiments with setal adhesive pads of the Colorado potato beetle (Russell and Johnson 2007; Voigt et al. 2008), whose spatula has a diameter of around 5 μm, was significantly reduced on substrates composed of particles of the size 0.3–1.0 μm.

Models describing individual spatula adhesion on a rough substrate were also presented by Sauer and Holl (2013) and Zhou et al. (2014). In those studies, rough substrates were modeled as a set of equidistantly distributed cylindrical pillars or Gaussians, respectively. Adhesion on those rough substrates was not determined.

There are other theories which assume that the spatula forms a full contact with the substrate. According to these theories, enhanced adhesion of setal pads on rough substrates results from a random distribution of the seta length and/or their hierarchical organization (Arzt et al. 2003; Hui et al. 2005). The adhesion of a rigid spatula on a rough surface was also modeled by Peressadko and Gorb (2004). Further improvements of the model presented here will be made below:

- by taking into account the interfacial fluid which is present in insect adhesive pads.
- by taking into account that the way in which a spatula comes into contact with the substrate could be modified by shear forces applied to the spatula along the substrate (Filippov et al. 2011).
- by taking into account the detachment process (most probably by peeling) by solving the problem in 3D and considering the real shape of the spatula which might provide more realistic results than those obtained with the present model.
- by extending the current model to a hierarchical organization of setae, which could be the next step in the generalization of the setal adhesion problem.

In addition, it would be interesting to study other boundary conditions, such as the presence of friction in the contact.

4.2 Shear-Induced Adhesion of Biological Spatula-Like Attachment Devices

Contact formation by shear force is one of the possible reasons why most biological hairy locomotory attachment systems rely on spatulate structures. Shear force applied to spatulae which are initially oriented at various angles results in the proper alignment of the spatulae and an increase of the contact area to smooth and rough substrates. This principle emphasizes the critical role of the terminal elements (spatulae) in fibrillar adhesion.

The various attachment systems of insects, arachnids and reptiles have been intensively studied (see review by Creton and Gorb 2007) in order to reveal the functional principles behind their amazing dynamical adhesive performance. Hairy (fibrillar) types of such systems consist of arrays of hairs called setae usually containing one or several levels of hierarchy (Hiller 1968; Autumn et al. 2000; Gorb 2001, 2005; Kim and Bhushan 2007). This structural complexity allows them to form a large contact area on flat and rough surfaces, thus effecting strong adhesion based on a combination of molecular interaction and capillary attractive forces (Autumn et al. 2000; Autumn and Peattie 2002; Langer et al. 2004; Huber et al. 2005b). The hierarchical level of the seta that is directly responsible for the formation of an intimate contact with the substrate is not punch-like but, as mentioned above, rather resembles a thin spatula, or just a thin film (Stork 1980a, 1983; Gorb 1998a, 2008; Persson and Gorb 2003; Spolenak et al. 2005; Tian et al. 2006).

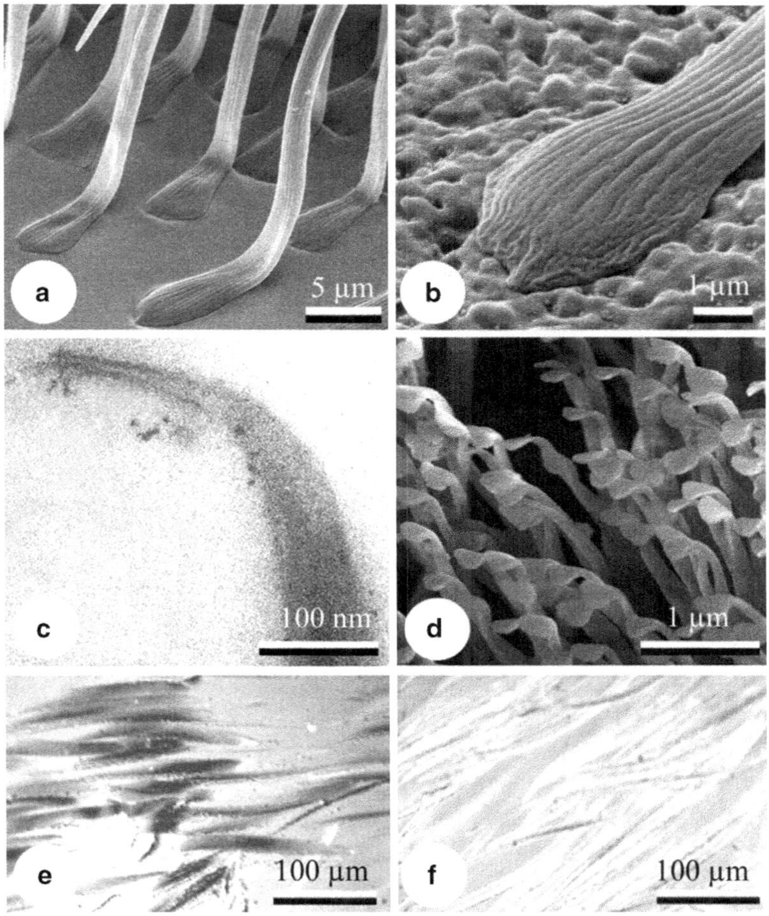

Fig. 4.10 (**a**, **b**) SEM, (**c**) TEM, (**d**) cryo-SEM and (**e**, **f**) light microscopy images of spatula-shaped thin-film terminal elements in hairy attachment pads of various animals. (**a**) Spatulae of the beetle *Gastrophysa viridula* in contact with a flat substrate. (**b**) Single spatula of the same species in contact with a substrate with an average asperity dimension in the range of 300 nm. (**c**) Longitudinal section of a single spatula of *Gekko gecko*. (**d**) Several spatulae of *G. gecko*. (**e**, **f**) Setae of the spider *Cupiennius salei* in the reflection light microscope during distal (**e**) and proximal (**f**) sliding on a glass substrate (black areas correspond to the sites of contact between spatulae and glass surface). (From Filippov et al. 2011)

Recently, the contact formed between an individual spatula-like terminal element of different animals and the substrate has been visualized by using cryo-SEM (Varenberg et al. 2010). Spatulae usually feature a gradient of thickness from their base to the tip (fly: Gorb 1998a; gecko: Persson and Gorb 2003; beetle: Eimüller et al. 2008) (Fig. 4.10b, c). In contact with a substrate, spatulae are aligned and orientated to the distal direction of the pad (Fig. 4.10a, b). Several hypotheses have been proposed for explaining the functional importance of such a contact geometry:

1. enhancement of adaptability to the rough substrate (Persson and Gorb 2003);
2. contact formation by shear force rather than by normal load (Autumn et al. 2000);
3. increase of the total peeling line by an array of multiple spatulae (Varenberg et al. 2010);
4. contact breakage by peeling off (Gao et al. 2005).

The Kendall peeling model (Kendall 1975) has recently been employed to explain the role of multiple spatulate tips in the enhancement of the total peeling line (Varenberg et al. 2010). It is well known that the application of a normal force can enhance adhesion (Popov 2010). However, for hairy attachment systems, the adhesion force always remains smaller than the initially applied normal force (Schargott et al. 2006). This would not be enough to enable walking on a ceiling. Another possibility of enhancing the adhesion force is application of a shear force. We have previously shown that the application of shear force to a particular direction of a spider's spatulate setae results in an increase of the real contact area (Niederegger and Gorb 2006) (Fig. 4.10e, f). Flies also employ shear movements with their attachment devices during contact formation (Niederegger and Gorb 2003). Autumn et al. (2000, 2006) revealed a strong shear dependence of the measured pull-off force in the gecko attachment system and even called this effect "*frictional adhesion*".

Since the effective elastic modulus of thin plates is very small, even for relatively stiff materials such as keratin or chitin-reinforced arthropod cuticle, peeling geometry is of fundamental importance for adhesion on rough substrates (Persson and Gorb 2003) due to the low deformation energy stored in the material during contact formation. We have previously shown that adhesion in such systems depends on the nature of the substrate roughness (Sect. 4.1; cf. Gorb 2001; Peressadko and Gorb 2004; Huber et al. 2007), and it is stronger in attachment devices containing predominantly spatulate tips (Voigt et al. 2008).

The important role of spatulate contact elements in the generation of adhesion has recently been experimentally demonstrated on artificial, bioinspired surfaces with similar geometry (Kim and Sitti 2006; Kim et al. 2007, 2008; Murphy et al. 2007). Recent theoretical considerations support the importance of applied shear force (pretension) for an increase of the peel-off force (Chen et al. 2009). However, it remains unclear how the contact on a rough substrate can be generated by shear, especially if the spatulae are initially not aligned in the plane of the substrate.

Below, we use a numerical approach to study the dynamics of spatulate tips during contact formation on rough substrates. The following questions are asked:

1. What is the role of the thickness gradient during contact formation?
2. Does applied shear force contribute to the enhancement of the contact area on the rough substrate?
3. Is there any optimal shear distance/force for a single spatula to form a proper contact?

4.2.1 Microscopical Examination of Various Spatulae

To visualize setal tips (spatulae), we examined the attachment pads of various representatives of insects, flies, spiders, and reptiles. Insect and spider tarsi were cut off from anesthetized animals with a fine razor blade. In the case of reptiles, molted toe skin was used. Some samples were brought in contact with a glass slide by using fine forceps. The contact was formed by applying a slight shear movement, as previously described to be the natural movement in contact formation in flies (Niederegger and Gorb 2003). Other samples were mounted on stubs, with their contact side up, for further examination by scanning electron microscopy (SEM), transmission electron microscopy (TEM) (for details of sample preparation for SEM and TEM see Gorb 1998a), cryo-SEM (for details of sample preparation see Gorb 2006) and light microscopy.

After air drying and sputter coating with gold–palladium, the samples were examined with a scanning electron microscope (SEM) (for further details of sample preparation, see Gorb 1998a). For cryo-SEM, the samples were mounted on metal holders, frozen in liquid nitrogen, and transferred to the Hitachi S-4800 cryo-SEM (Hitachi High-Technologies Corp., Tokyo, Japan) equipped with a Gatan ALTO 2500 cryo-preparation system (Gatan Inc., Abingdon, UK). Possible contamination by frozen crystals of condensed water was prevented by sublimating for 2 min (sample at −90 °C, cooler at −140 °C). After sublimation, the samples were sputter-coated with gold–palladium (to a thickness of 3–6 nm) in the preparation chamber and examined in the cryo-SEM at an accelerating voltage of 3 kV at −120 °C (for details of sample preparation, see Gorb 2006). Some samples were fixated, embedded in epoxy resin, dissected, mounted, contrasted, and examined in a transmission electron microscope (TEM) (for details of sample preparation, see Gorb 1998a, b).

Examples of the spatulate contact geometry of different animals are shown in Fig. 4.10. The SEM images demonstrate a similar contact geometry of the setal tips in different animal lineages. Independent of their dimensions, spatulae make flat contact with the substrate with their free ends oriented in the distal direction (Varenberg et al. 2010) (Fig. 4.10a, b). Spatulae have a gradient of thickness from the distal (thinner) to the basal (thicker) part (Fig. 4.10b, c) (Eimüller et al. 2008). On rough substrates, spatulae are partially able to adapt to the surface profile (Fig. 4.10b). In the non-contact state, spatulae are preferably, but not ideally, aligned (Fig. 4.10d). When shear force is applied in a certain direction (depending on the species and on the site where the leg is attached), spatulae make contact with the substrate (Fig. 4.10e). When applied in the opposite direction, spatulae lose the contact (Fig. 4.10f).

4.2.2 Numerical Modeling of the Shear-Induced Contact of Spatulae with Rough Surfaces

To simulate shear-caused adhesion in a typical spatula-like structure, we applied the following model: the terminal part of the spatula is treated as a flexible elastic plate with its width (and the corresponding flexural stiffness) gradually varying along the x-coordinate. The plate is brought into an initial contact with a rigid, rough horizontal surface ("ceiling"). Generally, the spatula is supposed to be initially inclined to the horizontal plane with a trial tilt angle α which varies in the interval $0 < \alpha < \pi/2$ from a horizontal to a vertical orientation.

The conceptual structure of the model is presented in Fig. 4.11, which shows a 2D projection of the system onto the x-z-plane. It is expected that the very thin end part of the plate is flexible enough to *be attracted* quite easily to the rigid rough surface by van der Waals forces. For the sake of simplicity, we simulate it by a gradient of Morse potential $U_{VdW}(r) = U_o(1 - \exp(-ar))^2$ with a physically reasonable amplitude $U_0 \simeq 10$ nN × nm and the minimum located at the distance $a \simeq 1$ nm from the surface.

The rigid ceiling surface may be simulated by the self-affine fractal surface given by the real part of $Z(x, y) = A \iint dq_x dq_y C(q) \exp(iq_x x + iq_y y + \zeta)$ with scaling spectral density $C(q)$. Here, A is the amplitude of surface roughness, i is an imaginary unit, $q_{x,\ y}$ are Fourier components along the x and y directions and ζ is a random phase. Details of how to generate the profile $Z(x, y)$ were described in Chap. 2 and in a number of previous papers (Filippov and Popov 2007a, b; Popov et al. 2007). It is well accepted in the current literature (Persson and Gorb 2003) that such a representation of the surface is suitable for the majority of physical surfaces at the scale-invariant spectrum $C(q) = 1/q^\beta$ and the exponent value $\beta \approx 0.9$.

We will now concentrate our attention on a single spatula. Accordingly, we restrict the amplitude A of the surface profile to $A = 1$ nm. With this amplitude, the procedure generates random realizations of a rigid surface with a typical roughness of up to 4 nm, which is typical for the considered scale.

Van der Waals attraction to the surface competes with the resistance of the spatula to bending. According to the theory of elasticity (Landau and Lifshits 1981), the elastic energy of the flexible plate is given by the following integral:

$$W_{elastic} = \frac{E}{24(1-\nu^2)} \iint dxdyh^3(x,y) \left\{ \left(\frac{\partial^2 z}{\partial x^2} + \frac{\partial^2 z}{\partial y^2} \right)^2 + 2(1-\nu) \left[\left(\frac{\partial^2 z}{\partial x \partial y} \right)^2 - \frac{\partial^2 z}{\partial x^2} \frac{\partial^2 z}{\partial y^2} \right] \right\}, \tag{4.6}$$

where E is the Young's modulus of the plate material and ν is the Poisson ratio which is typically equal to $\nu = 1/3$. In the considered systems, the adhesive force may be comparable with the weight of the whole animal. Thus, the gravitational force acting on the spatula is negligible. Furthermore, due to strong internal damping in the

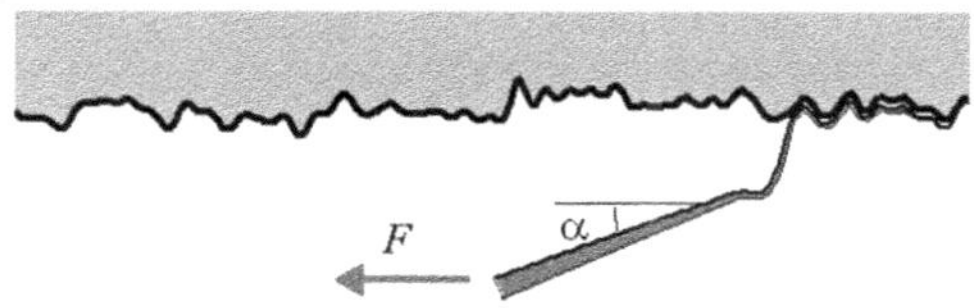

Fig. 4.11 Conceptual structure of the numerical model of an insect seta contacting a ceiling. We examine the terminal part of the "spatula" which is modeled by an elastic plate of variable thickness. It is brought into contact with the rough, rigid ceiling surface at an initial inclination angle α and pulled in the horizontal direction by an external force F. From Filippov et al. (2011)

spatula material, the system can be treated as over-damped. The over-damped dynamics of the system along the vertical z-coordinate is described by the equation of motion:

$$\gamma \frac{\partial z(x, y)}{\partial t} = -\frac{\partial W_{elastic}[z]}{\partial z} - \frac{\partial U_{VdW}[z]}{\partial z}, \tag{4.7}$$

where γ is the damping constant which determines a characteristic time scale of the process ($\gamma = 1$). Van der Waals bonding also produces a horizontal force $F^x_{VdW} = -\partial U_{VdW}[z(x)]/\partial x$. This force competes with an external shear force F^x. When F^x exceeds the total resistance of all instantly bonded segments $\int dxdyF^x_{VdW} > |F^x|$, the whole spatula moves along the x-direction according to the relation: $\gamma \partial x/\partial t = F^x - \int dxdyF^x_{VdW}$.

A typical numerically determined intermediate 3D configuration of the system described by Eq. (4.7) is shown in Fig. 4.12. The dynamic behavior leading to the picture in Fig. 4.12 is as follows: the spatula plate is initially attached to the surface with one of its end segments by the van der Waals force $F_{VdW} = -\frac{\partial U_{VdW}[z]}{\partial z}$. Over time, it relaxes to an equilibrium state in which it adheres to the surface by additional segments. The rate of attachment depends on the angle α between the plate and the surface, and it is normally faster for smaller angles α.

If the external force is nonzero, $F \neq 0$, it pulls the plate to the left and competes with the van der Waals attachment F_{VdW} of the "glued" segments. If the total attachment force is stronger than the horizontal component of the external force $\int dxdyF^x_{VdW} > |F^x|$, the spatula does not slide along the x-axis. However, the part which remains unattached can rotate and approach the surface $z(x, y) \rightarrow \langle Z \rangle$ (here, the symbol $\langle \ldots \rangle$ denotes a mean value) due to the action of the vertical component of the force $F^z > 0$; $F^z \sim z$. This rotation reduces the distances between the stiff and flexible surfaces and can greatly enhance the total adhesion. Generally, one can expect that a stronger shear force will cause faster attachment. However, if the shear force is too strong, exceeding the van der Waals bonding, it may even lead to the detachment of previously attached segments. In this case, the plate will start to slip along the surface, its rotation will stop and additional segments will not adhere to the ceiling. These qualitative considerations give rise to the following optimization

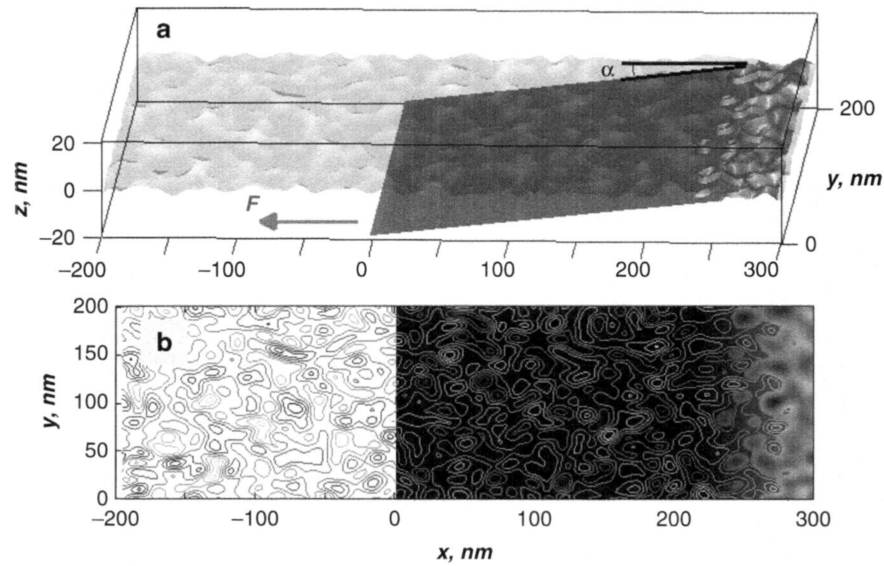

Fig. 4.12 (**a**) Typical 3D configuration of the elastic spatula (plate) at an intermediate stage of motion. The rigid ceiling is represented by the semi-transparent upper surface in the image. Instant configuration of the flexible surface is presented by the gray-scale map (the darker the color, the lower the value of the z-coordinate). (**b**) The image shows a contour plot of the rough surface where instant contact areas (shown in the gray-scale map) are directly visible. See also the two videos Movies 4.1 and 4.2 as electronic supplementary material. (From Filippov et al. 2011)

problem: up to what extent can the shear force be varied to stimulate attachment without rupturing the contact?

We performed two sets of numerical simulations: with a fixed initial inclination angle α and varying force F and with a fixed force F and varying angle α. The numerical experiment was organized as follows. First, in each experiment a rigid surface $Z(x, y)$ (with the same fractal properties in all setups) was generated as a 2D array. Each array had a size of [500, 200] cells (corresponding to an area of 500nm × 200nm), while a plate representing the terminal part of a spatula approaching the rigid surface had the size 200nm × 200nm. These sizes were found to be large enough to provide good statistics for substantial self-averaging of the integral values (like total force, fraction of the attached segments, etc.). Now we brought one end of the spatula into contact with the hard surface at some trial angle α and waited for a short period of time (about $t_0 = 2\mu s$ in dimensional units), during which the spatula starts to adapt spontaneously to the rigid surface (at zero external force $F_{t<t_0} = 0$). At $t = t_0$ we "turned on" a pulling force. This pulling process was continued up to a maximum time of $t_{max} = 10$ ms, which corresponds to the observed time for contact formation of the gecko toe (Gao et al. 2005).

Figure 4.13a shows how the fraction of the spatula in contact with the rigid surface (normalized to its total area) is growing at different values of the pulling

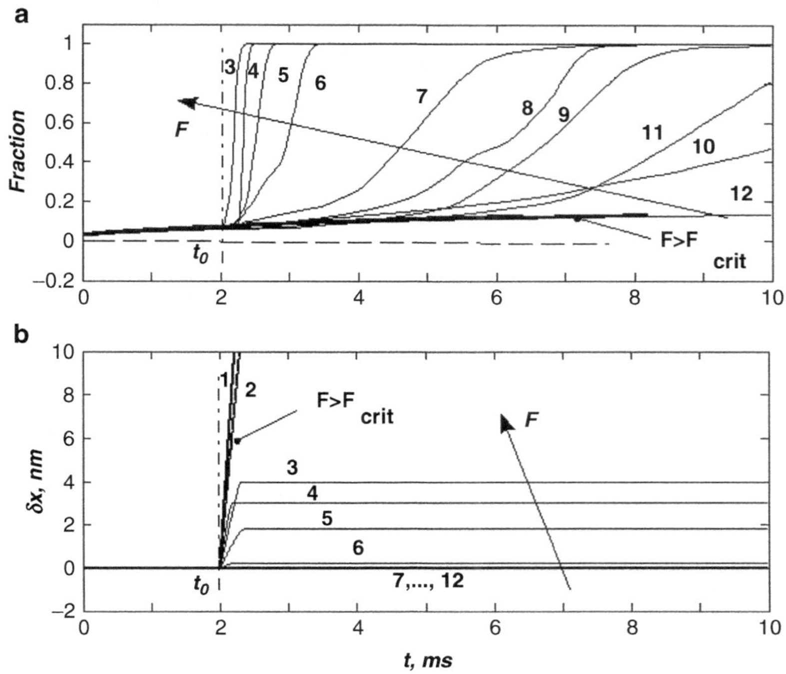

Fig. 4.13 Temporal development of (**a**) the attached fraction (contacting part of the spatula normalized to its total area) calculated at different values of the pulling force and (**b**) the corresponding drift δx of the attached end. The arrows in both panels mark the tendencies in fraction and drift changes at increasing force F. Bold curves 1 and 2 correspond to the forces $F = 22\,nN$ and $F = 20\,nN \geq F_{crit}$, i.e. higher than the critical value F_{crit}, which is sufficient to rupture the initial attachment and cause permanent slipping of the spatula. Curves 3–12 correspond to force F uniformly decreasing inside the interval $0 < F < F_{crit}$. All the curves are typical for an intermediate trial angle α (and obtained for a definiteness at fixed $\alpha = 0.2\pi$). (From Filippov et al. 2011)

force F and a fixed trial inclination angle $\alpha = const$ (here we used a representative intermediate angle $\alpha = 0.2\pi$). It can be seen that different random realizations of the substrate surface $Z(x, y)$ lead to small deviations as shown in Fig. 4.13a, but without affecting the general trend. This observation is important for ensuring that the obtained results are representative.

As expected, a higher force F leads to a faster decrease of the inclination angle. If the force is smaller than some critical value for detachment, $F < F_{crit} \approx 20\,nN$, which is in the range of forces previously experimentally measured for single gecko spatulae (Huber et al. 2005a), then the plate gradually tends to a horizontal orientation for $t \to \infty$. Whenever the force approaches the critical value F_{crit}, it becomes capable of breaking some of the already attached bonds and of causing a slight shift of the plate in a horizontal direction. Corresponding small horizontal displacements

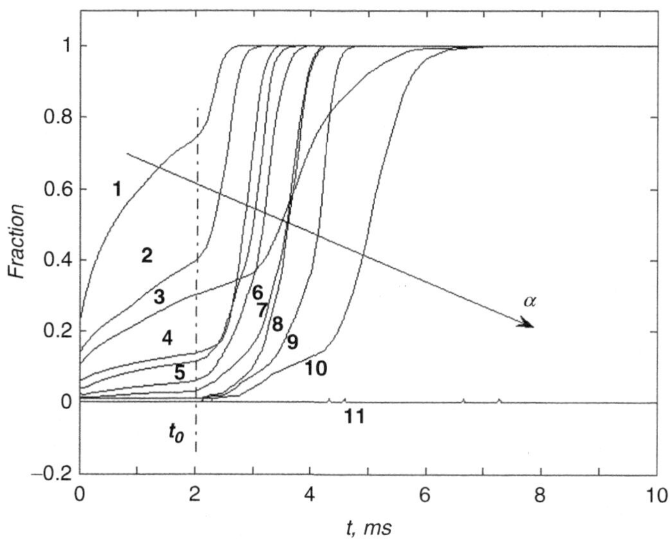

Fig. 4.14 Temporal development of the attached fraction of the spatula calculated at a fixed value of the pulling force $F = 10\,nN$ and an inclination angle which is gradually varied between an almost horizontal and a vertical orientation: $0.05 \cdot \frac{\pi}{2} \leq \alpha \leq 0.95 \cdot \frac{\pi}{2}$. Curves 1 and 11 correspond to the two extreme limits of this interval, respectively. The arrow marks a tendency of a decreasing fraction of attachment with a growing inclination angle. (From Filippov et al. 2011)

$\delta x = x - x_0$ of the system from its initial position x_0 are clearly seen in Fig. 4.13b (curves 3–6), where the temporal development of the shift $\delta x(t)$ is presented for different forces. The arrows in both panels (a and b) show how the temporal development changes with increasing external force F. Finally, if the force exceeds a critical threshold F_{crit} (curves 1 and 2), it completely disrupts the initial attachment and causes permanent sliding of the plate. In this case, the plate does not rotate and does not approach the horizontal orientation any further, thus not leading to any increase of adhesion. It is natural that in a biological system the animal cannot control the state (attachment and orientation) of each individual spatula, but presumably it can monitor the total resistance force of the entire spatula array, keeping it close to, but not exceeding, the critical value.

As mentioned above, the critical force F_{crit} depends on the initially attached area and thus on the inclination angle of a particular spatula. It is important to study variations of the time-dependent scenarios arising with changes of the initial angle α. The results of this variation are summarized in Fig. 4.14, where the temporal development of the attached fraction calculated at a fixed value of the pulling force $F = 10\,nN$ is shown for different trial angles α, which are varied in regular intervals from an almost horizontal to an almost vertical orientation: $0.05 \cdot \frac{\pi}{2} \leq \alpha \leq 0.95 \cdot \frac{\pi}{2}$. The two extreme variations of this angle are presented by curves 1 and 11, respectively. The arrow in the figure shows how the scenario varies with increasing angle in the range $0.05 \cdot \frac{\pi}{2} < \alpha < 0.95 \cdot \frac{\pi}{2}$. It can be clearly seen that the shear force enhances the attachment process, even at very acute angles (curves 1–3) since the fraction of attached segments spontaneously grows even at $t < t_0$.

4.2.3 Implications for Biological Systems

In contrast to the mushroom-shaped contact geometry adapted to long-term attachment, a spatulate contact geometry occurs widely in biological attachment devices used for locomotion (Gorb and Varenberg 2007). Locomotory function requires rapid and reliable contact formation and breakage, which can be achieved by the application of shear force (formation) and peeling (breakage) (Autumn et al. 2000, 2006; Niederegger and Gorb 2003; Gao et al. 2005; Tian et al. 2006; Chen et al. 2009).

Recently, Chen et al. (2009) generalized Kendall's (1975) model of an elastic film adhering to a substrate by incorporating the effect of pre-tension in the film. This has allowed these authors to investigate the effect of pre-tension on the orientation-dependent adhesion strength of a spatula on the substrate. The main result of Chen et al.'s (2009) study was that pre-tension can significantly enlarge the peel-off force at small peeling angles, while decreasing it at large peeling angles, resulting in strong reversible adhesion.

Our numerical model presented above demonstrates three additional functional aspects of spatulate geometry with regard to adhesion on smooth and rough surfaces:

- The contact area grows under applied shear force, especially when the spatulae are misaligned prior to contact formation, which is the case in the gecko system.
- The applied shear force has an optimum, which describes the situation when the maximal contact is formed, but no slip occurs. At such an equilibrium maximal adhesion can be generated.
- Due to the thickness gradient, the initial contact is formed by the spatula tip. This initial contact, together with applied shear, results in a further increase of the contact area, which is responsible for stronger adhesion.

These principles show the critical role of spatulate terminal elements in biological fibrillar adhesion.

However, even when the state of minimum free energy corresponds to complete contact, the elastic plate (spatula) may be trapped in a metastable state because of friction. In this case, because the kinetic friction is smaller than the static friction, sliding or vibrating the plate may increase the contact area (Niederegger and Gorb 2006). This effect is known experimentally: due to the sliding of spatulae for a short distance, the contact area (Niederegger and Gorb 2006) and friction/adhesion increase (Autumn et al. 2000, 2006). Thus, we can suggest that bio-inspired adhesives based on spatulate geometry have to be actuated according to the scheme preload–shear–peel, in contrast to terminal tips with a mushroom-shaped geometry (Varenberg and Gorb 2008).

4.3 Wet Attachment and Loss of the Fluid from the Adhesive Pads in Contact with the Substrate

4.3.1 Attraction Based on Liquid Bridges

The adhesive pads of some insects rely on wet adhesion caused by the pad fluid (Edwards and Tarkanian 1970; Bauchhenss 1979; Walker et al. 1985; Ishii 1987; Kosaki and Yamaoka 1996; Eisner and Aneshansley 2000) and have repeatedly been reported as having excellent attachment properties with high contact reliability (Gorb 2001, 2005; Federle 2006; Bullock and Federle 2009). In such wet adhesive systems, the liquid is squeezed out of tight contacts and builds bridges between two surfaces that are close to each other. The presence of pad secretion produced by specific epithelia is crucial for generating strong attractive forces and therefore a strong friction (Stork 1980a; Betz 2010). Aside from van der Waals and Coulomb forces, capillary forces mediated by the pad secretion are an important factor in the mechanism of those insects' attachment to substrates (Langer et al. 2004; Betz 2010).

Most insect pad secretions are probably emulsions containing water-based and lipid-based components (Gorb 2001; Federle et al. 2002; Vötsch et al. 2002; Betz 2010). This indicates that these secretions may serve various mechanical functions in making contact:

- Insect adhesive pad secretions form capillary bridges between setae and the substrate, increasing the contact area and hence both adhesion and friction forces.
- Non-Newtonian properties of the pad secretion are an additional mechanism preventing insects from slipping on smooth substrates (Gorb 2001; Vötsch et al. 2002; Betz 2010; Dirks et al. 2010). This mechanism probably results from the existence of yield stress when an emulsion droplet contains small enough sub-droplets.

The pad secretion of the beetle *Leptinotarsa decemlineata* was found to behave like a fluid with high viscosity (about a hundred times the viscosity of water) (Abou et al. 2010). Besides high viscosity, this beetle's pad secretion has a decreased evaporation rate, which is a crucial issue for micrometer-sized droplets (1–7 μm) because it ensures that the adhesion is robust under a variety of conditions (Abou et al. 2010).

In a series of experiments, it was shown that surface roughness strongly influences the attachment of insects. The adhesive pads of insects generate a much higher friction on either smooth surfaces or surfaces with large roughnesses (above 3 μm), whereas friction is lowest on substrates with a roughness ranging from 0.3 μm to 3 μm (Gorb 2001; Peressadko and Gorb 2004; Gorb et al. 2008). On rough substrates, wet attachment devices of insects are more effective than dry ones since the effective pad contact area is much larger when pad secretion covers substrate asperities (Drechsler and Federle 2006).

The microstructure of many plant surfaces strongly reduces insect attachment forces (Edwards 1982; Stork 1986; Edwards and Wanjura 1990; Eigenbrode 1996; Eigenbrode and Kabalo 1999; Brennan and Weinbaum 2001; Eigenbrode and Jetter 2002; Gaume et al. 2002; Gorb et al. 2005, 2008). To explain the anti-frictional properties of plant substrates covered with microscopic wax crystals, it was suggested that micro- and nanostructured surfaces may absorb the fluid from adhesive pads (Gorb and Gorb 2002). It is intuitively clear that fluid loss will be higher on rough surfaces. Indeed, it has been shown that on porous substrates, insects may produce significantly reduced forces, partially due to absorption of the fluid from the pads (Gorb et al. 2010). In this case, the reduced friction may be explained by the reduction of the fluid contact area if the rate of fluid loss is significantly higher than the rate of fluid production (Dirks and Federle 2011).

However, it remains unclear how the fluid moves on substrates with different roughnesses and different surface energies. To study how the fluid moves from the pad to rough substrates during contact formation we formulated a fluid loss model (see Sect. 4.3.3). The following questions were asked:

(i) Is fluid production on rough substrates stronger as compared to smooth substrates?
(ii) Does the amount of fluid remaining on the insect pad after the insect has moved on a substrate depend on the density of the substrate asperities?
(iii) How do the aspect ratio of the substrate asperities and the surface energy of the substrate (substrate affinity to the fluid) influence the amount of fluid lost during subsequent contact formation?

4.3.2 Microscopic Examination of Insect Prints with Wet Adhesion

To visualize the behavior of pad fluids during contact formation, we examined footprints of the attachment pads of the beetle *Gastrophysa viridula* (Coleoptera, Chrysomelidae) and the flies *Episyrphus balteatus* (Diptera, Syrphidae) and *Calliphora vicina* (Diptera, Calliphoridae) under a phase contrast light microscope and by cryo-SEM. Insect tarsi were cut off from anesthetized animals using a fine razor blade and were subsequently brought into contact with a glass slide using fine forceps. Contact was formed by applying a slight shear movement, as this has previously been described to be the natural mode of contact formation in flies (Niederegger and Gorb 2003). The samples were mounted on metal holders, frozen in liquid nitrogen, and transferred to a Hitachi S-4800 cryo-SEM (Hitachi High-Technologies Corp., Tokyo, Japan) equipped with a Gatan ALTO 2500 cryo-preparation system (Gatan Inc., Abingdon, UK). Possible contamination by frozen crystals of condensed water was eliminated by sublimating samples for 2 min (sample at −90 °C, cooler at −140 °C). After sublimation, samples were sputter-coated with gold–palladium (thickness of 3–6 nm) in the preparation chamber and examined in the SEM at an accelerating voltage of 3 kV at −120 °C (for more details of sample preparation see Gorb (2006)).

Artificial micropatterned polymer samples of the insect pads (Gorb et al. 2007), which were initially covered with a thin layer of oil (Mobile DTE oil ISO VG 46; ExxonMobil Lubricants & Specialties Europe, Antwerpen, Belgium), were used to make prints on various (smooth and rough) substrates. The prints were visualized using a stereo microscope M 205A (Leica, Wetzlar, Germany).

Examples of insect footprints on smooth substrates are shown in Fig. 4.15a–d. The phase contrast light microscopy images show that fly footprints have a hexagonal pattern of discrete fluid droplets, corresponding to the distribution of the tenent hairs of the attachment pad (Fig. 4.15a). Cryo-SEM images demonstrate that the shape of droplets that remain on the substrate and are frozen after breaking the contact between the insect foot and the substrate is convex (Fig. 4.15b). In contrast, when the foot in contact with the substrate is first frozen and then the contact between the foot and the substrate is broken, the frozen droplets remaining on the substrate are concave (Fig. 4.15c, d). This concave shape might be due to the fact that the largest portion of the fluid has been pressed out of the contact, thus leading to a meniscus-like appearance of the fluid at the circumference of the contact (Fig. 4.15c, d). If the fluid is not renewed, its amount will decrease in a series of sequential prints/steps. Such a decrease was observed to be greater on rough substrates (Fig. 4.15e, f).

4.3.3 Fluid Loss Model

The following 2D discrete model was proposed for elucidating the fluid transfer from elastic setae to a rough substrate in a series of footprints (Kovalev et al. 2013). In this model, fluid is initially only present on the surface of the setae. This surface is then brought into contact with the surface of a substrate and the distribution of the fluid between the two interacting surfaces is examined. After some time of equilibration, the surfaces, which all have a different affinity to the fluid, are separated again. The fluid fraction that remains on the elastic substrate surface (seta) during ten contact cycles may be analyzed. The fluid distribution between the two interacting surfaces can be described by time-dependent mesoscopic fluid density variations. The model includes a term that stabilizes the fluid density. The model seta is elastic and can conform to the topography of the substrate surface.

The utilized model is based on a standard approach of physical kinetics (Pitaevskii and Lifshitz 1981). It includes two contacting surfaces with a liquid layer between them. The surfaces are defined on discrete numeric grid which is uniform in the **x**-direction (Fig. 4.16).

The lower surface is assumed to be non-deformable and represents a typical, rough, rigid substrate. Generally, it has a complex, even fractal structure (Persson and Gorb 2003; Peressadko et al. 2005), which, for the purposes of the present

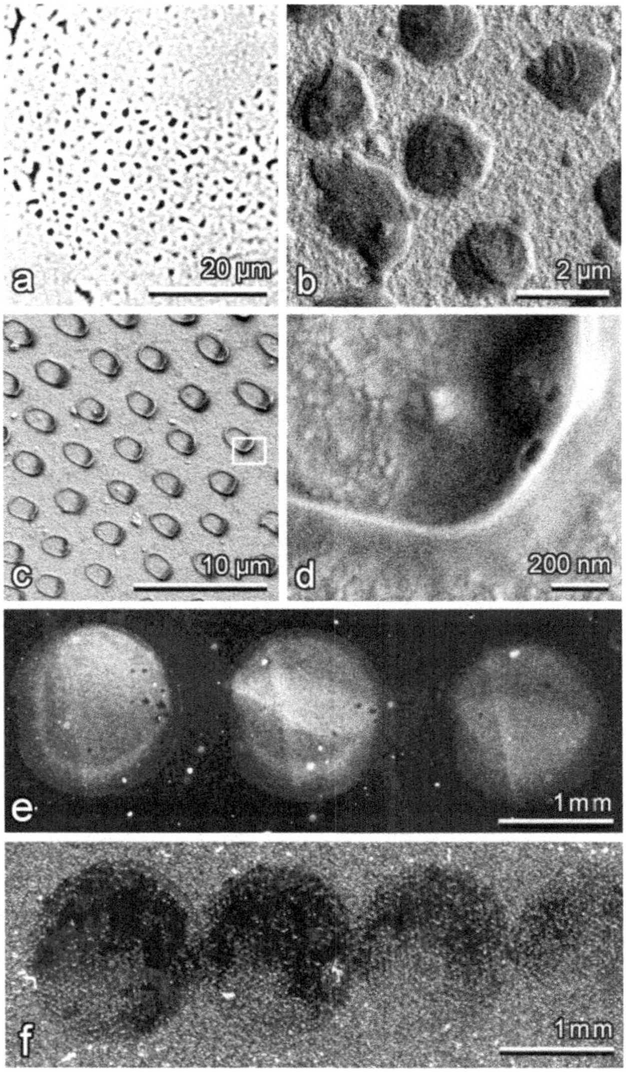

Fig. 4.15 Footprints of insects (**a–d**) and fluid residues from oil-coated artificial microstructured pads (**e**, **f**) on various substrates. (**a**) Footprints of the fly *Episyrphus balteatus*, phase contrast microscopy. (**b**) Cryo-SEM images of footprints of the fly *Calliphora vicina*. (**c** and **d**) The liquid residue after freezing the *C. vicina* foot with the liquid interface adhering to the surface after contact breakage. (**e** and **f**) Serial prints by artificially produced microstructured and oiled polyvinyl siloxane samples on a smooth (**e**) and a rough (R_a = 1.31 μm) (**f**) substrate. Note the different amounts of oil left on the substrate in the subsequent prints. (From Kovalev et al. 2013)

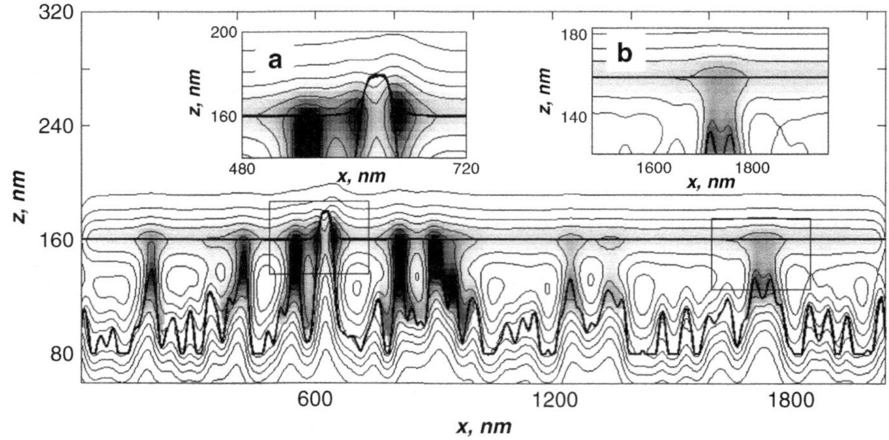

Fig. 4.16 Conceptual scheme of the experiment modeling the fluid distribution between an adhesive pad and a rough surface. Two potential minima, which correspond to the surfaces of an elastic seta (upper surface) and a rough substrate (lower surface), respectively, are shown by the bold line. The potential relief is represented by the thin lines of the contour plot. The fluid layer is shown as a grayscale area, with darker hues corresponding to higher densities. Two different meniscus types (described in the text) are shown in magnified representations in insets (**a**) and (**b**). (From Kovalev et al. 2013)

model, was generated by accumulation of the modes with different wave numbers (Filippov and Popov 2007a, b; Popov et al. 2007).

The upper surface corresponds to the adhesive pad of the insect or to that part of it which is normally flexible and adaptable to the substrate profile (Gorb and Beutel 2001; Eimüller et al. 2008; Filippov et al. 2011). As the simplest approximation, the adjusted geometry of the upper surface is represented by the following equation:

$$\frac{d^2 z_u}{dx^2} = -K(z_{0,u} - z_u), \tag{4.9}$$

where K is the effective stiffness of the upper surface, ($3{\cdot}10^7\ \mathrm{m}^{-1}$), $z_u(x)$ is a function that describes this upper surface, $z_{0,u}$ is zero level of z_u and corresponds to the effective loading.

Several individual cases of the fluid layer distribution (trial distributions) near the upper surface were modeled as uniform along the surface and a Gaussian function in a perpendicular direction, with a half maximum width of ~50 nm (Schuppert and Gorb 2006), which corresponds to the fluid layer thickness and was estimated from Fig. 4.15b, d.

The chemical potential of the fluid at an arbitrary point i between the surfaces is modeled by the Morse potential: $U_{i,\,j,\,k}(x,\,z) = U_{0,\,k}(1 - \exp\{-r_{ij}/r_0\})^2$. Here, r_{ij} represents the distance from point i to a point j on a surface $k = 1,2$ (upper or lower), $r_0 = 25$ nm. The impact of each discrete segment used in numerical simulation on the total potential is proportional to its length. Due to non-uniform discretization along the z-axis this length is proportional to the factor $\sqrt{1+(dz_k/dx)^2}$. As a result, the total surface potential takes the following form:

$$U_i = \sum_{j,k} U_{0,k}\left(1 - \exp\left\{-r_{ij}/r_0\right\}\right)^2 \sqrt{1+\left(\frac{dz_k}{dx}\right)^2}\Bigg|_{x=x_j}. \tag{4.10}$$

During the simulation, the fluid flow from the upper to the lower surface is studied. Initially, the fluid is present only on the upper surface. Then, the two surfaces are brought into contact for 100 ms. This time interval was selected as approximately corresponding to the time interval for which a single insect foot is in contact with a substrate during walking (A. Kovalev and J. Langowski unpublished data). During this interval, the fluid is allowed to redistribute between the surfaces according to the following kinetic equation:

$$\gamma^{-1}\frac{d\rho}{dt} = \eta_{eff}\Delta\rho - \rho\left[U_i - (\rho - 1)(2\rho - 1)\right], \tag{4.11}$$

where $\rho(x, y)$ is the effective particle density of the liquid, $\Delta = \partial^2/\partial x^2 + \partial^2/\partial y^2$ is a Laplace operator, η_{eff} is the diffusion coefficient or effective surface tension ($6{\cdot}10^{-4}$ m^2 s^{-1}), γ^{-1} is the characteristic time necessary for forming a liquid bridge between seta and substrate (10 ms, the time needed for the wet spatula of a Colorado beetle attachment pad to establish the equilibrium contact, Abou et al. 2010). This time is proportional to the viscosity of the pad fluid (Abou et al. 2010). The local term $\rho[U - (\rho - 1)(2\rho - 1)]$ in Eq. 4.11 takes a standard form for the chemical potential $F(\rho) = [-U\rho^2/2 - \rho^3 + \rho^4/2]$. To fulfill the requirement of mass conservation in the absence of liquid sources or drains, we included an additional operation into the numerical procedure which keeps the total amount of liquid fixed at each iteration step.

After a contact time of 100 ms, the surfaces are separated. At each point, the fluid now redistributes between the two surfaces. The amount of liquid must be divided according to the mutual relation between the local surface potentials at any given point. In other words, the fluid will remain on the surface with the deeper local potential. The next cycle starts with the remaining fluid homogeneously redistributed on the upper surface (attachment pad), which is brought into contact with a new dry and rough surface. This procedure is repeated 10 times.

The upper surface loses fluid at each cycle. It is convenient to characterize this process by the fraction of the fluid remaining on the upper surface normalized to the initial amount of the fluid. We will use this '*fraction*' as a main parameter in the next section below.

4.3.4 Influence of Various Factors on the Fluid Distribution

The chemical potential of the substrate in the sense of affinity to the fluid (which is analogous to the surface energy) was analyzed as a second factor influencing the fluid redistribution between the two contacting surfaces. The ratio of the affinities of the two surfaces was chosen as a characteristic parameter: $\mu = U_{0,\ up}/U_{0,\ down}$. Here, $U_{0,\ up}$ and $U_{0,\ down}$ correspond to the two different coefficients U in Eq. (4.10) for the upper (= seta) and lower (= substrate) surfaces, respectively.

Figure 4.16 presents a typical experimental configuration. Potential minima for the upper and lower surfaces are shown by the thick contour line. The shape of the liquid layer is displayed as a grayscale map. One can see how the fluid is redistributed in the total potential formed by both surfaces. The fluid fills the potential minima and forms clearly visible liquid bridges. Two basic types of such bridges can be easily distinguished in the picture. One type, which appears where the surfaces are in direct contact, is shown in inset (a) of Fig. 4.16. In such contacts, a liquid meniscus with high density surrounds the contacting surfaces. The second type (inset b in Fig. 4.16) is formed in areas of close proximity between the surfaces. In this case, both the fluid density and the flow rate decrease monotonously with increasing distance between the surfaces.

We will now separately examine the dependence of the draining rate on the chemical and geometrical properties of the rough surface. Keeping this in mind, it is useful to first use a regular hard surface with just one periodic roughness component. In this case, the liquid loss depends on the structure density n_s, which equals the number of structures (roughness components) per 1 μm. All the menisci in this case are of the type shown in inset (a) of Fig. 4.16.

The fluid fraction remaining on the upper surface (seta) during consecutive contact cycles is presented at a logarithmic scale in Fig. 4.17. The set of curves here displays a dependence of the fluid fraction on the structure's density n_s. It is

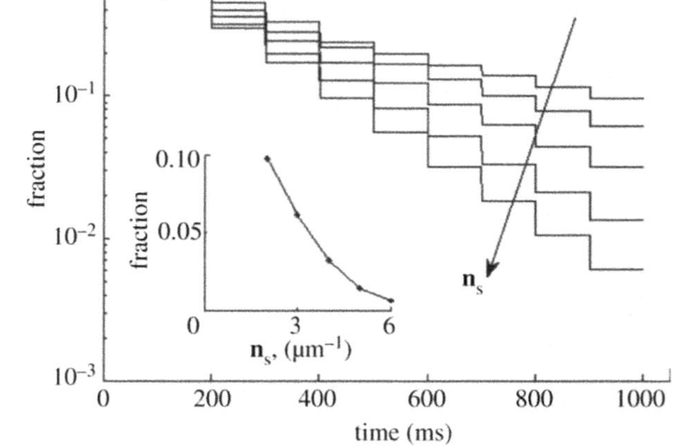

Fig. 4.17 Fluid fraction remaining on the elastic setae (upper surface) in sequential contact cycles on a rough substrate with regular roughness structures. The tendency of growing structure density, n_s, is indicated by the arrow. Dependence of the fraction of liquid remaining on the seta after the final contact cycle on n_s is shown in the inset. (From Kovalev et al. 2013)

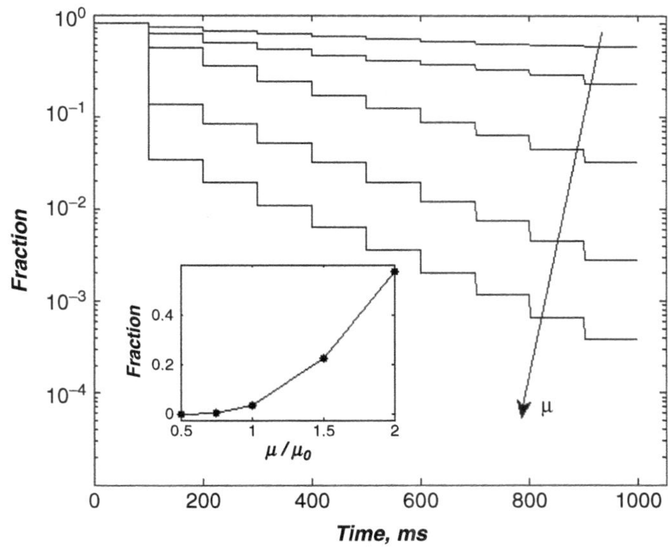

Fig. 4.18 Fluid fraction remaining on the elastic setae (upper surface) in sequential contact cycles on substrates with different affinities to the liquid (μ). The tendency with growing μ is indicated by the arrow. The dependence of the fraction after the final contact cycle on μ/μ_0 is shown in the inset. (From Kovalev et al. 2013)

important to note that the curves here are close to the exponent (linear in a logarithmic scale) at high n_s only, but differ from it when n_s decreases, due to the non-linear nature of Eq. (4.11). The fraction of the fluid left on the upper surface after all the contact cycles for different values of n_s is summarized by the inset in Fig. 4.17. The graph shows the fast non-hyperbolic decrease of the remaining fluid fraction depending on n_s.

The reason for such dependence is that the fluid loss is controlled by two factors: the number of menisci and the flow path length (the average distance the fluid has moved during the contact time). Both these factors are proportional to n_s. It is important to note that about 10% of the fluid remains on the upper surface after 10 contact cycles for $n_s = 2\mu\text{m}^{-1}$.

Now we modeled the influence of the different affinity of the surfaces to the fluid on the fraction of the fluid remaining on the seta. The set of curves in Fig. 4.18 represents the fraction kinetics at a logarithmic scale for different values of μ when $n_s = 4\mu\text{m}^{-1}$. Because of the non-linearity of Eq. 4.11, the shape of the kinetic curves differs from the naively expected exponent. This difference is especially pronounced at $\mu = U_{0,\ up}/U_{0,\ down}$.

The dependence of the liquid fraction remaining on the upper surface after 10 contact cycles on μ/μ_0 is shown in the inset in Fig. 4.18. It is visible in the figure that the ratio of the affinities, μ/μ_0, has a stronger influence on the fraction than the period of roughness components. If the upper surface has a two times higher affinity to the fluid than the substrate, it will lose just 43% of the fluid after 10 contact cycles.

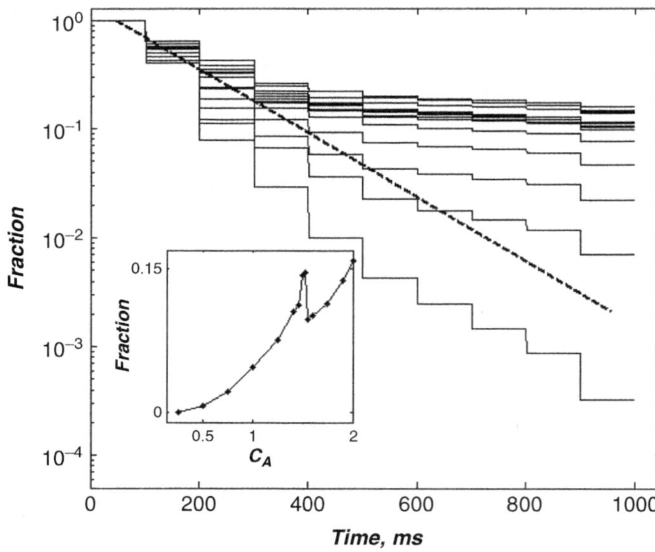

Fig. 4.19 Fluid fraction remaining on the elastic setae (upper surface) in sequential contact cycles on the substrate with different effective aspect ratios of roughness. The expected fluid loss on a completely flat substrate is shown by the dashed line. Dependence of the fluid fraction remaining on the seta after the final contact cycle on the roughness is shown in the inset. (From Kovalev et al. 2013)

Next, we modeled the influence of the amplitude of the roughness components on the surface on the fraction of the remaining fluid. The set of kinetic curves for different roughness amplitudes with fixed parameters of $\mu = 1$ and $n_s = 4\ \mu m^{-1}$ is shown in Fig. 4.19. The inset in Fig. 4.19 summarizes the fraction values after 10 contact cycles for different roughness amplitudes. It is directly visible in the figure that the remaining fluid fraction is higher for a surface with random size of the asperities than for one with periodic asperities. Two factors may be responsible for this phenomenon. The average number of menisci is the same for both fractal and periodic surfaces, but some of them are of the type shown in inset (b) in Fig. 4.16, which have a different impact on the fluid flow because, according to the kinetic Eq. (4.11), the flow rate is lower in more remote contacts. Additionally, when the distance between neighboring peaks is irregular, the mean flow time is longer.

A simple geometric sequence is expected for the contact between absolutely flat surfaces. Such a sequence corresponds to an exponential decrease as indicated by the dashed line in Fig. 4.19. The fluid loss rate is proportional to the original amount of liquid. That is why, for some ranges of roughness, the rate of fluid redistribution is initially faster than the expected exponent, but becomes slower in later approaches. For a substrate with small roughnesses, the initial rate of fluid loss is typically higher than for a smooth substrate. Since soft setae surface fits the smooth structures better, the fluid can directly reach the pits in the hard surface. Besides, the effective area of the lower surface is bigger than that of the flat one. At a roughness amplitude of

$C_A = 1.5$, the aspect ratio of the basic structures is close to 1. This changes the impact of the surface tension (represented by the Laplacian term in Eq. 4.11) on the kinetics and leads to better conservation of the fluid fraction on the upper surface. This effect manifests itself by a local peak seen in the inset in Fig. 4.19. In the main part of the figure, this peak corresponds to an area densely filled by the majority of the curves.

4.3.5 *Discussion of the Numerically Obtained Results and Biological Consequences*

The aim of this model was to estimate the fluid expended by an insect running on surfaces of varying roughness, since the fluid contributes to the attachment on rough surfaces. There is an optimal volume of pad secretion at which the pad adhesion reaches its maximum. When secretion accumulates for some minutes, this may lead to the flooding of the attachment pads (Fig. 4.15c, d). In the flooded state, both the capillary forces and pad adhesion are reduced. Yet after an insect has taken several steps (seven steps for a cockroach on a smooth substrate), the pad secretion volume approaches a stationary state in which the accumulated secretion is largely transferred to the substrate and the fluid lóss is compensated for by a rather slow influx of newly secreted fluid. High fluid loss on rough substrates (Fig. 4.15e, f) causes a reduction in adhesion, due to the reduced fluid contact area and therefore reduced capillary forces.

The ability to adhere to rough substrates provides some important benefits to insects, such as exploiting new food sources, reducing competition with other species, and escaping parasites/predators that are unable to hold on to such substrates. The influence of surface roughness on insect attachment has been experimentally examined in several studies (Gorb 2001; Peressadko and Gorb 2004; Drechsler and Federle 2006; Gorb and Gorb 2008; Voigt et al. 2008; Bullock and Federle 2009). In the following, we will discuss the model presented above and the results with regard to the influence of the structure's density, the surface energy, and the roughness amplitude on the fluid loss.

We presented a simplified, 2D model of the fluid dynamics for the theoretical description of the fluid loss from an insect foot during locomotion. In this context, we were interested in the long-term fluid redistribution between the contacting surfaces of insect pad and substrate. The model's simplified fluid representation includes a term that stabilizes the fluid density. Due to its simplicity, the model lacks the precision of an exact physical model, but it does allow a semi-quantitative description of the system behavior in a large parameter space.

In the presented numerical study, a higher density of substrate irregularities leads to higher fluid loss from the adhesive pad. This effect may be responsible for the reduced friction of insect adhesive pads on substrates with fine roughnesses (0.3–1.0 μm). This reduction in contact forces has earlier been explained by the specific geometry of the spatula-like terminal elements of insect tenent setae

(Peressadko and Gorb 2004; Gorb and Gorb 2008) and their thickness (Filippov et al. 2011); they are better able to generate sufficient contact with larger surface irregularities. Interestingly, many plant surfaces to which insects fail to adhere have such a critical roughness due to the presence of tiny wax crystals on their surfaces (Gorb and Gorb 2002).

Recently, it has been experimentally demonstrated that the behavior of fluid drops on solid, smooth surfaces differs essentially from that on nanoporous ones (Gorb et al. 2010). On smooth surfaces, the contact angle (CA) of an oil drop remains almost the same over a certain period of time, whereas on nanoporous surfaces, CA–time curves show a fast decrease of the CA during the first 20 s after drop deposition. This effect was explained by rapid absorption of the fluid by the porous medium just after drop deposition. It is later followed by stabilization of the CA due to the "saturation" of the porous surface by the absorbed fluid (Gorb et al. 2010). These experimental findings and the results of the present numerical simulation lead us to an important conclusion: the use of wet adhesive pads for attachment purposes may be much more costly during locomotion on micro- and nanorough surfaces, especially if the fluid is lipid-like.

The results of the present study show that liquid loss can be minimized if the surface energy of the pads (their affinity to the fluid) is high enough. This conclusion is based on the existence of an optimal amount of fluid between the pad and the substrate which enhances adhesion (Kovalev et al. 2012). Therefore, the quality of wet adhesion is dependent on the duration of the contact, especially if fluid is continuously secreted by the pad into the contact area: if the contact lasts too long, this will lead to reduced adhesion due to the accumulation of a thick fluid layer (Drechsler and Federle 2006; Bullock et al. 2008).

On the other hand, the pad affinity to the fluid should not be too high, because in this case the substrate will not be sufficiently wetted. If the substrate affinity to the fluid is very high, the fluid can wet the substrate well, but will be quickly drained from the pad. This may also lead to an adhesion reduction due to the thick fluid layer causing a hydroplaning effect, as previously observed for the superhydrophilic surface of the peristome in the carnivorous plants of the genus *Nepenthes* (Bohn and Federle 2004). The hitherto observed biphasic fluids of insects (Gorb 2001; Federle et al. 2002; Vötsch et al. 2002; Betz 2010) exhibit a sufficient, but average affinity to almost any natural substrate (presumably, $\mu > 1$). Since hydrophilic surfaces have a high affinity to the water-based component of the pad secretion, and non-polar surfaces have a high affinity to the lipid-based component, the wetting of the substrate will always be sufficient, and the fluid will not be rapidly absorbed from the pad (if the pad's affinity to the pad secretion is higher than that of the substrate).

According to the numerical simulations performed above, an abundant fluid loss from the adhesive pad takes place on a substrate with fine roughness (high density and low amplitude of the substrate irregularities). The fraction of fluid loss decreases with decreasing roughness amplitude; however, the contact area between the fluid and the substrate decreases for roughness amplitudes smaller than 1.

Presumably, different amounts of fluid are produced by the pad glands of insects on different substrates, but this hypothesis has no experimental evidence yet. Fluid production "on demand" may compensate for its loss on rather smooth surfaces or on those with a fine roughness.

One can conclude that a numerical approach to studying fluid dynamics during the contact formation of an adhesive wet insect pad as applied here shows that an increase in the periodicity length of the substrate leads to a decrease in the fluid loss from the pad. In other words, substrates with a fine roughness take up pad fluid faster. An increased affinity of a solid substrate to the fluid leads to an increase of fluid loss from the pad. With an increasing aspect ratio of the substrate irregularities (porosity), the fluid loss also increases. The numerical results obtained here agree well with previous observations on insects and with experimental results on fluid absorption on nanoporous substrata.

4.4 Self-Alignment System of an Adhesive Fruit

4.4.1 The Plant **Commicarpus helenas** *in Nature*

The plant genus *Commicarpus* consists of about 30–35 species distributed in arid areas of Africa and western Asia (Meikle 1978; Douglas and Spellenberg 2010). The funnel-shaped fruit is 10-ribbed, with 5–10 viscid and 5 mucilaginous glands at its distal part. As described by Struwig and Siebert (2013), after fertilization the upper, petaloid part of the flower falls off, whereas the lower part enlarges and develops into a protective structure around the fruit (Fig. 4.20a), called the *anthocarp* (Joshi and Rao 1934; Vanvinckenroye et al. 1993; Hickey and King 2000). The shape of the anthocarp and the arrangement of the glands are species-specific for *Commicarpus* (Struwig et al. 2011a, 2011b), varying from cylindrical, fusiform and clavate to elliptic clavate. The apex is surrounded by either 5 or 10 glands, which can be stalked or sessile. Additionally, sessile wart-like glands are scattered across the surface of the anthocarp (called fruit throughout the text) below the apex (Fig. 4.20b).

The fruits of *Commicarpus helenas* plants growing on Fuerteventura, Canary Islands, Spain, form a quick and robust adhesive contact with almost any surface (Brandes 2015), with five radially arranged apical glands (Fig. 4.20b). More than 80% of these fruits have been observed to adhere with their full set of apical adhesive glands even to uneven and corrugated surfaces (Fig. 4.20e–h).

In ripe fruits, the adhesive secretion is discharged and the fruit generates an adhesive contact whenever an apical gland is touched by the surface of a moving object (Fig. 4.20c). If some pulling force is applied, the ripe fruit will be easily detached from the plant stem and adhere to the substrate (Fig. 4.20d). Our field observations showed that applying some pulling force to a fruit which has formed an initial contact with one apical gland leads to the self-alignment of the fruit on the

Fig. 4.20 Fruits (anthocarps) of *Commicarpus helenas*. (**a**) Intact fruits on the plant ready for adhesion to the surface of a potential dispersal vector. (**b**) Single fruit with five radially arranged apical glands (indicated by arrows) delivering adhesive secretion as soon as a contact is formed. (**c**) Single, not yet completely aligned fruit adhering with only two glands (on the right-hand side) to a glass surface (view from below through the transparent glass). (**d**) Single, completely aligned fruit adhering with five glands to the glass surface (view from above, cryo-SEM image). (**e**)–(**h**) Fruits adhering to human skin. Please note that most of them adhere to the surface with all five radially arranged apical glands. (From Filippov et al. 2017)

substrate and thus to additional apical glands adhering to the surface (Fig. 4.20e–h), providing stronger adhesion. Since the radial arrangement of 5–10 apical adhesive glands is very characteristic of *Commicarpus* fruits, the question arises whether this specific shape of the fruit and the radial arrangement of apical glands are adaptations for self-alignment and adhesion enhancement during the initial contact of the fruit with a potential dispersal vector.

Numerical modeling was undertaken to answer the following questions (Filippov et al. 2017):

1. Does the specific shape of the fruit and the specific apical and radial arrangement of adhesive glands provide a self-alignment mechanism to various substrate geometries during contact formation?
2. What is the minimal/optimal number of apical glands necessary for self-alignment at contact formation?
3. Does the self-alignment effectiveness depend in any way on the direction of the external pulling force relative to the position of the first gland in contact?

4.4.2 Numerical Model of Commicarpus *Adhesion to Rough Surfaces*

To simulate the adhesive contact of the radially arranged apical glands, we constructed the following simplified model (Fig. 4.21). First of all, we reduced the complex structure of the fruit with their adhesive contact glands to a "pyramid" shape with N adhesive contact points and one central point elastically connecting all of them. In the simulation, the number of glands, $N = 2, 3, \ldots$, is not restricted by our observations in nature. It can vary, thus allowing us to model different systems and to extract information for further possible optimization.

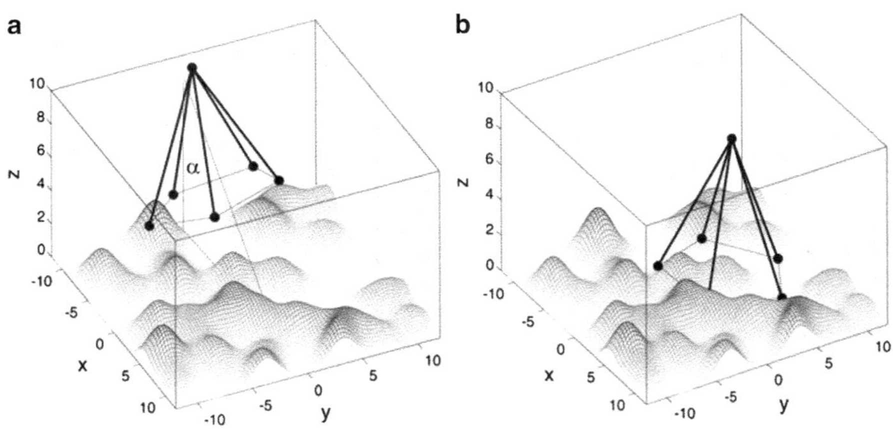

Fig. 4.21 Conceptual structure of the model. (**a**) Pyramidal configuration with N apical contact points. In this particular case, $N = 5$ corresponds to the real configuration of the *Commicarpus* fruit. The pyramid is initially placed at the angle α to the vertical line of the pyramid near a rough substrate surface. It is attracted to the surface by its apical contact points which correspond to the apical glands of the real fruit. (**b**) After some time, depending on the relative configuration of apical points to the surface profile, the configuration achieves complete contact with the surface. (From Filippov et al. 2017)

In the particular case illustrated in Fig. 4.21, we used the same number of glands, $N = 5$, as observed in the real system depicted in Fig. 4.20. Before contact formation with some rough surface, the pyramid is placed at some arbitrary angle α to the vertical line as shown in Fig. 4.21a.

It is supposed that, due to adhesion forces, an apical gland of the fruit that is close enough to the surface will be attracted to the substrate. Since adhesion is a short-range force, the pyramidal configuration can for a long time remain attached by the initially contacting points only. The probability and the stability of the completely "attached" configuration and the progress of the attraction process are determined by a compromise between a number of competing forces and parameters. Some of these parameters are:

1. roughness of the surface,
2. adhesive strength generated by the glands,
3. external forces acting on the fruit,
4. angle at which any external force is applied, and
5. fluctuations of the external force with time.

To avoid time-consuming calculations in modeling the role of these different parameters, we minimized the number of mechanical degrees of freedom of the problem and reproduced the system to $N + 1$ movable points, where $n = 1, 2, \ldots,$ N basal adhesive points are elastically connected with each other as well as with the one in the geometrical center of the apical construction. The latter point is numbered as point number zero, $n = 0$.

The real fruit is relatively rigid, but the apical glands are slightly flexible. To maintain a more or less fixed structure with very few movable segments in three-dimensional space, the elastic interaction is modeled in the following way, as has been done in previous studies (for a review see Popov et al. (2016) and references therein). The shape is provided with strong longitudinal stiffness of the 2-valley potential $k^{\|}$, preventing any extension and compression of the practically rigid segments connecting the neighboring nodes with each other $\vec{r}_j$ and with the central one ($j = 0, 1, \ldots, N$). The corresponding force $\vec{f}^{\|}_{jk} = k^{\|}\left(\vec{r}_j - \vec{r}_k\right) \times \left[1 - \left(\left(\vec{r}_j - \vec{r}_k\right)/r_{0jk}\right)^2\right]$ tends to keep the distance between the nodes $\vec{r}_j$ and $\vec{r}_k$ close to the equilibrium r_{0jk} due to the attraction factor $\left[1 - \left(\left(\vec{r}_j - \vec{r}_k\right)/r_{0jk}\right)^2\right]$. The equilibrium array r_{0jk} is calculated from the initial distribution of the segments reproducing the realistic configuration.

For this particular problem, it is convenient to model the rough surface $Z(x, y)$ by a random deposition of Gaussians with varied positions, amplitudes and widths $Z(x,y) = \sum_n G_n(x,y,\{x_n,y_n\}) = \sum_n a_n \exp\left[-\left((x-x_n)^2 + (y-y_n)^2\right)/w_n^2\right]$. The characteristic scale of the surface irregularities is adjusted to be smaller to, or comparable with, the scale of the pyramid. This is regulated by the number of Gaussians, i.e. the typical distance between the hills and valleys of the randomly accumulated $Z(x, y) = \sum_n G_n$ surface. Finally, the amplitude A of roughness after accumulation is regulated by the normalization $Z(x, y) \rightarrow A(Z(x, y) - \min(Z))/(\max(Z) - \min(Z))$ and can be anywhere between a flat surface ($A = 0$) to values comparable to the length of the pyramid segments.

Here and below, all the lengths and forces of the system under consideration are measured in units of the characteristic strength f_0 and distance of the adhesion force r_0, respectively. For definiteness and simplicity, the adhesion force is generated by the Morse potential $U_{VdW} = f_0 r_0[1 - \exp(-(r - r_0)/r_0)^2]$, which is often used for such studies (see examples above). In other words, $f_0 = 1$ and $r_0 = 1$.

It is supposed that every apical point (gland) interacts via adhesion with each segment of the numerically generated discrete array of the surface $Z(x, y)$. At the same time, the central point on top of the pyramid (which is not adhesive) is affected by a constant external force $\vec{F}_{ext}$. In general, this force can be directed at some (arbitrary) angle β to the vertical plane (x, z). Due to the elastic connection between this point and all the other ones, this force is transferred to the motion of the whole pyramid.

It is generally accepted that for the scale of the problem under consideration, one can neglect inertial terms in the Newtonian equations and reduce the problem to an over-damped one. Then, motion takes the following form: $\tau \partial \vec{r}_n / \partial t = \vec{f}_{elastic} + \vec{f}_{VdW} + \vec{F}_{ext}$. Here, the forces are of a different nature, depending on the particular point ($n = 0$ and $n = 1, .., N$) and must be numerically accumulated from all the sources described above. As is typical for over-damped equations, the multiplicative constant τ defines the characteristic time of the process and can be used as a unit to measure all time-dependant values: $\tau = 1$.

Affected by the combined adhesion, elastic and external forces, the pyramid will be moving and rotating in three-dimensional space. Adhesion gradually turns its basal plane to the surface and, at the same time, the system tends to shift along the substrate in some direction found as a compromise between the relief of U_{VdW} and the external force $\vec{F}_{ext}$. A projection of this motion to the horizontal plane (x, y) for the same particular realization of the surface $Z(x, y) = \sum_n G_n$ as in Fig. 4.21 is shown in Fig. 4.22.

Intuitively, one can expect that if the external force is weak enough, the pyramid will contact the surface with its base, stick to it due to the adhesion force and stop. At the opposite limit, a strong force pulling at the pyramid top will overcome adhesion at the base and continue to rotate the system, i.e., the apical point (gland) can be detached from the surface after some temporary contact with it. This qualitative speculation can be quantitatively modeled by a calculation of the temporal development of the angle $\alpha(t)$ and the velocity of the central point $\vec{V}_0(t)$. To reduce the number of degrees of freedom, we first study the case of the flat surface $A = 0$ and motion along the x-axis only at $\beta = 0$ and $\vec{V}_0(t) \rightarrow Vx_0(t)$.

The results of this study are shown in Fig. 4.23, where the temporal development of the angle α and of the velocity of the central point Vx_0 are calculated for different

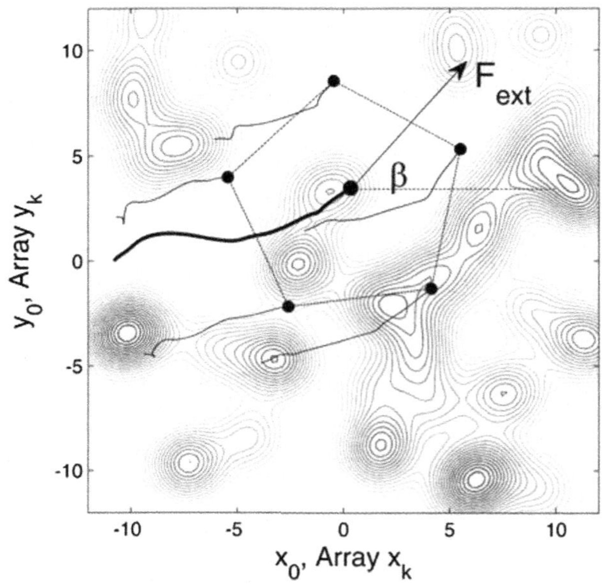

Fig. 4.22 Vertical projection of the system shown in Fig. 4.21. The projections of the trajectories of the apical points (glands) and the central point between the configurations shown in panels (**a**) and (**b**) of Fig. 4.21 are plotted by thin and bold curves, respectively. The contour plot reproduces the same random rough surface as presented in Fig. 4.21. A constant external force F_{ext} is applied to the central point of the pyramid at the azimuth angle β to the axis x. The dynamic process described in Figs. 4.21 and 4.22 is also shown in the supplementary Movies 4.3, 4.4, 4.5, 4.6 and 4.7 for different numbers of contact glands. (From Filippov et al. 2017)

values of the external force. Here, the curves are plotted at a logarithmic scale along the time axis in order to extend the short time interval of fast rotation at the very beginning and to compress the much longer asymptotic part with only small changes of the variables. The main reason for this difference in the rates of the process is the strong motion of the central point caused by the rotation of the pyramid. This is clearly visible in panels (a) and (b) in Fig. 4.23. Even small deflections in the curves of angle $\alpha = \alpha(t)$ are accompanied by maxima of the velocity $Vx_0(t)$. As an example, two local maxima of $Vx_0(t)$ are marked by vertical lines crossing both panels of Fig. 4.23.

Rotation of the pyramid is also accompanied by relatively fast motions of all adhesive points. This motion is presented in Fig. 4.24 by a set of thin curves for the array $vx_k(t)$, where $k = 1, 2, \ldots, N$. The maxima of these curves are much lower that the main maximum of $Vx_0(t)$ (bold line), but still well pronounced. It is also important to note the strong difference between the spatial and temporal development of the velocities in panels (a) and (b), respectively. The reason for this is that the contact points are almost completely attached to the substrate, and the variable x_0 practically does not change, after $x_0 \approx 2$ so the displacements along the x-axis stop rather quickly. However, some slow drift of all the components of the system still occurs as can be seen in the logarithmic plot shown in the inset in Fig. 4.24b.

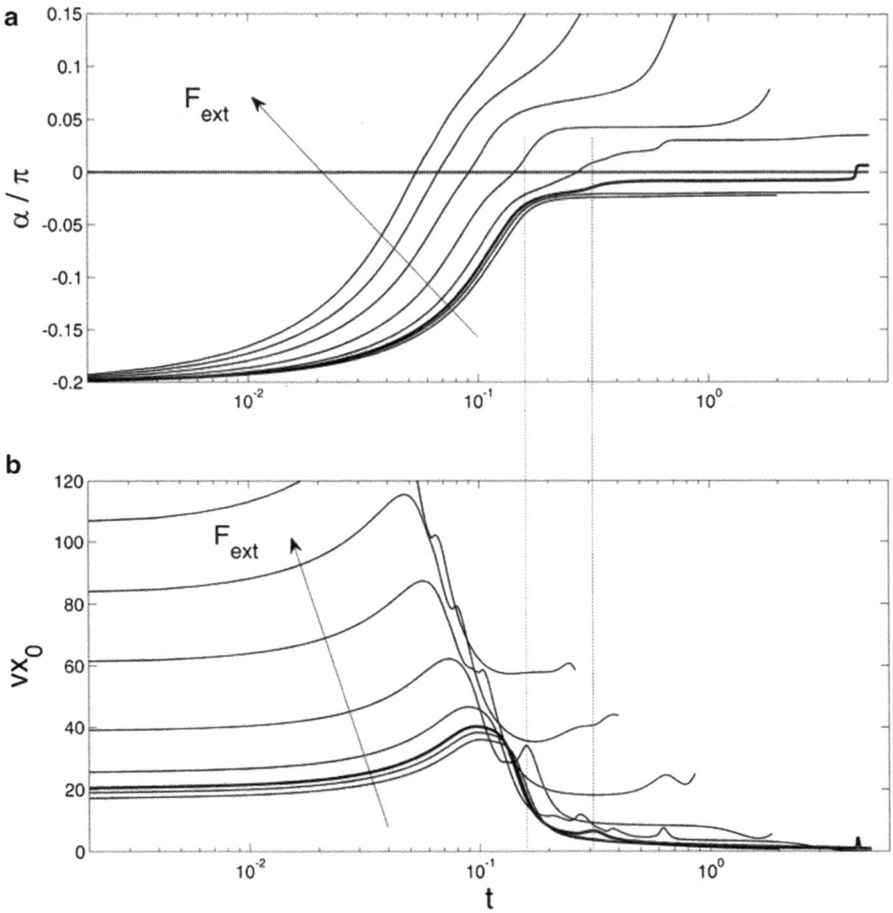

Fig. 4.23 Temporal development of angle α (**a**) and the velocity of the central point Vx_0 (**b**) calculated at different values of the external force for a flat contact surface. The saddle-point curve corresponding to the critical value of the force F_{ext} at which the system stops exactly in the vertical position is plotted with a bold line. For more explanations see the text. (From Filippov et al. 2017)

Let's now take a look at the situation with an arbitrary orientation of the external force, $\beta \neq 0$. If the pyramid starts to form the contact from an arbitrary inclined position with an angle $\alpha \neq 0$, one can expect that its rotation will essentially depend on the direction of the external force. Indeed, when an external influence is absent ($F_{ext} = 0$), the pyramid will spontaneously fall to the surface in the direction of x. If the external force ($F_{ext} \neq 0$) is parallel to this direction, it supports the rotation. In all other cases, the result of the interplay between this force and adhesion is not so clear and one needs to perform further numerical simulations.

The results of the solution for a number of different angles $\beta = 0, \pi/8, \pi/4, .. \pi$ are presented in Fig. 4.25. This figure demonstrates that up to the perpendicular force

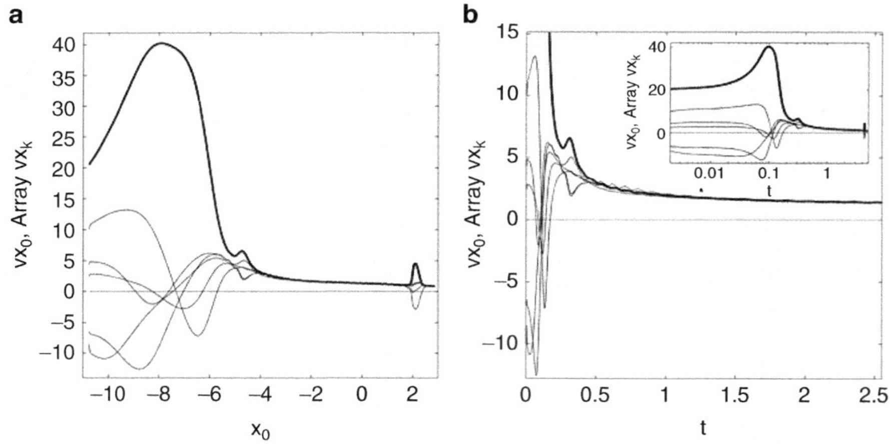

Fig. 4.24 Spatial (**a**) and temporal (**b**) development of the velocities. Panels (**a**) and (**b**) show the velocities of the central point (bold line) and the array of the contact points vx_k (where $k = 1, 2, .., N$) plotted against distance and time, respectively. The inset shows the same temporal development as in (**b**) in logarithmic scale, in order to extend the initial interval of fast motion at the beginning of the process. Distance and time are given in arbitrary units. (From Filippov et al. 2017)

$\beta = \pi/2$, the absolute velocity $|V_0| = \sqrt{Vx_0^2 + Vy_0^2}$ generally increases: this tendency is indicated by the bold arrow in Fig. 4.25. Beyond this value, the tendency changes and is now represented by the fine arrow in the opposite direction. The reason for this is that the x-projection of the external force is now opposite to the x-projection of the adhesion force. These two forces are competing at the beginning of the process, reducing the pyramid rotation and, as a result, the total velocity $|V_0|$. The minimum of the horizontal velocity is reached for the exactly negative orientation of the external force at $\beta = \pi$.

To determine the minimal/optimal number of apical glands necessary for contact formation, we will now examine the simplest case of the fixed angle $\beta = 0$ and the external force $F_{ext} = 0.25$, varying the number of glands, $N = 2, 3, 4, \ldots$. Figure 4.26 presents the results of the simulation based on these constants, showing the set of time-dependent absolute velocities $|v_0|$ at different numbers of contact points $N = 2$, 3, 4, ... The main plot of this figure shows that for the numbers $N = 5$, $N = 6$ and $N = 7$, the curves practically coincide at the beginning. After some time, small differences appear. The inset in Fig. 4.26, which magnifies the vertical scale for a fragment of the complete image, illustrates the monotonous variation of $|v_0|$ for different values of N.

Above, we artificially reduced the roughness of the substrate to zero, $A = 0$, in order to make the dependence on different parameters of the problem more transparent. However, for any real contact problem, the amplitude of roughness is a very important property of the surface. To extract information related to this property in the same systematic manner as we did for other parameters, we will now again use a

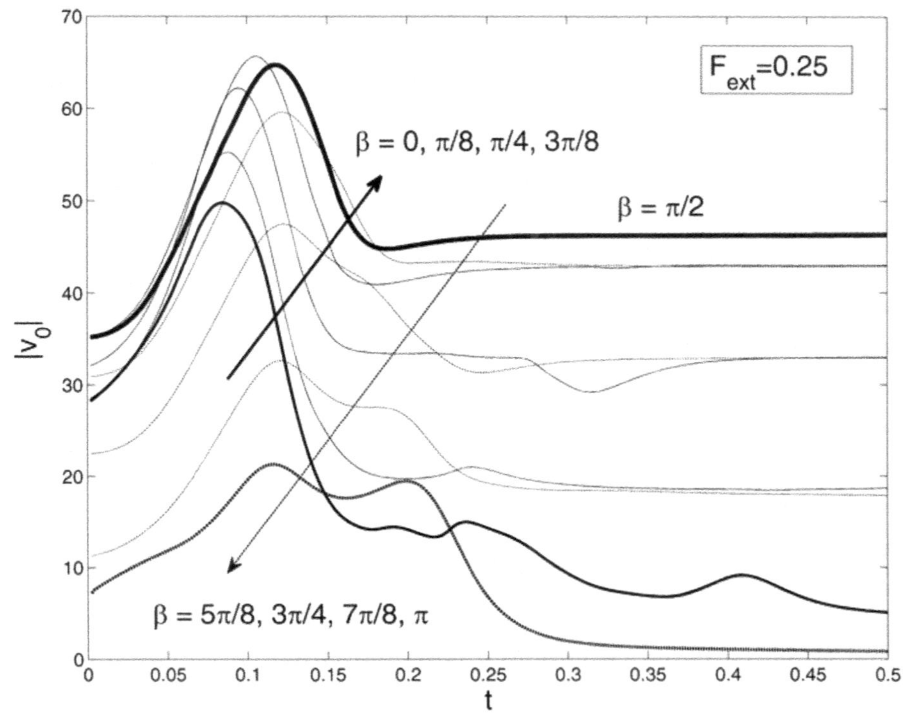

Fig. 4.25 Influence of the azimuth angle β on the temporal development of the absolute velocity of the central point $|v_0|$ at the fixed intermediate external force $F_{ext} = 0.25$. The curves corresponding to the symmetrically important directions $\beta = 0, \pi/2, \pi$ are represented by bold lines. (From Filippov et al. 2017)

fixed number of glands, namely $N = 5$, as it represents the actual biological case. Besides, we will return to the trivial angle $\beta = 0$ and again use a fixed intermediate value of the external force, $F_{ext} = 0.25$.

It is important to note that the chosen value $F_{ext} = 0.25$ makes it possible to study all stages of fruit motion in each numerical experiment because this force is not so strong as to immediately turn the fruit (and thus quickly remove it from contact), nor so weak as to allow the apical glands to stick to the substrate from the very beginning of contact formation. Intuitively, one can expect that varying roughness of the contact surface, $A \neq 0$, will cause a contest between the following two factors: On the one hand, if the surface is strongly corrugated, it is difficult to achieve a good spatial configuration of the glands that allows simultaneous contact of all or the majority of them with the surface. On the other hand, if the surface consists of deep "hills and valleys", every gland, once attached, will be strongly trapped in narrow spaces between the "hills" of the roughness. In this case, it may be difficult to remove it from such a trapped position.

Which factor prevails in the contact formation process certainly depends on the spatial distribution of the contact points. One can expect that a simple planar "contact

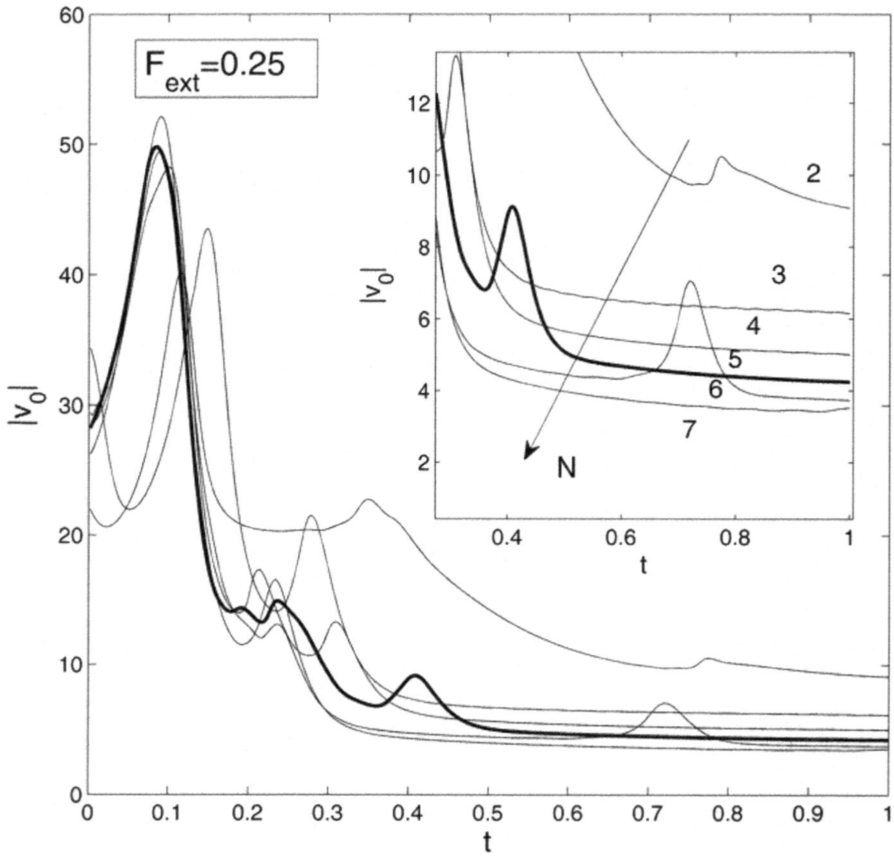

Fig. 4.26 Temporal development of the absolute velocity $|v_0|$ for different numbers of contact points, $N = 2, 3, 4, \ldots$. Inset magnifies the $|v_0|$ values for the range of $0.3 < t < 1$, illustrating the monotonous tendency of the velocity $|v_0|$ with the number of contact points N. Velocity and time are given in arbitrary units. (From Filippov et al. 2017)

pad" may conflict with the random asperities on a complex surface because an initially formed contact may prevent the formation of new ones. From our previous studies of analogous contact problems, we know that adhesion can be strongly enhanced by a "hairy" configuration of the contact device, which allows every fragment of the system to turn and find a better contact configuration. In order to test this idea for the adhesive fruit system under consideration, we calculate the temporal development of angle $\alpha(t)$ and absolute velocity $|v_0|$ for a set of different amplitudes of roughness A.

Figure 4.27 shows that, while the behavior is relatively complex, the contact ability of the adhesive fruit system is generally stronger on rough substrates. The value of A varies, starting from a completely flat surface ($A = 0$) up to $A_{crit} \approx 0.5$. The latter value corresponds to the roughness at which the previously used

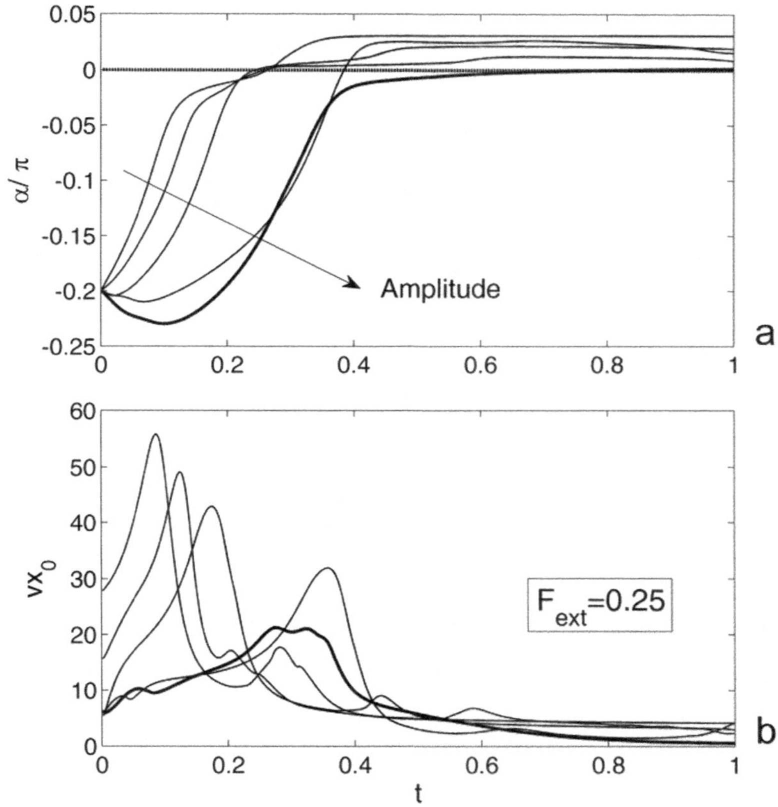

Fig. 4.27 Temporal development of the angle α (**a**) and the absolute velocity $|v_0|$ (**b**) at different amplitudes of roughness ($A = 0$, 0.125, 0.25, 0.375, 0.5). A varies from a completely flat surface ($A = 0$) to the roughness $A_{crit} \approx 0.5$, at which the external force $F_{ext} = 0.25$ becomes critical; the corresponding curves are represented by bold lines. (From Filippov et al. 2017)

"intermediate" external force $F_{ext} = 0.25$ becomes critical. This means that in contrast to the flat substrate surface, at $F_{ext} = 0.25$ the system completely adheres to the substrate. Both the angle α and the absolute velocity $|v_0|$ simultaneously tend to zero: the corresponding curves in Fig. 4.27a, b are represented by bold lines.

In the real system, another important source for randomness and possible instability is the randomly fluctuating external force F_{ext}. While we cannot account for all the possible sources of fluctuations, it may be assumed that the force F_{ext} caused by potential dispersal agents, such as vertebrates, is never fixed in a real situation. Additionally, the angle β of the force acting on the fruit continuously varies. Below, we will model the fluctuations of the angle β.

Mathematically, these fluctuations mean that even at a more or less fixed strength of the external force ($F_{ext} \approx const.$), the direction of the force F_{ext} (direction of the wind, for example) permanently changes. To simulate this, an additional equation of

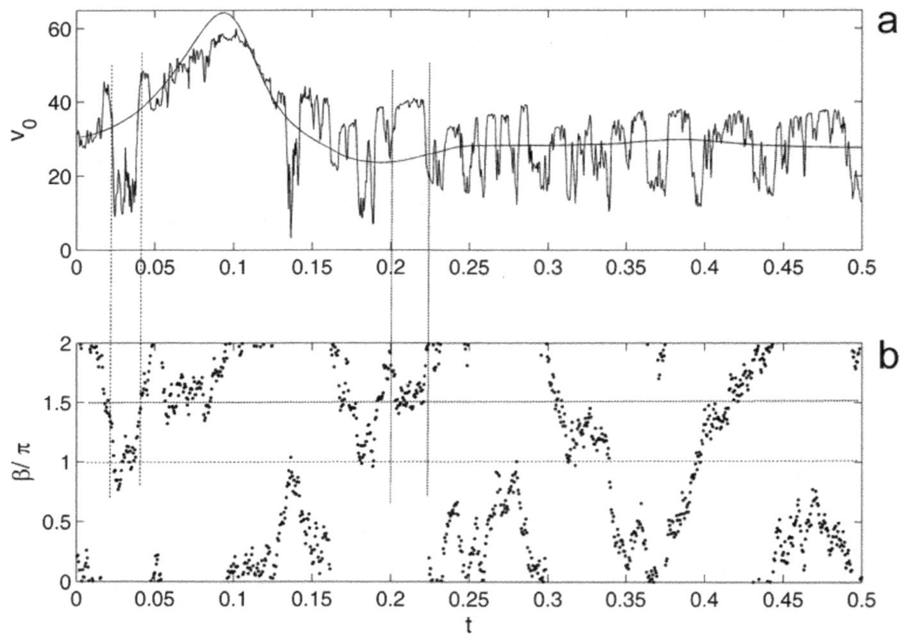

Fig. 4.28 (**a**) Time-dependent absolute velocity $|v_0|$ at random walk of the azimuth angle β. (**b**) The rotation of the angle $\beta(t)$ in the plot is reduced to the physical interval $[0, 2\pi]$. The vertical straight lines mark two typical regions, where the angle is close to the values of maximal and minimal velocities (near $\beta = \pi$ and $\beta = 3\pi/2$, respectively). (From Filippov et al. 2017)

motion (providing some kind of chaotic component) is required for the angle $\partial\beta/\partial t = \xi(t)$. It describes the so-called „random walk"of the angle β. The time-dependent source of fluctuations $\xi(t)$ here is δ-correlated Gaussian noise with a mean value of zero, $<\xi(t)> = 0$, and the strength $<\xi(t)\xi(t')> = \sigma\delta(t - t')$, which is defined by the intensity σ of the fluctuations.

Above (cf. Figure 4.25), we already saw that different angles β led to different dynamic scenarios, changing the absolute velocity $|V_0|$ in particular temporal development. Now, the angle β of the external force F_{ext} permanently changes, $\partial\beta/\partial t = \xi(t)$. Depending on the relationship between the characteristic times and forces of the attachment on the one hand and their fluctuations on the other hand, different scenarios are possible. If the fluctuations are too weak (or if their oscillations are too fast), they will not affect the motion at all. At the opposite limit of sufficient force F_{ext} and slow rotation of its direction $\partial\beta/\partial t = \xi(t)$, we return to the already discussed case of a strong and regular force F_{ext}. The most interesting intermediate case is presented in Fig. 4.28 which shows a complicated temporal development. It corresponds to the situation when the direction of F_{ext} changes a few times during the characteristic rotation time of the angle α and, as a result, essentially varies the absolute velocity $|V_0|$.

In this case, despite the extremely strong effect of the fluctuations and the wide variations of the velocity $|V_0|$, the behavior remains stable and the fruit will attach to the surface. It is interesting to note the well pronounced correlations between the instant angles β in Fig. 4.28a and the resulting velocities $|V_0|$ in Fig. 4.28b. These correlations are found to be in perfect agreement with the results found earlier for the array of regular fixed angles β as presented in Fig. 4.25. It is expected from Fig. 4.25 that when the randomly walking angle $\beta(t)$ remains close to $\beta = \pi$ or $\beta = \pi/2$ ($\beta = 3\pi/2$) for a relatively long time period, the velocity will be attracted to its maximal or minimal values at this stage of motion. To elucidate this case, we plot the rotation of the angle $\beta(t)$ in Fig. 4.28a as reduced to the physically meaningful interval $[0, 2\pi]$. Two typical regions where the angle is close to the values of maximal and minimal velocities (near $\beta = \pi$ and $\beta = 3\pi/2$, respectively) are indicated by the vertical straight lines connecting both panels.

Referring to a "desirable stability" of the behavior even under relatively strong fluctuation means that, despite of strong variations, absolute velocity $|V_0|$ still remains close to the function obtained at an intermediate constant angle β. From a symmetrical point of view, one can predict that in the case of permanent rotation, the curve will be close to that at the angle $\beta = \pi/4$. However, Fig. 4.28 shows that the curves are slightly lower in the interval between $\beta = 0$ and $\beta = \pi$. This leads to some shift of the mathematical expectation for the average $|V_0|$. As a result, the value appears to be around the angle $\beta = \pi/5$; the corresponding (smooth) curve is added in Fig. 4.28a.

4.4.3 Biological Significance of the Obtained Results

After coming into contact with the surface of any surrounding object, such as stones, other plants, soil, or potential dispersal agents, the *Commicarpus* fruit may experience an external force caused by wind, substrate vibrations, touch, etc. The results of our numerical modeling demonstrate that this force, if it exceeds some critical value, can either enhance or reduce the contact area (and consequently, adhesion) and even remove the fruit from the surface, depending on its direction. If the force falls short of the critical value, the contact (and consequently, adhesion) will be almost always enhanced.

As mentioned in the previous Section (4.4.2), the effect of this force depends on its angle. If the angle tends to rotate the fruit away from the other adhesive glands, the chance that rotation continues and will remove the fruit from the substrate is rather high. If the angle is in the opposite direction, the chance for the fruit to be removed is low and the chance of enhanced contact/adhesion is high. Any intermediate angles would effect an intermediate result with different degrees of probability.

However, in reality the external force is seldom oriented in one particular direction. It usually changes direction, and this, in turn, will change the force rate/speed. So how does the variability of the external forces influence the fruit's anchoring to the substrate? We found that if the external force changes its direction at times corresponding to instances of fruit rotation, then in spite of non-constant force directionality, the fruit will remain in stable contact and sometimes the probability of adhering firmly to the substrate with the maximum number of its apical adhesive glands will even be enhanced.

From an evolutionary point of view, the question of the optimal number of apical adhesive glands is probably the most important one. Representatives of the genus *Commicarpus* typically have 5 or 10 glands (one or two circles of glands, respectively). The results of our simulations show that having only one or two contact points might not be enough for maintaining the contact in the face of external forces. With three contact points, however, the self-stabilization mechanism will start, although still rather weak. Furthermore, we did not obtain any stronger contact formation with more than five individual apical contacts. This particular geometry of adhesive contacts at the pyramid base reaches its saturation point with approximately five individual contacts. In other words, more than five apical glands would be redundant for the effects discussed here.

Natural substrates to which *Commicarpus* fruits usually adhere are not smooth and flat, but rather strongly corrugated. The specific geometry of *Commicarpus* fruits with 5 or 10 adhesive glands situated apically at the perimeter of the anthocarp represents an adaptation to enhance the self-alignment after an initial adhesive contact formed with just one individual gland. Such a distribution of adhesive glands increases adhesion to rough substrates and in the face of an external force with variable directions. Three glands are sufficient for this effect to occur. This geometrical adaptation, in combination with the properties of the adhesive secretion (which we did not model), enables the fruit to generate strong adhesion forces after the formation of the first discrete adhesive contact. From our field observations, we conclude that the fruits adhere well to rough surfaces.

Furthermore, the results of our numerical simulations demonstrate that an increase of the substrate's roughness amplitude leads to an increase in the adhesive ability of the fruit, because on a rough surface, the action of any external forces will almost always be redirected into a proper direction. That is why this situation will lead to a stronger anchoring of the fruit to the 3D surface, due to the recruiting of a higher number of individual apical glands in contact. All the mentioned features contribute to the success of the epizoochorous dispersal of *Commicarpus* fruits and to their anchoring and stabilization between stones and in the soil.

References

Abou B, Gay C, Laurent B, Cardoso O, Voigt D, Peisker H, Gorb SN (2010) Extensive collection of femtolitre pad secretion droplets in the beetle *Leptinotarsa decemlineata* allows nanolitre microrheology. J R Soc Interface 7(53):1745–1752

Arzt E, Gorb S, Spolenak R (2003) From micro to nano contacts in biological attachment devices. Proc Natl Acad Sci U S A 100:10603–10606

Autumn K, Peattie AM (2002) Mechanisms of adhesion in geckos. Integr Comp Biol 42:1081–1090

Autumn K, Liang YA, Hsieh ST, Zesch W, Chan WP, Kenny TW, Fearing R, Full RJ (2000) Adhesive force of a single gecko foot-hair. Nature 405:681–685
Autumn K, Dittmore A, Santos D, Spenko M, Cutkosky M (2006) Frictional adhesion: a new angle on gecko attachment. J Exp Biol 209:3569–3579
Bauchhenss E (1979) Die Pulvillen von *Calliphora erythrocephala* Meig. (Diptera, Brachycera) als Adhäsionsorgane. Zoomorphologie 93:99–123
Betz O (2010) Adhesive exocrine glands in insects: morphology, ultrastructure, and adhesive secretion. In: Byern J, Grunwald I (eds) Biological adhesive systems part 1. Springer, pp 111–152
Bohn HF, Federle W (2004) Insect aquaplaning: *Nepenthes* pitcher plants capture prey with the peristome, a fully wettable water-lubricated anisotropic surface. Proc Natl Acad Sci U S A 101 (39):14138–14143
Brandes D (2015) Some contributions on *Commicarpus helenae* (Roemer and Schultes) Meikle on Fuerteventura (Canary Islands, Spain). Braunschweiger Geobotanische Arbeiten 10:67–75
Brennan EB, Weinbaum SA (2001) Effect of epicuticular wax on adhesion of psyllids to glaucous juvenile and glossy adult leaves of *Eucalyptus globulus* Labillardière. Aust J Entomol 40:270–277
Bullock JMR, Federle W (2009) Division of labour and sex differences between fibrillar, tarsal adhesive pads in beetles: effective elastic modulus and attachment performance. J Exp Biol 212:1876–1888
Bullock J, Federle W (2011) Beetle adhesive hairs differ in stiffness and stickiness: in vivo adhesion measurements on individual setae. Naturwissenschaften 98:381–387
Bullock JMR, Drechsler P, Federle W (2008) Comparison of smooth and hairy attachment pads in insects: friction, adhesion and mechanisms for direction-dependence. J Exp Biol 211:3333–3343
Chen B, Wu PD, Gao H (2009) Pre-tension generates strongly reversible adhesion of a spatula pad on a substrate. J R Soc Interface 6:529–537
Creton C, Gorb SN (2007) Sticky feet: from animals to materials. MRS Bull 32:466–472
Diens G, Klemm H (1946) Theory and application of the parallel plate plastometer. J Appl Phys 17:458–471
Dirks J-H, Federle W (2011) Mechanisms of fluid production in smooth adhesive pads of insects. J R Soc Interface 8:952–960
Dirks J-H, Clemente CJ, Federle W (2010) Insect tricks: two-phasic foot pad secretion prevents slipping. J R Soc Interface 7:587–593
Douglas NA, Spellenberg R (2010) A new tribal classification of Nyctaginaceae. Taxon 59:905–910
Drechsler P, Federle W (2006) Biomechanics of smooth adhesive pads in insects: influence of tarsal secretion on attachment performance. J Comp Physiol A 192:1213–1222
Edwards PB (1982) Do waxes of juvenile *Eucalyptus* leaves provide protection from grazing insects? Aust J Ecol 7:347–352
Edwards JS, Tarkanian M (1970) The adhesive pads of Heteroptera: a re-examination. Proc R Entomol Soc Lond A 45(1–3):1–5
Edwards PB, Wanjura WJ (1990) Physical attributes of eucalypt leaves and the host range of chrysomelid beetles. Symp Biol Hung 39:227–236
Eigenbrode SD (1996) Plant surface waxes and insect behaviour. In: Kerstiens G (ed) Plant cuticles – an integral functional approach. BIOS, Oxford, pp 201–222
Eigenbrode SD, Jetter R (2002) Attachment to plant surface waxes by an insect predator. Integr Comp Biol 42(6):1091–1099
Eigenbrode SD, Kabalo NN (1999) Effects of *Brassica oleracea* waxblooms on predation and attachment by *Hippodamia convergens*. Entomol Exp Appl 91:125–130
Eimüller T, Guttmann P, Gorb S (2008) Terminal contact elements of insect attachment devices studied by transmission X-ray microscopy. J Exp Biol 211:1958–1963

Eisner T, Aneshansley DJ (2000) Defense by foot adhesion in a beetle (*Hemisphaerota cyanea*). Proc Natl Acad Sci U S A 97(12):6568–6573
Federle W (2006) Why are so many adhesive pads hairy. J Exp Biol 209:2611–2621
Federle W, Riehle M, Curtis ASG, Full RJ (2002) An integrative study of insect adhesion: mechanics and wet adhesion of pretarsal pads in ants. Integr Comp Biol 42:1100–1106
Filippov AE, Popov VL (2007a) Fractal Tomlinson model for mesoscopic friction: from microscopic velocity-dependent damping to macroscopic Coulomb friction. Phys Rev E 75:027103
Filippov AE, Popov V (2007b) Flexible tissue with fibres interacting with an adhesive surface. J Phys Condens Matter 19:096012
Filippov A, Popov V, Gorb S (2011) Shear induced adhesion: contact mechanics of biological spatula-like attachment devices. J Theor Biol 276:126–131
Filippov AE, Gorb EV, Gorb SN (2017) Radial arrangement of apical adhesive sites promotes contact self-alignment of fruits in *Commicarpus* plants (Nyctaginaceae). Sci Rep 7:10956
Gao H, Wang X, Yao H, Gorb SN, Arzt E (2005) Mechanics of hierarchical adhesion structures of geckos. Mech Mater 37:275–285
Gaume L, Gorb SN, Rowe N (2002) Function of epidermal surfaces in the trapping efficiency of *Nepenthes alata* pitchers. New Phytol 156:479–489
Gilett J, Wigglesworth V (1932) The climbing organ of an insect, *Rhodnius prolixus* (Hemiptera; Reduviidae). Proc R Soc Lond B 111:364–376
Gorb SN (1998a) The design of the fly adhesive pad: distal tenent setae are adapted to the delivery of an adhesive secretion. Proc R Soc Lond B 265:747–752
Gorb SN (1998b) Frictional surfaces of the elytra to body arresting mechanism in tenebrionid beetles (Coleoptera: Tenebrionidae): design of co-opted fields of microtrichia and cuticle ultrastructure. Int J Insect Morphol Embryol 27:205–225
Gorb SN (2001) Attachment devices of insect cuticle. Kluwer Academic Publishers, Dordrecht
Gorb S (2005) Uncovering insect stickiness: structure and properties of hairy attachment devices. Am Entomol 51:31–35
Gorb SN (2006) Fly microdroplets viewed big: a Cryo-SEM approach. Microsc Today 14(5):38–39
Gorb SN (2008) Smooth attachment devices in insects. In: by Casas J, Simpson SJ (eds) Advances in insect physiology, vol. 34: insect mechanics and control. Elsevier Ltd./Academic, London, pp 81–116
Gorb SN, Beutel RG (2001) Evolution of locomotory attachment pads of hexapods. Naturwissenschaften 88:530–534
Gorb EV, Gorb SN (2002) Attachment ability of the beetle *Chrysolina fastuosa* on various plant surfaces. Entomol Exp Appl 105:13–28
Gorb EV, Gorb SN (2008) Effects of surface topography and chemistry of *Rumex obtusifolius* leaves on the attachment of the beetle *Gastrophysa viridula*. Entomol Exp Appl 130:222–228
Gorb SN, Varenberg M (2007) Mushroom-shaped geometry of contact elements in biological adhesive systems. J Adhes Sci Technol 21(12–13):1175–1183
Gorb EV, Haas K, Henrich A, Enders S, Barbakadze N, Gorb SN (2005) Composite structure of the crystalline epicuticular wax layer of the slippery zone in the pitchers of the carnivorous plant *Nepenthes alata* and its effect on insect attachment. J Exp Biol 208:4651–4662
Gorb SN, Sinha M, Peressadko A, Daltorio KA, Quinn RD (2007) Insects did it first: a micropatterned adhesive tape for robotic applications. Bioinspir Biomim 2:S117–S125
Gorb EV, Voigt D, Eigenbrode SD, Gorb SN (2008) Attachment force of the beetle *Cryptolaemus montrouzieri* (Coleoptera, Coccinellidae) on leaflet surfaces of mutants of the pea *Pisum sativum* (Fabaceae) with regular and reduced wax coverage. Arthropod Plant Interact 2:247–259
Gorb EV, Hosoda N, Miksch C, Gorb SN (2010) Slippery pores: anti-adhesive effect of nanoporous substrates on the beetle attachment system. J R Soc Interface 7:1571–1579
Heepe L, Gorb SN (2014) Biologically inspired mushroom-shaped adhesive microstructures. Annu Rev Mater Res 44:173–203

Hickey M, King C (2000) The Cambridge illustrated glossary of botanical terms. Cambridge University Press, Cambridge

Hiller U (1968) Untersuchungen zum Feinbau und zur Funktion der Haftborsten von Reptilien. Z Morphol Tiere 62:307–362

Hosoda N, Gorb SN (2011) Friction force reduction triggers feet grooming behaviour in beetles. Proc R Soc B 278:1748–1752

Huber G, Gorb SN, Spolenak R, Arzt E (2005a) Resolving the nanoscale adhesion of individual gecko spatulae by atomic force microscopy. Biol Lett 1:2–4

Huber G, Mantz H, Spolenak R, Mecke K, Jacobs K, Gorb SN, Arzt E (2005b) Evidence for capillarity contributions to gecko adhesion from single spatula nanomechanical measurements. Proc Natl Acad Sci U S A 102(45):16293–16296

Huber G, Gorb SN, Hosoda N, Spolenak R, Arzt E (2007) Influence of surface roughness on gecko adhesion. Acta Biomater 3:607–610

Hui C-Y, Glassmaker N, Jagota A (2005) How compliance compensates for surface roughness in fibrillar adhesion. J Adhes 81:699–721

Ishii S (1987) Adhesion of a leaf feeding ladybird *Epilachna vigintioctomaculata* (Coleoptera: Coccinellidae) on a vertically smooth surface. Appl Entomol Zool 22(2):222–228

Joshi AC, Rao VSR (1934) Vascular anatomy of the flowers of four Nyctaginaceae. J Indian Bot Soc 13:169–186

Kendall K (1975) Thin-film peeling – the elastic term. J Phys D Appl Phys 8:1449–1452

Kim TW, Bhushan B (2007) Adhesion analysis of multi-level hierarchical attachment system contacting with a rough surface. J Adhes Sci Technol 21(1):1–20

Kim S, Sitti M (2006) Biologically inspired polymer microfibers with spatulate tips as repeatable fibrillar adhesives. Appl Phys Lett 89:261911

Kim S, Aksak B, Sitti M (2007) Enhanced friction of elastomer microfiber adhesives with spatulate tips. Appl Phys Lett 91:221913

Kim S, Sitti M, Jang J-H, Thomas EL (2008) Fabrication of bio-inspired elastomer nanofiber arrays with spatulate tips using notching effect. IEEE Nano 8:780–782

Kosaki A, Yamaoka R (1996) Chemical composition of footprints and cuticula lipids of three species of lady beetles. Jpn J Appl Entomol Zool 40:47–53

Kovalev AE, Varenberg M, Gorb SN (2012) Wet versus dry adhesion of biomimetic mushroom-shaped microstructures. Soft Matter 8:7560–7566

Kovalev AE, Filippov AE, Gorb SN (2013) Insect wet steps: loss of fluid from insect feet adhering to a substrate. J R Soc Interface 10:20120639

Kovalev AE, Filippov AE, Gorb SN (2018) Critical roughness in animal hairy adhesive pads: a numerical modeling approach. Bioinspir Biomim 13:066004

Landau LD, Lifshits EM (1981) Theory of elasticity, 2nd edn. Pergamon Press, Oxford

Langer MG, Ruppersberg JP, Gorb S (2004) Adhesion forces measured at the level of a terminal plate of the fly's seta. Proc R Soc Lond B 271:2209–2215

Meikle RD (1978) A key to *Commicarpus*. Notes Roy Bot Gard Edinb 36:235–249

Murphy MP, Aksak B, Sitti M (2007) Adhesion and anisotropic friction enhancements of angled heterogeneous micro-fiber arrays with spherical and spatula tips. J Adhes Sci Technol 21 (12–13):1281–1296

Niederegger S, Gorb SN (2003) Tarsal movements in flies during leg attachment and detachment on a smoothe substrate. J Insect Physiol 49:611–620

Niederegger S, Gorb SN (2006) Friction and adhesion in the tarsal and metatarsal scopulae of spiders. J Comp Physiol A 192:1223–1232

Peisker H, Michels J, Gorb SN (2013) Evidence for a material gradient in the adhesive tarsal setae of the ladybird beetle *Coccinella septempunctata*. Nat Commun 4(1661):1–7

Peressadko A, Gorb SN (2004) Surface profile and friction force generated by insects. In: Boblan I, Bannasch R (eds) Fortschritt-Berichte VDI 15, Nr. 249. VDI Verlag, Düsseldorf, pp 257–263

Peressadko A, Hosoda N, Persson B (2005) Influence of surface roughness on adhesion between elastic bodies. Phys Rev Lett 95(12):124301
Persson B, Gorb S (2003) The effect of surface roughness on the adhesion of elastic plates with application to biological systems. J Chem Phys 119:11437–11444
Pitaevskii L, Lifshitz EM (1981) Physical kinetics. Pergamon Press, London
Popov VL (2010) Contact mechanics and friction. Physical principles and applications. Springer, Berlin
Popov VL, Starcevic J, Filippov AE (2007) Reconstruction of potential from dynamic experiments. Phys Rev E 75(6 Pt 2):066104
Popov V, Filippov A, Gorb SN (2016) Biological microstructures with high adhesion and friction. Numerical approach. Physics-Uspekhi 59:829–845
Russell A, Johnson M (2007) Real-world challenges to, and capabilities of, the gekkotan adhesive system: contrasting the rough and the smooth. Can J Zool 85:1228–1238
Sauer R, Holl M (2013) A detailed 3D finite element analysis of peeling behaviour of gecko spatula. Comput Methods Biomech Biomed Eng 16:577–591
Schargott M, Popov V, Gorb SN (2006) Spring model of biological attachment pads. J Theor Biol 243:48–53
Scherge M, Gorb SN (2001) Biological micro- and nanotribology. Springer, Berlin et al
Schuppert J, Gorb SN (2006) The wet step: visualization of the liquid bridges in the attachment pads of the beetle *Gastrophysa viridula*. Comp Biochem Physiol A 143:97
Spolenak R, Gorb SN, Gao H, Arzt E (2005) Effects of contact shape on the scaling of biological attachments. Proc R Soc A 461:305–319
Stork NE (1980a) Experimental analysis of adhesion of *Chrysolina polita* (Chrysomelidae, Coleoptera) on a variety of surfaces. J Exp Biol 88:91–107
Stork NE (1980b) A scanning electron microscope study of tarsal adhesive setae in the Coleoptera. Zool J Linnean Soc 68:173–306
Stork NE (1983) The adherence of beetle tarsal setae to glass. J Nat Hist 17:583–597
Stork NE (1986) The form of plant waxes: a means preventing insect attachment? In: Juniper BE, Southwood TRE (eds) Insects and the plant surface. Edward Arnold, London, pp 346–347
Struwig M, Siebert SJ (2013) A taxonomic revision of *Commicarpus* (Nyctaginaceae) in southern Africa. S Afr J Bot 84:44–64
Struwig M, Jordaan A, Siebert SJ (2011a) Anatomical adaptations of *Boerhavia* L. and *Commicarpus* Standl. (Nyctaginaceae) for survival in arid environments of Namibia. Acta Biol Cracov Ser Bot 53(2):50–58
Struwig M, Siebert SJ, Klaassen ES (2011b) Nyctaginaceae. Notes on *Commicarpus* Standl. in Southern Africa, including a new record for Namibia. Bothalia 41:289–292
Tian Y, Pesika N, Zeng H, Rosenberg K, Zhao B, McGuiggan P, Autumn K, Israelachvili J (2006) Adhesion and friction in gecko attachment and detachment. Proc Natl Acad Sci U S A 103 (51):19320–19325
Vanvinckenroye P, Cresens E, Ronse Decraene L-P, Smets E (1993) A comparative floral developmental study in *Pisonia, Bougainvillea* and *Mirabilis* (Nyctaginaceae) with special emphasis on the gynoecium and floral nectaries. Bull Jard Bot Natl Belg 62:69–96
Varenberg M, Gorb SN (2008) Shearing of fibrillar adhesive microstructure: friction and shear-related changes in pull-off force. J R Soc Interface 4:721–725
Varenberg M, Pugno N, Gorb S (2010) Spatulate structures in biological fibrillar adhesion. Soft Matter 6:3269–3272
Voigt D, Schuppert J, Dattinger S, Gorb SN (2008) Sexual dimorphism in the attachment ability of the Colorado potato beetle *Leptinotarsa decemlineata* (Coleoptera: Chrysomelidae) to rough substrates. J Insect Physiol 54:765–776
Vötsch W, Nicholson G, Müller R, Stierhof Y-D, Gorb SN, Schwarz U (2002) Chemical composition of the attachment pad secretion of the locust *Locusta migratoria*. Insect Biochem Mol Biol 32:1605–1613

Walker G, Yule AB, Ratcliffe J (1985) The adhesive organ of the blowfly, *Calliphora vomitoria*: a functional approach (Diptera: Calliphoridae). J Zool (Lond) 205:297–307
Wolff J, Gorb SN (2012) Surface roughness effects on attachment ability of the spider *Philodromus dispar* (Araneae, Philodromidae). J Exp Biol 215:178–184
Zhou Y, Robinson A, Steiner U, Federle W (2014) Insect adhesion on rough surfaces: analysis of adhesive contact of smooth and hairy pads on transparent microstructured substrates. J R Soc Interface 11:20140499

Chapter 5
Anisotropic Friction in Biological Systems

Abstract Biological surfaces covered with micro- and nanostructures, oriented at some angle to the plain may cause strong mechanical anisotropy. Some of them also exhibit pronounced flexibility due to the material of the supporting layer or due to flexible connecting joints. Flexible systems have a wide range of functions including the transport of particles in insect cleaning devices and the propulsion generation during slithering locomotion of snakes. In this chapter, we study the dependence of the anisotropic friction on the slope of the structures, rigidity of their joints, and sliding speed. A system of this kind is the snake skin consisting of stiff scales embedded in a flexible supporting layer. Additionally, there is also microstructure with strongly anisotropic orientation on these scales, which provides frictional anisotropy of the skin. The main function of such hierarchical anisotropic structures is to generate low sliding friction in the forward sliding direction, and high propulsive force along the substrate. Snakes are also able to dynamically adapt their friction interactions by redistributing their local pressures and changing their winding angles, when either friction anisotropy is suppressed by the low friction substrate, or when the external force displacing snake overcomes friction resistance on inclines. In order to understand these biotribology problems, we develop a set of corresponding numerical models.

Anisotropic surfaces are widespread in the non-biological world, ranging from the molecular level (Liley 1998) to the macroscopic level. The crystal structure of solids leads to the anisotropy of their surfaces on an atomic level. In engineering, anisotropy of certain texture patterns of polycrystalline materials is produced naturally or

Electronic supplementary material The online version of this chapter (https://doi.org/10.1007/978-3-030-41528-0_5) contains supplementary material, which is available to authorized users.

A. E. Filippov, S. N. Gorb, *Combined Discrete and Continual Approaches in Biological Modeling*, Biologically-Inspired Systems 16,
https://doi.org/10.1007/978-3-030-41528-0_5

artificially during manufacturing of the materials. Surface anisotropy manifests itself even on geological scales where, due to tectonics, a majority of structures have well-developed anisotropy.

There is also a huge variety of biological surfaces covered with micro- and nanostructures oriented at some angle to the supporting surface (Nachtigall 1974; Gorb 2001; Schege and Gorb 2001). Such structures cause mechanical anisotropy due to different effects, such as friction and/or mechanical interlocking, e. g. during sliding in contact with another surface or during propulsion generation on (or within) a substrate for the purpose of locomotion or for transporting items. These structures have been previously described in a variety of mechanical systems belonging to different organisms such as the insect unguitractor plate (Dashman 1953; Goel 1972; Conde-Boytel et al. 1989; Seifert and Heinzeller 1989; Gorb 1996), interlocking mechanisms of joints in insect legs and antennae (Gorb 2001), insect ovipositor valvulae (Müller 1941; Smith 1972; Mickoleit 1973; Austin and Browning 1981; Gorb 2001), animal attachment pads (Autumn et al. 2000; Gorb and Scherge 2000; Bohn and Federle 2004; Huber et al. 2005; Niederegger and Gorb 2006; Gorb et al. 2007), the inner surface of pitcher plants (Clemente et al. 2009; Gorb and Gorb 2009, 2011), wheat awns (Elbaum et al. 2007), fluid-guiding systems of plants (Roth-Nebelsick et al. 2012), butterfly wings (Zheng et al. 2007), etc.

5.1 Frictional-Anisotropy-Based Mechanical Systems in Biology

The surface outgrowths of most such anisotropic surfaces in biology are connected to a supporting layer which is rather rigid and relies on the ratchet principle in its mechanical behavior (Fig. 5.1). However, some systems exhibit a pronounced flexibility of surface structures due to a specialized flexible joint connecting the surface structures to the rigid supporting layer or due to the flexible material of the supporting layer. A typical macroscale system of the latter kind is snake skin which consists of rather stiff scales (Klein et al. 2010) embedded in a rather flexible supporting layer (Fig. 5.2). The preferred orientation of both the scales themselves and the surface microstructures has been discussed to be the key feature responsible for the frictional anisotropy in this particular system (Hazel et al. 1999; Niitsuma et al. 2005; Berthé et al. 2009; Hu et al. 2009). In addition, there is another microstructure with a strongly anisotropic orientation which has recently been demonstrated to provide frictional anisotropy to the snake skin (Berthé et al. 2009; Benz et al. 2012).

Shark skin exhibits a similar arrangement of stiff surface denticles embedded in a flexible collagenous supportive layer (Reif and Dinkelacker 1982). Another example

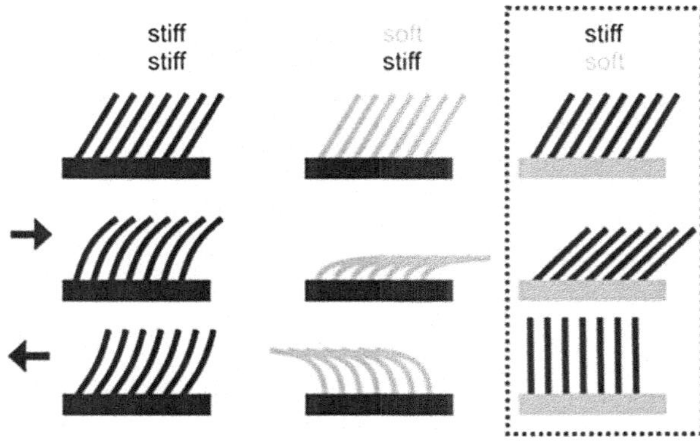

Fig. 5.1 Diagram of ratchet-like frictional anisotropic systems. Left-hand column: stiff protuberances on a stiff supporting layer. Middle column: soft protuberances on a stiff supporting layer. Right-hand column: stiff protuberances on a soft supporting layer. The framed column represents the system considered here. First row: systems in non-deformed state. Second row: deformation caused by sliding in the direction of the protuberance slope. Third row: deformation caused by sliding in the direction opposite to the protuberance slope. (From Filippov and Gorb 2013)

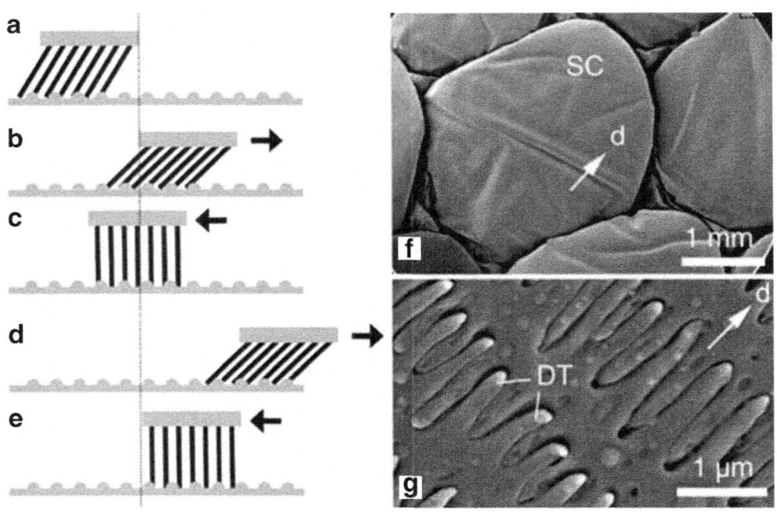

Fig. 5.2 Anisotropic system that assists body propulsion for locomotion. (**a–e**) Diagrams showing how a soft-embedded sloped stiff array of protuberances can generate propulsion long a non-smooth substrate. (**f**, **g**) Lateral scales of the snake *Python regius* at different magnifications (Scanning Electron Microscopy). Abbreviations: *d* direction toward the tail (caudal), *DT* denticulations, *SC* scales. (From Filippov and Gorb 2013)

is provided by the burr-covered *Galium aparine* plant leaves, stems and fruits (Bauer et al. 2010), where the burrs are connected to the supporting layer with a flexible joint (Gorb and Gorb 2002). Cleaning devices of insects also consist of rigid setae

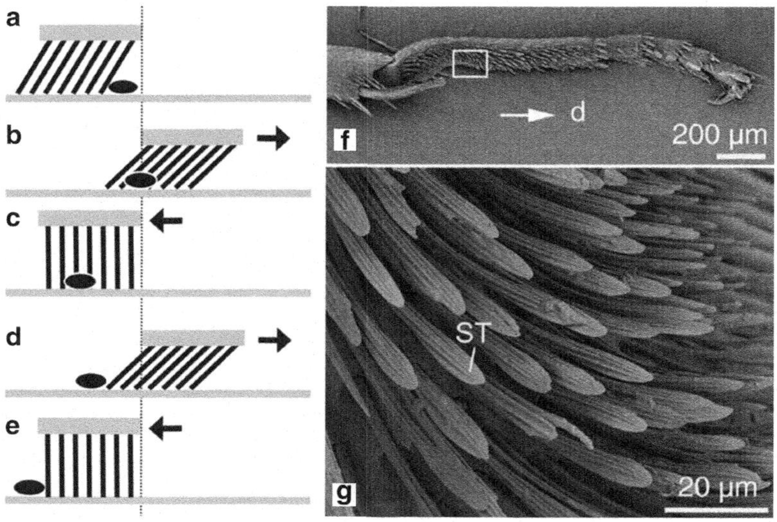

Fig. 5.3 System that generates particle movement for cleaning. (**a–e**) Diagram showing how the soft-embedded sloped stiff array of protuberances can generate unidirectional particle motion along a substrate. (**f**, **g**) Foreleg of the ant *Formica polyctena* with a specialized cleaning device at different magnifications (SEM). Abbreviations: *d* distal direction, *ST* setae. (From Filippov and Gorb 2013)

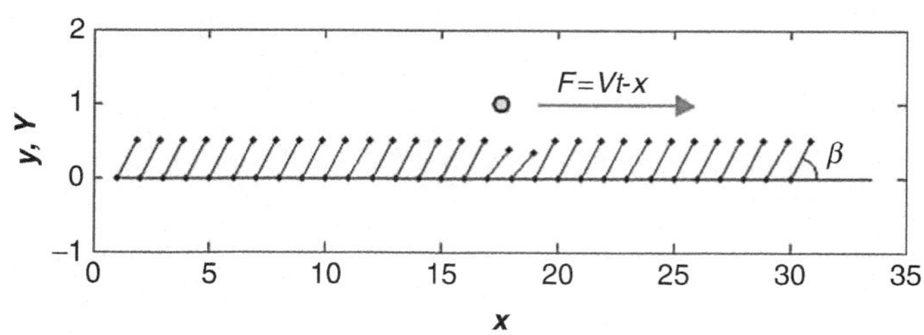

Fig. 5.4 Conceptual diagram of the model. The motion of the system is determined by Eq. (5.1) at equal damping constants, $\gamma = \gamma_\beta = 1$, fixed elasticity $K = 1$ of the external spring and an interaction $f_0 = 1$ between the probe and the hairs. Two other parameters, external velocity V and rigidity against rotations K_β, remain varied. (From Filippov and Gorb 2013)

connected to the surface by flexible joints (Schönitzer and Penner 1984; Schönitzer 1986; Schönitzer and Lawitzky 1987) (Fig. 5.3).

These numerous examples have a wide range of functions, such as the transport of particles (cleaning devices), the positioning of a leaf on top of another leaf or propulsion during slithering locomotion (snakes).

Since the rigidity of the support must have an influence on the mechanical behavior of these systems (as recently shown for snake skin by Benz et al. 2012), we developed a model for studying their mechanical behavior (Fig. 5.4), in particular the dependence of the anisotropic friction efficiency on (1) the slope of the surface structures, (2) the rigidity of the joining sites between the neighboring scales, and

(3) the sliding speed. Based on the proposed model, we suggest a generalized optimal set of variables for maximizing functional efficiency of anisotropic systems of this type. Finally, we will discuss the optimal combination of such parameters from the perspective of biological systems.

5.1.1 *Numerical Model of Anisotropic Friction in Propulsion and Particle Transport*

The model presented here explains the behaviors associated with two distinct mechanical functions of anisotropic surfaces with soft-embedded surface protuberances: propulsion/locomotion and transport of items/particles. In such systems, the interaction between the tips of the protuberances and the substrate asperities or particles during a sliding motion affects the non-symmetric resistance of the anisotropic surface differently, depending on (1) the direction of movement and (2) the orientation of the hairs. However, this result is evident and similar for other types of anisotropic surfaces with stiff or soft hairs embedded in a stiff matrix.

To construct a numerical model of stiff hairs embedded in a soft matrix, we considered a simple configuration of a probe connected to a spring, driven at a fixed velocity V and interacting with a 1-dimensional periodic array of stiff "hairs". The probe interacts with the nearest hair with a force that depends on the inclination angle β of the hair to the substrate plane: $f(\beta) = f_0 \sin(\beta) + const$. This force has a maximum $f(\beta) = f_0$ for the hair standing perpendicular to the surface and $f(\beta) \to const$ at $\beta \to 0$. A conceptual diagram of the model is presented in Fig. 5.4. In the real system, the interaction between the probe and the hair in horizontal orientation is much smaller than at non-zero angles $const < < f_0$. In the picture of the basic model below, for the sake of simplicity we will neglect this impact on the interaction and suppose that $f(\beta) \to 0$ at $\beta \to 0$.

At any moment in time t, the orientation of each hair in that instant depends on the relation between its place in the array $x_j, j = 1, 2, \ldots N$ and the current position of the probe x. Normally, the hairs will tend to keep the angle $\beta \to \beta_0$ close to its equilibrium β_0 by means of some elastic force $f_{elastic} = K_\beta(\beta_0 - \beta)$. In biological structures, $K_{elastic}$ typically is in the range of 1–100 μN/rad. The inclination angle at a given instant, $\beta = \beta(x_j, t)$, is dynamically determined by a balance between the pressure of the probe $f(\beta)$ and the elastic force $f_{elastic} = K_\beta(\beta_0 - \beta)$. In turn, the probe registers a reaction of the hair $-f(\beta)$ and is driven by the external force $K(Vt - x)$. The following model equations result:

$$\begin{aligned} \frac{\partial^2 x}{\partial t^2} &= K(Vt - x) - f(\beta) - \gamma \frac{\partial x}{\partial t}; \\ \frac{\partial \beta}{\partial t} &= \gamma_\beta^{-1} \left[f(\beta) + K_\beta(\beta_0 - \beta) \right]. \end{aligned} \tag{5.1}$$

Here, rotation of the hair inside a viscous substrate is supposed to be strongly over-damped, and the material properties of the substrate are assumed to be fully described by the elastic constant K_β and the damping constant γ_β. The damping constant γ_β of the substrate defines the characteristic time scale of the process and may be experimentally restored for a particular system. To study the general properties of the model below, it is convenient to measure time in the units of γ_β^{-1} and assume $\gamma_\beta = 1$. A preliminary study of the model revealed that the clearest manifestation of friction anisotropy appears when (a) the damping constant of the external device is comparable to the internal damping $\gamma \simeq \gamma_\beta = 1$ and (b) the external elastic force $K\Delta x$ at $\Delta x \simeq x_{j+1} - x_j = 1$ is also comparable to the maximum level of interaction, $\max\{f(\beta)\} = f_0$ between the probe and the hairs ($K\Delta x \simeq f_0$). The period of real structures in different biological systems has a wide range (from 1 to 1000 μm), but typically is between 1 and 10 μm. Thus, our unit of length corresponds to 1–10 μm. Below, for the sake of clarity, we will use the equation $K\Delta x = f_0 = 1$ and characterize the system by means of the remaining free parameters, relative elasticity $K_\beta/K \equiv K_\beta$ and external velocity V. The system of differential equations is simple enough to be easily solved mathematically using any standard mathematics software. In particular, we used the MatLab program and applied the time step $\Delta t = 10^{-4} \div 10^{-2}$, which is small enough to achieve stable behavior across the entire interval of variations of both parameters K_β and V.

Even before reaching a numerical solution, one can intuitively predict that in the cases of either an extremely rigid $K_\beta \to \infty$ or an extremely soft $K_\beta \to 0$ the probe actually interacts either with an almost fixed periodic substrate potential or with a practically flat surface, where the hair is easily pushed into a lying position under the influence of the external body (probe) sliding along the hair array. In both cases, the probe effectively interacts with a static potential. The inclination angle $\beta \to \beta_0$ may still make friction anisotropy possible, but one can expect that the friction force depends weakly on the direction of motion and its values are very similar at positive ($V > 0$) and negative ($V < 0$) velocities. Another situation is created by some intermediate constant K_β, when the probe, moving against the inclination of the hairs $V < 0$, raises them almost vertically, provoking strong resistance to the motion. In this case, resistance should essentially differ for the two directions of motion. From a mathematical point of view, this is an example of movable systems providing directed transport as a result of the anisotropic organization of the motion (Fleishman et al. 2004; Filippov and Popov 2008).

5.1.2 *Typical Temporal Development and Mean Values of Forces*

The typical temporal development of the friction force $F_{friction}(t)$ at positive ($V > 0$) and negative ($V < 0$) velocities is presented in panels (a) and (b) of Fig. 5.5, respectively. Bold lines in both cases indicate the mean friction force

$$< F_{friction} >= \frac{1}{t} \int_{t=0}^{t} F_{friction}(t), \quad (5.2)$$

accumulated between the starting moment $t = 0$ and the current time t. The mean force $<F_{friction}>$ averaged over sufficiently long periods of time can be used to characterize the differences in the system properties at varied elasticities K_β and velocities V.

It is important to note that the model is designed to be robust against particular choices of the elastic force $f_{elastic}(\beta)$. Above, this force was assumed to follow the linear function $f_{elastic} = K_\beta(\beta_0 - \beta)$, in part because we believe that the elastic force uniformly increases with the deviation of the angle $\beta_0 - \beta$ from its equilibrium value β_0, and in part because currently we do not know its actual behavior. An alternative model might be $f_{elastic} = K_\beta \sin(\beta_0 - \beta)$. This equation reflects the fact that distortions of the skin, which tend to effectively function as springs preventing a rotation of the fibers, are proportional to the cosine: $\delta l \sim \cos(\beta_0 - \beta)$. The corresponding force $f_{elastic} = K_\beta \sin(\beta_0 - \beta)$ degenerates into a linear force $f_{elastic} = K_\beta(\beta_0 - \beta)$ when the angle deviations are small, $(\beta_0 - \beta) \rightarrow 0$. The results obtained from this variant of the model must coincide with the linear one in the limit $(\beta_0 - \beta) \rightarrow 0$. In principle, a greater difference is expected for larger deviations. These can appear at negative velocities $V < 0$ and a weak elastic constant $K_\beta \ll 1$, where the external force is able to rotate the fibers powerfully. However, the equation $f_{elastic} = K_\beta \sin(\beta_0 - \beta)$ at $K_\beta \ll 1$ also becomes unrealistic, given that the force $f_{elastic}$ must grow uniformly at large deviations.

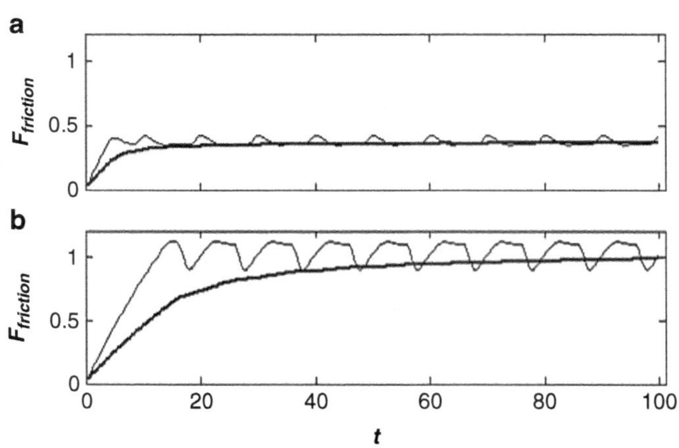

Fig. 5.5 Typical temporal development of the friction force for (**a**) positive velocities $V > 0$ and (**b**) negative velocities $V < 0$. Bold lines in both cases indicate the mean friction force ($F_{friction}$) starting from $t = 0$ to a current moment t. The parameters are the same as in Fig. 5.4. (From Filippov and Gorb 2013)

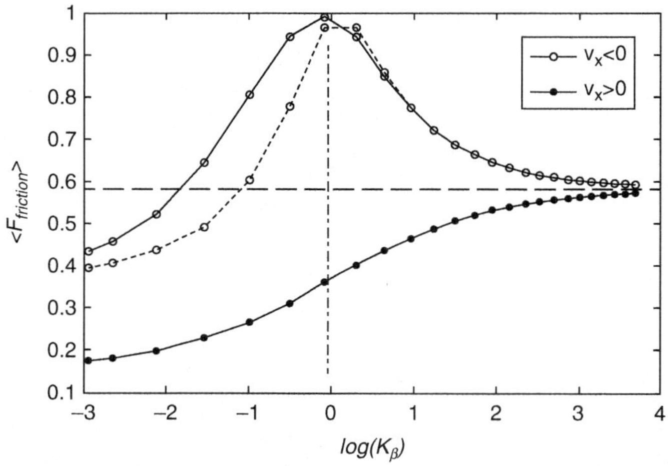

Fig. 5.6 Mean friction forces for fixed positive and negative velocities $V = \pm 1$, accumulated for a few orders of the absolute value $10^{-3} \leq K_\beta < 10^4$. The optimal elasticity at $K_\beta \approx 1$, which corresponds to the maximal anisotropy of motion, is indicated by the dash-dotted line. To compare two variants of the model, numerical simulation with $f_{elastic} = K_\beta \sin(\beta_0 - \beta)$ was also performed (dotted curve). Different variants of the kinetics of the described process, with diverse relationships between elasticity and the direction of motion, are illustrated in the supplementary movies 5.1–5.4. (From Filippov and Gorb 2013)

To compare two variants of the model, we also performed numerical simulations with $f_{elastic} = K_\beta \sin(\beta_0 - \beta)$. The results are shown as a dotted curve in Fig. 5.6. As expected, a deviation is observable only for a combination of negative velocity $V < 0$ and a weak elastic constant $K_\beta \ll 1$. For a stronger $K_\beta \geq 1$ and for all positive velocities $V > 0$, two variants coincide almost perfectly (for positive V, the difference is practically invisible in the figure). Also, it has to be taken into account that the majority of the results below are obtained for an elastic constant close to the optimal value $K_\beta \simeq 1$. This allows us to limit ourselves to the simplest linear model, $f_{elastic} = K_\beta(\beta_0 - \beta)$. It is also important to acknowledge that this basic, minimalistic model neglects to consider many important parameters, such as the density and geometry of the stiff fibers, as well as the fiber length, thickness, etc., which may also affect the results. All these questions remain open for more targeted further studies of concrete biological systems. The main advantage of the present model is its simplicity and transparency, allowing us to consider the importance of structural flexibility for tuning frictional anisotropy.

It is also interesting to study the inverse problem and apply friction anisotropy (which is, in fact, caused by the anisotropy of interactions between substrate and probe) to produce the directed drift of some "cargo". To do this, we put the probe on top of the substrate and perform periodic oscillations of the substrate. It is possible to set up a mathematical model to confirm that this anisotropy of interactions truly leads to a directed motion. Typical cases of the transport of cargo are illustrated in the supplementary movies 5.5 and 5.6.

We found directed drift even when applying gently colored random noise of fluctuations with slightly dominating frequency Ω instead of strictly periodic oscillations. Such stability in the face of perturbations, and even the ability of the system to produce directed motion at weakly pronounced direction, may be assumed to be extremely important for the biological applications of the effect. However, below we will limit ourselves to strictly periodic oscillations with defined frequency Ω.

In this case, the numerical experiment runs as follows. We use Eq. (5.1), but change the external force to zero: $K(Vt - x) = 0$. Instead of applying external forces, we move the substrate periodically to find out how the probe position changes during a sufficiently lengthy run. Directed drift is characterized by a non-zero mean velocity $<V_x>$:

$$< V_x >= \frac{1}{t} \int_{t=0}^{t} V_x(t) \neq 0. \tag{5.3}$$

Figure 5.7 shows the dependence of the mean drift velocity $<V_x>$ on the elastic constant K_β at some representative frequencies Ω. The larger $<V_x>$, the more pronounced the effect. The dot-dashed line in Fig. 5.7 corresponds to the optimal elasticity $K_\beta \simeq 1$.

At high frequencies, the curves of $<V_x>$ start to shift down monotonously. We do not show all the curves here so as not to overcomplicate the figure. Instead, the direction of the shift is indicated by an arrow. It is possible to collect optimal values of the drift velocity (found on each curve near $K_\beta \simeq 1$) and plot them as a function of Ω (see Fig. 5.8). Starting at a frequency slightly higher than $\Omega = 0.5$ (around $\Omega = 2\pi/10$, shown in Fig. 5.7), the drift decreases exponentially, especially at $\Omega \rightarrow \infty$. This exponential decay is clearly confirmed by the logarithmic plot in the inset in Fig. 5.8. At low frequencies of $\Omega \leq 0.5$, the drift velocity is almost constant. Thus, at small frequencies of $\Omega \leq 0.5$, the probe always is captured by the substrate motion and tends to move in one direction. From a practical point of view, it is also necessary to record how long each cycle of the motion lasts. This period is inversely proportional to the frequency $T \sim 1/\Omega$ and becomes very long for $\Omega \rightarrow 0$. Thus, from the biological point of view, we observe that a frequency near the bending point is optimal.

5.1.3 Main Results and Biological Implications

The surprising – and not at all trivial – result of the study of the proposed model is this: the degree of anisotropy, defined here as the relationship between forces that generate resistance to sliding motions in both directions, is strongly dependent on the stiffness of the joints between the individual sloped protuberances (hairs) and their supporting layer. Quite unexpectedly, the anisotropy is maximal at an intermediate

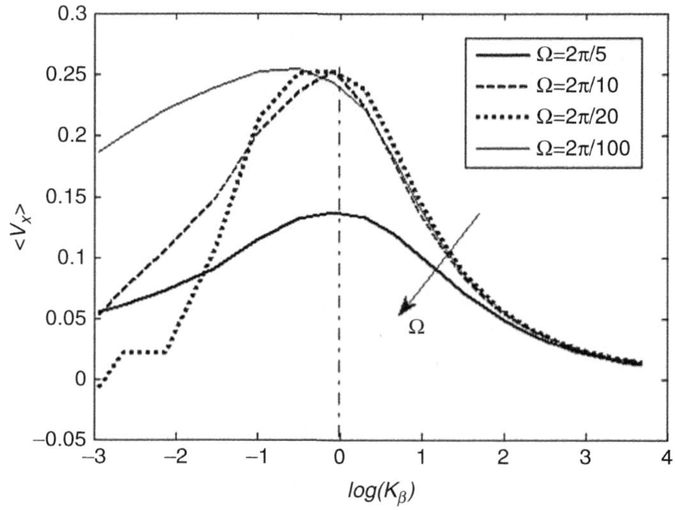

Fig. 5.7 Dependence of mean drift velocity $<V_x>$ on the elasticity of the surface K_β, calculated for some representative frequencies Ω. Dot-dashed line corresponds to the optimal elasticity $K_\beta \approx 1$, at which drift velocity is maximal. The arrow indicates the monotonous shift of the curves $<V_x>$ at frequencies higher than $\Omega = 2\pi/10$. Most of these curves (except for $\Omega = 2\pi/5$) are not shown here to do not overload the figure. (From Filippov and Gorb 2013)

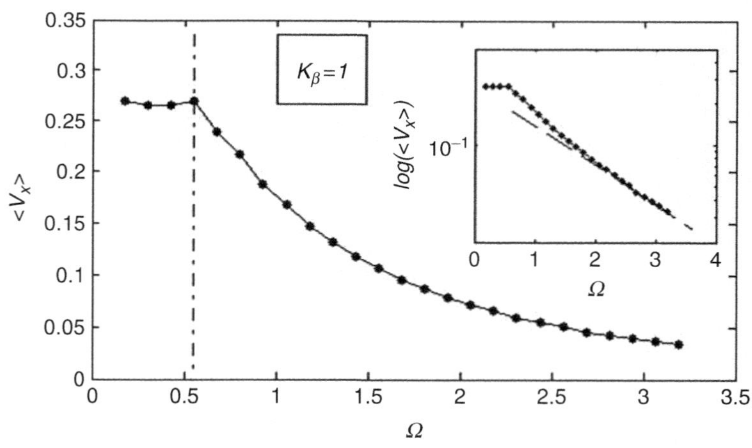

Fig. 5.8 Frequency dependence of drift velocity $<V_x>$ calculated at fixed optimal elasticity $K_\beta \approx 1$. Logarithmic plot (inset) confirms an exponential decay of the drift velocity $<V_x>$ at high Ω. (From Filippov and Gorb 2013)

stiffness of these joints. Presumably, this is the reason why frictional systems based on softly-embedded stiff sloped hairs occur widely in biological mechanical devices that deal with propulsion/locomotion or transport of items/particles.

Another important conclusion that arises from this model is that the degree of anisotropy depends on the frequency/velocity of sliding movements. There is a frequency at which the cumulative directed displacement, caused by oscillations in opposite directions, is maximal. The use of such a frequency in combination with a particular stiffness of the joints leads to an enhancement of the propulsive/carrying performance by several orders of magnitude.

It can be assumed that in biological evolution, oscillation frequency and joint stiffness were important variables for the optimization of concrete mechanical systems. Systems requiring a particular oscillating frequency to function properly for some reasons, such as characteristic velocity of muscle contraction, might be tuned to optimum performance by tuning the mechanical properties of their hair joints. On the other hand, systems with joints with constant mechanical properties might adapt their performance by tuning their specific oscillating frequency within the range of mathematically available frequencies. An optimal stiffness of joints presumably corresponds to specific force values at which the particular mechanical system usually operates. Stiff joints will not improve performance at very low operating forces, and vice versa, soft joints will not maintain strong frictional anisotropy at very high forces.

Some biological anisotropic systems, such as snake skin, have an even higher degree of hierarchical organization of their anisotropically oriented structures (such as scales and denticulations), which may allow for an additional optimization of anisotropic frictional behavior at different macro- and microscopic levels of organization.

One very interesting observation regarding the behavior of the proposed model is that in "cargo"-transporting devices, the transporting function completely fails at a certain minimal oscillating frequency. Nevertheless, both types of systems may properly operate at quite a wide range of joint stiffnesses and sliding velocities – which in biological evolution might be under the control of selective pressure. The model allows for a wide mathematical range of mechanical properties in combination with oscillating frequencies. However, in biological systems, the spectra of these variables must be much narrower as a result of limitations caused by specific biological factors. For example, transport that lasts very long – even though the transporting process is effective – may not be adequate for specific physiological processes within a living organism, or may even take too long as compared to the organism's lifetime. On the other hand, the oscillating frequency cannot be extremely high, since the effectiveness of transport would decrease. Typical frequencies of sliding in various biological systems are in the range of 0.5–25.0 Hz, and sliding velocities tend to be in the range of 0.01–50.00 m/s. Mutual relations between these parameters are well within the range of the optimum predicted by the model.

Recently, bio-inspired systems with anisotropic frictional properties have been developed (Murphy et al. 2007; Manoonpong et al. 2017; Tramsen et al. 2018). We believe that our model might aid in the further optimization of such systems for technological purposes. The verification of the model within different biological systems is not a simple task, and it is currently still in progress. Some preliminary results showing the influence of the flexibility of supporting tissues on the degree of frictional anisotropy were published recently (Benz et al. 2012). However, the verification of our model requires a much broader range of parameters than the range used in the paper by Benz et al. (2012).

5.2 Anisotropic Surface Nanostructures of Snake Skin

Locomotion without extremities has important tribological consequences in snakes, as their ventral body surface is in almost continuous contact with the substrate. In order to facilitate locomotion, the surface of the snake skin must generate low friction – to support sliding in the forward direction – and simultaneously produce high friction – to enable propulsive force generation along the substrate (Renous et al. 1985; Baum et al. 2014a, b). It is well known that the frictional behavior of two materials in contact with each other depends on various factors, such as surface energy and the material properties of both surfaces, but one of the most important parameters is the surface roughness of the two bodies in contact (Bowden and Tabor 1986; Scherge and Gorb 2001). Thus, in the case of a snake, the specific roughness of the substrate on which the snake moves must play a very important role in the generation of frictional forces. The ventral surface of the snake is not smooth either, but rather covered with a very specific kind of micro- and nanostructure known as microdermatoglifics (Picado 1931; Hoge and Santos 1953; Maderson 1972; Price 1982; Renous et al. 1985; Irish et al. 1988; Chiasson and Lowe 1989; Price and Kelly 1989; Hazel et al. 1999; Gower 2003; Berthé et al. 2009; Abdel-Aal et al. 2012; Schmidt and Gorb 2012; Baum et al. 2014a, b), and these previous authors have suggested that the specific ventral surface of the snake skin is of high relevance for the generation of anisotropic friction and thus the facilitation of snake locomotion. Previous atomic force microscopy and confocal laser scanning microscopy studies revealed non-symmetric but regular denticle-like nanostructures on the ventral scales of the vast majority of snake species. The structures are 2.46 ± 0.45 µm long, 0.60 ± 0.11 µm wide, and oriented caudally and parallel to the longitudinal body axis (Baum et al. 2014a, b) (Fig. 5.9). Detailed data regarding the morphology of skin surface nanostructures were provided in earlier publications (Schmidt and Gorb 2012; Baum et al. 2014a, b).

Previous experimental data using various tribological approaches clearly revealed anisotropic friction on the ventral scale surface of snakes (Hazel et al. 1999; Berthé et al. 2009; Hu et al. 2009; Abdel-Aal et al. 2012; Marvi and Hu 2012; Baum et al. 2014a, b). Meanwhile, there are biomimetic surface structures with anisotropic friction (Filippov and Gorb 2013; Baum et al. 2014a; Greiner and Schäfer 2015; Mühlberger et al. 2015) inspired by the micro- (Greiner and Schäfer 2015) and

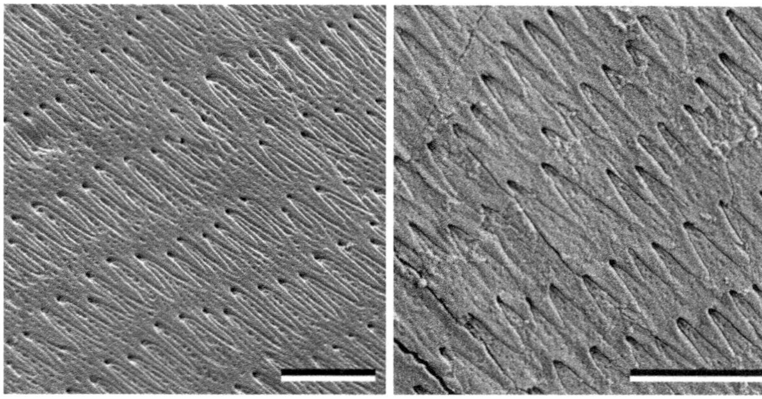

Fig. 5.9 Scanning electron micrographs of ventral snake skin. Left: rattlesnake (*Crotalus* sp.). Right: cobra (*Naja nigricollis*). Left upper corner of both images points toward the head. Denticles are caudally oriented. Scale bars = 5 μm. (From Filippov and Gorb 2016)

nanostructures (Filippov and Gorb 2013; Baum et al. 2014b; Greiner and Schäfer 2015; Mühlberger et al. 2015) of snake skin. However, it is known that the ventral surface of snake skin has a very broad range of frictional properties, and the degree of anisotropy ranges widely as well. This might be due to the variety of species studied, the diversity of approaches used for friction characterization (AFM, microtribometer, sliding test on a slope) and/or the variety of substrates used in the experiments.

The study presented here was undertaken to understand and perhaps even predict the interactions between the nanostructure arrays of the ventral surface of snake skin and various substrates and it aims to numerically model the frictional properties of the structurally anisotropic surfaces that are in contact with various sizes of asperities. Above (Sect. 5.1.1), we showed the effect of the stiffness of surface structures on frictional anisotropy (Filippov and Gorb 2013), whereas in this model we concentrate on the role played by relative dimensions in connection with skin structures and substrate roughness in friction generated in different sliding directions (Filippov and Gorb 2016).

5.2.1 *Modeling of the Frictional Behavior of Snake Skin*

To simulate the friction anisotropy of skin covered with anisotropic microstructures, we used an appropriate modification of the Tomlinson-Prandtl (TP) model. In accordance with the above observations, the skin is covered with a slightly randomized periodic structure of asymmetric holes with short, relatively deep slopes on one side and long, shallow slopes on another. One of the simplest ways to mimic such a structure in numerical simulation is to use an array of almost periodically placed

Gaussian curves with slightly randomized (negative) amplitudes and positions that have different widths in two opposite directions:

$$U(x) = \sum_j G(x - x_j). \tag{5.4}$$

Here, index $j = 1, 2, \ldots, N$ numerates the positions of the minima. It runs along the entire system. The total length of the system is defined by the condition $L = \sum_{j=1}^{N} dx_j$. The distances between minima of nearest Gaussians are determined by the array $dx_j = dx_0(1 + \zeta_j)$. The statement that the system is almost periodic means the distance between minima varies around the average value $dx_0 = const = L/N$ with δ-correlated random deviations ζ_j, and these variations are relatively small:

$$< \zeta_j >= 0; \; < \zeta_j \zeta_k >= \Delta_x \delta_{jk}; \quad \Delta_x \ll 1. \tag{5.5}$$

The anisotropic Gaussians are defined by the following formula:

$$G(x - x_j) = G_j \exp\left\{-\left[(x - x_j)/\Lambda_j\right]^2\right\}, \tag{5.6}$$

where G_j and Λ_j are randomized in the same manner as dx_j with corresponding parameters $\Delta_{G,\,\Lambda} < < 1$. Furthermore, to realistically reproduce observable anisotropic forms of the skin surface (Eq. 5.4), we use different widths in positive and negative directions:

$$\begin{cases} \Lambda^+ = 0.5dx_0; & (x - x_j) > 0; \\ \Lambda^- = 0.1dx_0; & (x - x_j) < 0. \end{cases} \tag{5.7}$$

The typical form of the randomized effective potential $U(x)$, obtained after accumulation of the Gaussians in Eq. 5.1 with all the conditions of Eqs. 5.5, 5.6 and 5.7 taken into account, is shown in the conceptual image (Fig. 5.10). If the probe used in the standard TP friction model is a point of zero size (shown as a black ball in Fig. 5.10), this potential causes an effective force $f_{surf}(X) = -\partial U(x)/\partial x|_{x=X}$, which acts in the equations of motion:

$$\partial^2 X/\partial t^2 = f_{surf}(X) - \eta \partial X/\partial t + K(Vt - X), \tag{5.8}$$

where V, K and η are the velocity, the elastic constant and the damping constant of an external spring, respectively, driving the probe with an instant coordinate X. Typical temporal development of friction forces is shown in Fig. 5.11.

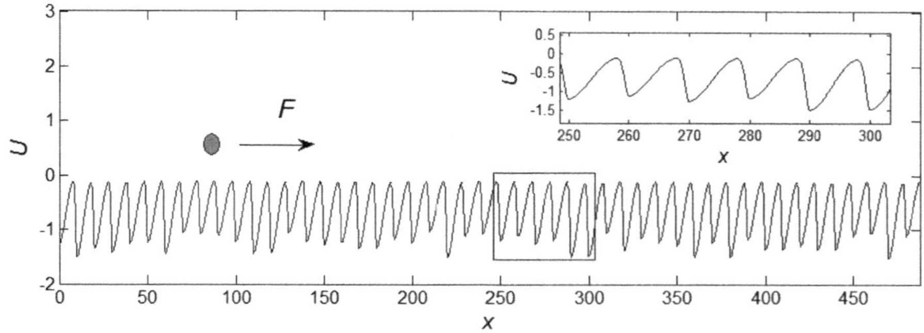

Fig. 5.10 Conceptual structure of the model. The potential $U(x) = \sum_j G(x - x_j)$ is constructed using an array of anisotropic Gaussian curves. The anisotropy of the Gaussian curves is shown for a small fragment of the skin surface (marked by the rectangle) and enlarged in the inset. The probe (which represents an asperity of the substrate) is driven by an external force and shown as a dark circle. It can be either a point of zero size or a body of a different size, ranging up to sizes comparable to the period of the potential. (From Filippov and Gorb 2016)

A more realistic variant of the model corresponds to a limited (non-zero) size θ of the probe, which simulates a characteristic size of the substrate asperities interacting with the skin's surface potential $U(x)$. In this case, total surface force $f_{surf}(X)$ is equal to an integral of all impacts accumulated along all segments of the surface taken in interval $|X - x| < \theta$:

$$f_{surf}(X) = -\frac{\int_{X-\theta}^{X+\theta} \Phi(X-x)\partial U(x)/\partial x dx}{\int_{X-\theta}^{X+\theta} \Phi(X-x)\, dx} \tag{5.9}$$

with a kernel which monotonously decreases with the distance from the center of the probe body $|X - x|$. For the characteristic size θ, it is quite self-consistent to simply take $\Phi(X - x) = \exp(-|X - x|/\theta)$.

It can be expected that at the limit $\theta \ll dx_0$, the model will reduce itself to the case of a body of zero size, where anisotropy of the surface is most pronounced. At the opposite limit, $\theta > dx_0$, the body will cover some number of the periods of the system. As a result, the anisotropy will become less pronounced, and at the extreme limit of $\theta \gg dx_0$, the body will treat the surface as practically flat and symmetric in both directions.

5.2.2 *Mean Friction Forces of Snake Skin and Their Variations*

Our simulations generally confirm these expectations. The results are summarized in Fig. 5.11 which presents friction forces for the five representative values

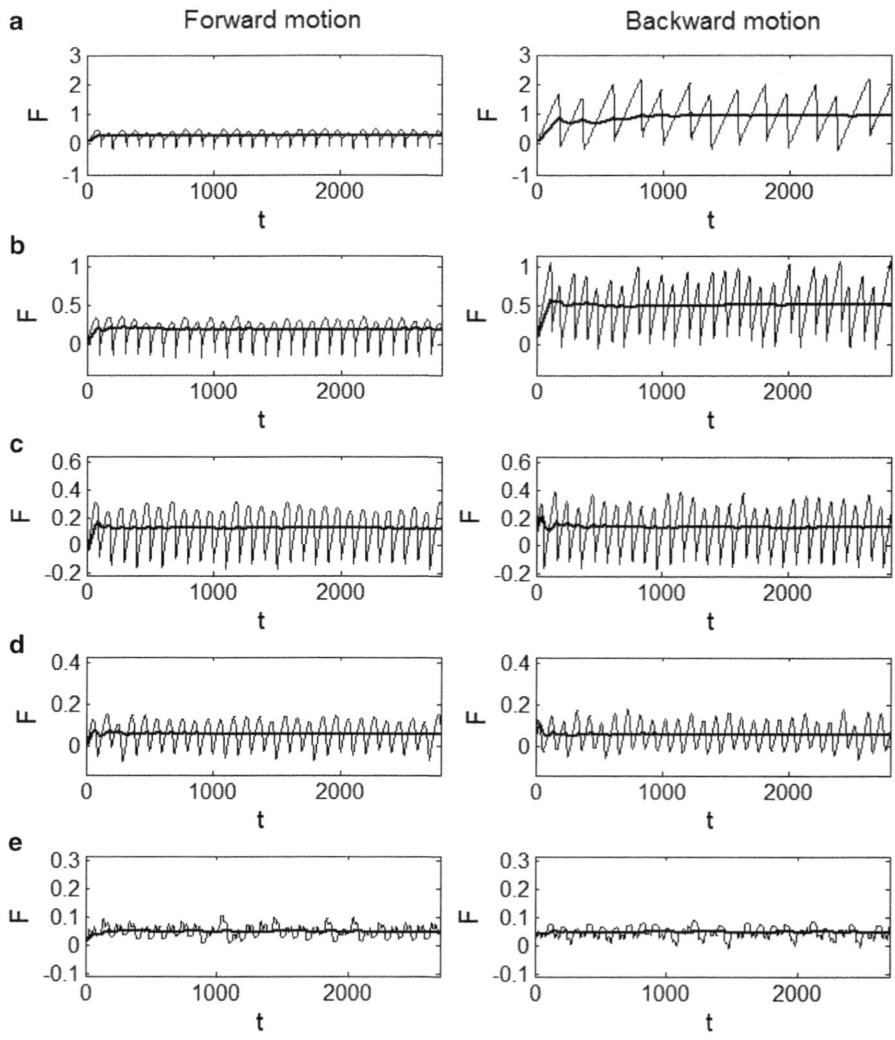

Fig. 5.11 Typical temporal development of friction forces for the forward and backward directions of motion, shown in the left-hand and right-hand panels of subplots (**a**)–(**e**) for 5 representative sizes of the probe $\theta = \{.01\Lambda^-, \Lambda^-, (\Lambda^- + \Lambda^+)/2, \Lambda^+, 3\Lambda^+/2\}$, respectively. Bold lines in all the plots correspond to the time-averaged friction forces. Note the different vertical axes, which monotonously decrease from subplots (**a**) to (**e**) but coincide in each pair of left and right plots. (From Filippov and Gorb (2016)). Formation of these plots in dynamics is shown in supplementary movie 5.7

$$\theta = \{.01\Lambda^-, \Lambda^-, (\Lambda^- + \Lambda^+)/2, \Lambda^+, 3\Lambda^+/2\} \tag{5.10}$$

of the width θ in subplots (a–e), respectively. Left-hand and right-hand panels for each of the cases (a–e) reproduce the force for positive and negative directions of

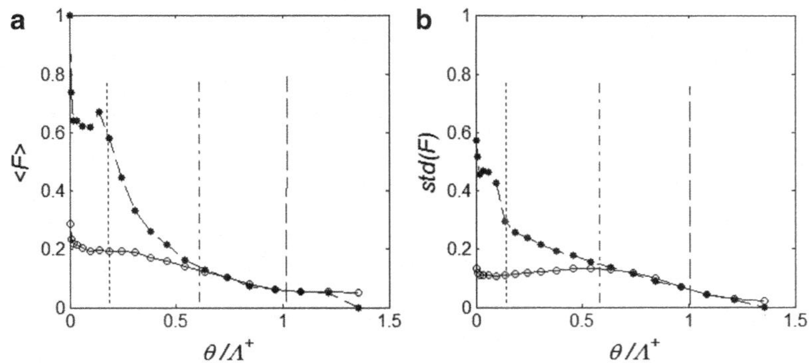

Fig. 5.12 Dependence of the mean friction forces (**a**) and the standard deviations (**b**) for forward (white circles) and backward (black circles) directions of motion, calculated in the intervals between probe sizes that correspond to the representative values shown in Fig. 5.11. Dotted, dash-dotted and dashed lines mark the cases $\theta = \Lambda^-$, $\theta = (\Lambda^- + \Lambda^+)/2$ and $\theta = \Lambda^+$, respectively. (From Filippov and Gorb (2016)). The dynamic formation of these curves is illustrated in supplementary movie 5.8

motion, respectively. Well-pronounced qualitative differences between these two directions are immediately apparent in the first three cases: strong stick-slip behavior in one direction and smooth motion in the other direction.

We also calculated instant time-averaged mean friction:

$$< F(t) >= \frac{1}{t} \int_0^t F_{friction}(t) dt'.$$

This is presented as bold lines in all panels (Fig. 5.11). For the stationary process here, when time is headed toward infinity, $t \to \infty$, the friction tends toward a constant value, $<F(t)> \; \to \; < F > \; = const.$, which is dependent on the size of the probe/asperity. For relatively small probes, <*F*> is considerably higher for a positive direction of motion than in the opposite direction (Fig. 5.12).

It is interesting to note that a relatively medium-sized probe, which in reality corresponds to a surface asperity from $\theta = \Lambda^+$ to $\theta = 1.5\Lambda^+$ (with an order of period dx_0), leads to a practically complete smoothening of the friction curves in both directions, as preliminarily expected for the larger $\theta \gg dx_0$. This means that frictional anisotropy only appears on those substrates of the ventral surface of the snake skin that have a characteristic range of roughness either less than or comparable to the dimensions of the skin microstructure. In other words, scale relief should reflect an adaptation to the particular range of surface asperities of the substrate. However, in cases where many substrate asperities interact simultaneously with the skin surface, the stick-slip behavior might not be as strongly pronounced as in the present model. It can be weaker due to the random distribution of denticle tips on the snake scales (Liley 1998).

In the literature, there are only a few studies that report on frictional properties of snake skins on different roughnesses (Berthé et al. 2009; Hu et al. 2009; Abdel-Aal et al. 2012; Marvi and Hu 2012). Berthe et al. (2009) performed frictional experiments with different scales of *Corallus hortulanus* in three directions on nine different rough surfaces and showed frictional anisotropy along both the rostro-caudal and the medio-lateral body axes on all tested substrate roughnesses. In another study, rough spheres with $R_a = 4$ μm (Abdel-Aal et al. 2012) and 2.4 μm (Baum et al. 2014a, b) were used as a sliding probe. The friction coefficient obtained in the cranial direction was always significantly lower than that obtained in the caudal direction. An estimation of the frictional behavior of snake skin on rigid styrofoam material (Marvi and Hu 2012) also showed anisotropic frictional properties in both directions (forward and backward).

Some enhancement of frictional anisotropy was also found in our previous experiments on cushioned skin (the soft underlying layer) versus uncushioned skin (the rigid underlying layer) of the snake *Lampropeltis getula californiae* when in contact with a rough rigid substrate. In frictional experiments with anesthetized snakes on relatively smooth and rough surfaces ($R_a = 20$ and 200 μm, respectively), the rough surfaces demonstrated frictional anisotropy that almost completely disappeared on the smooth surfaces (Mühlberger et al. 2015). However, these latter experiments presumably show the effect of the interlocking of individual scales on surfaces with coarse roughnesses. Thus, based on data from previous studies and the results of our numerical modeling presented here, we can assume that the particular dimensions of the nanostructures on the ventral scales were adapted to enhance frictional anisotropy at nanoscale substrate roughnesses. Frictional anisotropy at the micro- and nanoscales is provided by the macro–/nanoscopic patterns on the ventral scales. It can therefore be concluded that the frictional anisotropy of the ventral surface is provided by two hierarchical levels of structures: scales and denticles.

This is why snakes, whose locomotory ability is greatly decreased on smooth substrates, always rely on a certain dimension of roughness (even nanoscale roughness, where scales cannot be used, might be sufficient for generating propulsion) – a fact that perfectly agrees with the numerical model presented in this study.

5.3 Snake Locomotion with Change of Body Shape Based on the Friction Anisotropy of the Ventral Skin

Snakes are able to dynamically change their frictional interactions with a surface by means of at least three different methods: (1) adjusting the angle of their scales (Marvi and Hu 2012), (2) redistributing their weight throughout various points of contact with the substrate (Hu et al. 2009; Marvi and Hu 2012), and (3) changing their winding angles (Alben 2013). We observed that snakes change their winding angles either when friction anisotropy is suppressed by a particular roughness of the substrate or when the external force displacing the snake surpasses friction resistance during their locomotion on inclines (Filippov et al. 2018). Numerical modeling was undertaken to understand this behavior and perhaps even predict the snake's specific means of locomotion, given the interactions between the ventral surface of the snake skin and the substrate. The adaptation of the winding curvature considered here represents an enhancement of friction anisotropy in critical behavioral situations, such as movement on low-friction substrates or while moving up and down a slope.

5.3.1 Dynamic Change of Frictional Interactions

In order to facilitate slithering or serpentine locomotion, snakes keep their ventral body surfaces in almost continuous contact with the substrate (Abdel-Aal 2018). The friction forces generated by this contact are of crucial importance for propulsion generation. Due to their specific surface microstructure (Picado 1931; Hoge and Santos 1953; Maderson 1972; Irish et al. 1988; Chiasson and Lowe 1989; Price and Kelly 1989; Hazel et al. 1999; Gower 2003; Abdel-Aal et al. 2012; Schmidt and Gorb 2012), the ventral scales of snakes generate lower friction in the forward direction, which supports sliding, and higher friction in the lateral direction, enabling propulsive force generation during lateral winding (Renous et al. 1985; Berthé et al. 2009; Baum et al. 2014b). Since friction depends on the surface energy, material properties and surface roughness of both bodies in contact (Bowden and Tabor 1986; Scherge and Gorb 2001), the slithering behavior of snakes should change on substrates with different surface properties.

The friction anisotropy required for propulsion generation (Hu et al. 2009; Marvi and Hu 2012) may be rather sensitive to the roughness of the substrate on which the snake moves (Filippov and Gorb 2016). At specific relative dimensions of the snake skin microstructure and the substrate asperities, friction anisotropy may be very low. Previously, we numerically studied interactions between the microstructure of the ventral surface of the snake skin and various sizes of substrate asperities as they were in contact with each other (Filippov and Gorb 2016). Our model showed that the ventral surface of the snake skin demonstrates friction anisotropy only on those substrates that have a characteristic range of roughness either less than or comparable to the dimensions of the skin microstructure. This has an important tribological consequence for snake locomotion: at some substrate roughnesses, friction anisotropy may not support propulsion generation by normal slithering. In this case, a snake tries to adapt its body shape in a particular manner that enables the generation of propulsion. Here, we aim at modeling the behavioral adaptation of the snake to maintain friction anisotropy during locomotion on substrates with low friction or on inclines. Some previous snake locomotion models (Alben 2013; Wang et al. 2014) analyzed the functional minimizing locomotion energy expenses. This approach deepened our understanding of snake locomotion in general and is well suited for robotic applications. Our approach here is based on the analysis of a particular locomotion pattern of a real snake with friction parameters characteristic of the ventral snake skin.

5.3.2 Experimental Observations

We performed an experiment in which we video-recorded snakes (Schokari Sand Racer, *Psammophis schokari*) moving on substrates with different roughnesses. For a control experiment, we used a horizontally oriented epoxy resin replica of P150 polishing paper (3M Deutschland GmbH, Neuss, Germany), $R_a = 17$ μm, in lieu of a rough substrate, following a previous study by Hu et al. (2009). The snakes moved forward easily on the control substrate, demonstrating a body shape similar to a sine function. We used an epoxy resin replica of a fine polishing paper (Buehler, Lake Bluff, IL, USA) with 300 nm asperity size ($R_a = 0.35$ μm), which is known to decrease friction anisotropy (Filippov and Gorb 2016), as a low-friction substrate. A maximum incline of 6° was chosen for the experiments, which prevented the snakes from slipping down. The specific configurations of the body shape of the snake during slithering locomotion on the low-friction substrate are shown in Fig. 5.13. It is quite striking that the snake tries to compensate for low friction anisotropy with stronger curvatures of its windings (see supplementary movie 5.9).

Numerical modeling was undertaken to analyze this behavior and perhaps even predict the snake's specific means of locomotion, given the interactions between the ventral surface of the snake skin and the substrate. We aimed to model the changes in body shape that a real snake undergoes during slithering locomotion, taking the role of friction anisotropy into account. Previously, in another numerical experiment, we showed the general effect of the stiffness of surface structures on friction anisotropy (Filippov and Gorb 2013), whereas here we concentrate on the role of the combination of friction anisotropy and winding behavior in snake locomotion. In particular, we attempted to model behavioral adaptations aimed at overcoming locomotion problems encountered on inclines and/or on substrates with fine levels of roughness, where friction anisotropy is suppressed.

5.3.3 Numerical Model of Snake-Like Motion

To understand which friction parameters are important in the slithering snake's locomotion, we constructed and investigated a numerical model of the snake-like motion. Our goal here is to build a simple model of the locomotion that accounts for the anisotropic friction properties of the snake body within the narrow interval of values reported for natural systems. In contrast to other snake locomotion models, where the snake body shape was presented as a sinusoidal or triangular wave (Wang et al. 2014), or some arbitrary smooth function minimizing the energy expenses functional (Alben 2013), we aimed at modeling the shape of a real snake during locomotion over substrates with low friction. Our snake model is organized as follows.

The body of the animal is represented by an array of elastically connected segments, which are allowed to move in two directions, $\{x, y\}$. Each segment is

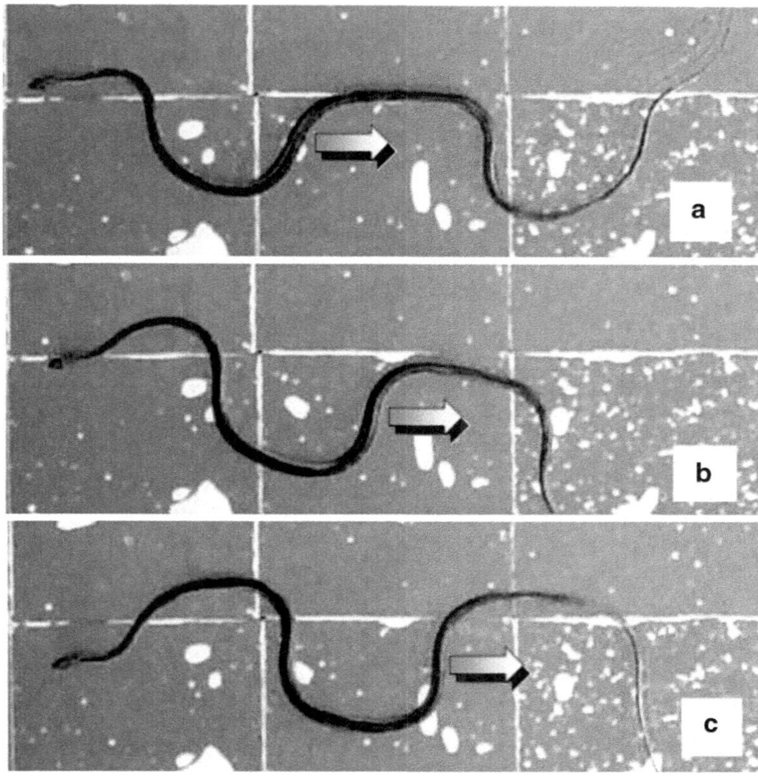

Fig. 5.13 Specific configurations of the body shape of the snake *Psammophis schokari* during slithering locomotion on an inclined substrate with low friction. Note the typical shape of the snake body while moving on such a low-friction surface. It is moving almost without (or with minimal) forward propulsion. This can be easily recognized due to the snake's position relative to the markings on the substrate. Additionally, it is clear that typical soliton-like waves (arrow) propagate along the body against the direction of motion and new solitons at the cranial part of the body are generated as old ones at the caudal end of the body are annihilated. (From Filippov et al. (2018)). See also supplementary movie 5.9

numbered as $j \equiv \{1 : N\}$, where N is the total number of the segments and defined by two end points $\{x_j, y_j\}$ and $\{x_{j+1}, y_{j+1}\}$. Considering that the length L of the studied animal is equal to $L = 105$ cm, and taking into account that numerical simulations should not be too time-consuming, it is convenient to use $N = 36$ points of the numerical chain. In this case, for the straight line $\{x_j(t = 0), y_j(t = 0) = 0\}$ applied with the initial condition and equidistant segments $dx_j = x_{j+1} - x_j$, each value corresponds to a 5 cm long part of the real body:

$$dx_j(t = 0) \equiv dl_0 = 5\,\text{cm}. \tag{5.11}$$

At an arbitrary point in time t, the coordinates y_j of the moving segment are not equal to zero. Given the real body of the animal, the length of each segment, however, must be conserved:

$$dl_j = \sqrt{dx_j^2 + dy_j^2} = const \equiv dl_0. \tag{5.12}$$

This condition is tricky to maintain exactly in a simple model based on elastic force alone. One way to satisfy this condition is to use a strong two valley interaction between two neighboring points in both directions, $\{x_{j-1}, y_{j-1}\}$,$\{x_j, y_j\}$ and $\{x_{j+1}, y_{j+1}\}$, produced by the isotopic Higgs-like potential $U_j^{eff} = U_o dl_j^2\left(dl_0^2 - dl_j^2/2\right)/2$. This potential leads to x and y components of the corresponding force:

$$f_{j;x}^{eff} = U_o dx_j\left(dl_0^2 - dl_j^2\right), \quad f_{j;y}^{eff} = U_o dy_j\left(dl_0^2 - dl_j^2\right). \tag{5.13}$$

At a strong interaction constant U_o, such forces maintain (isotropically) all the values dl_j as close as possible to the trial length of the segments $dl_j \simeq dl_0$. Additionally, we assume that every point of the array $\{x_j, y_j\}$ is subjected to the elastic force $f_j^{elastic}$, which returns the body of our modeled snake to its unperturbed straight form:

$$f_{j,x}^{elastic} = K\left(x_j(t=0) - x_j\right), \quad f_{j,y}^{elastic} = K\left(y_j(t=0) - y_j\right). \tag{5.14}$$

Here, K is the corresponding elastic constant that controls the transversal rigidity of the snake's body. For given initial conditions, an unperturbed configuration corresponds to the straight line along the x coordinate: $x_j = x_j(t=0) \neq 0$, $y_j(t=0) = 0$.

All the perturbations in the frame of the model come from *excitations* produced – in the real biological system – by the snake's muscles and manifest in a consistent wave-form driven along the snake body. These perturbations alternatively displace the y coordinate from its unperturbed configuration $\{y_j\} = 0$ in both directions $y > 0$ and $y < 0$. Our numerical model is expected to reproduce the observed configurations of the real snake body as naturally as possible.

As can be seen directly in videos of a real snake and in Fig. 5.13, while it is moving on a low-friction surface, the typical shape of the snake body has a form of almost planar waves in positive $y > 0$ and negative $y < 0$ directions propagating along the body against the direction of motion and separated by regions that are quite sharp and perpendicular to the direction of motion. Each elementary wave can be treated as a smooth step function with well-defined positive and negative plateaus and regulated widths of the transitions between them. It is convenient to generate such a wave by application of so-called Fermi-Dirac functions:

$$f_k(x;t) = \pm\sqrt{F_k}/\{1 + \exp\left[-(x - X_k(t))/\Delta\right]\}. \tag{5.15}$$

It can be easily proven that each such function results in zero or a unit situated far from the moving bending point $X_k(t)$. The interval of variation of Eq. 5.15 is determined by the width Δ, and its amplitude can be regulated by the factor $\sqrt{F_k}$. If two such functions, having different signs of the exponentials and shifted to the values $\pm\frac{\Delta X}{2}$ in different directions from the position $X_k(t)$, are multiplied as

$$F_k(x;t) = \pm F_k/\left\{1 + \exp\left[-\left(x - X_k(t) + \frac{\Delta X}{2}\right)/\Delta\right]\right\}/\left\{1 + \exp\left[\left(x - X_k(t) - \frac{\Delta X}{2}\right)/\Delta\right]\right\}, \tag{5.16}$$

we will obtain a localized (positive or negative) plateau, which has a total width of ΔX and moves together with $X_k(t)$, as seen in Fig. 5.14a. Below, we use $X_k(t) = Vt$ with a constant dimensionless velocity of $V = 1$, which normalizes other units of the problem.

Now this equation is ready to be applied as a wave of force dynamically producing deformations along the snake's body. We will take sequential waves acting alternatively in both directions, $y > 0$ and $y < 0$, and adjust the factors F_k and width ΔX 'a posteriori' in the course of the numerical experiment to create realistic shapes of the snake that correspond to different regimes of its motion.

In particular, let us note that when watching the locomotion of real snakes, we observed as a rule two pairs $\{k\} = 1, ...4$ of alternating positive $F_1 = F_3 = +F$, and negative waves $F_2 = F_4 = +F$ moving along the body. This fact, as well as the observed relationship between the length of each wave and the length of the region where its line turns, were originally applied to obtain a suitable zero approximation for the widths Δ and ΔX. For the length $L = 105$ cm and $\{k\} = 1, ...4$, the widths are approximately equal to $\Delta \approx 2.5$ cm, $\Delta X \approx 47.5$ cm, and for typical relations $\Psi = L_x/L_y$ between maximum and minimum in the x and y directions, $3 < \Psi < 9$. The factor F varies in the interval 5.8 cm < F < 17.5 cm.

The forces that remain unclear are the longitudinal and transversal components of the friction, which are crucial for generating locomotion with waves produced by the muscular forces $F_k(x;t)$. According to the main goal of this study, we assume (1) the existence of an anisotropic friction $F^{\|}_{friction}$ along the longitudinal body axis and (2) another component of the force $F^{\perp}_{friction}$ that is perpendicular to the body axis (lateral direction) and symmetric (the same friction coefficient for sinistral and dextral directions). Therefore, we define the longitudinal component of the force as:

$$F^{\|}_{friction} = \begin{cases} F^{\|}_{v>0}, & sign\left(v_{\|}\right) = +1; \\ F^{\|}_{v<0}, & sign\left(v_{\|}\right) = -1, \end{cases} \tag{5.17}$$

where $F^{\|}_{v>0}$ and $F^{\|}_{v<0}$ are two different specific friction forces corresponding to different friction force values per segment when the projection of the velocity along the given segment is parallel $sign(v_{\|}) = +1$ (forward motion of the stretched snake, caudad motion of the substrate) or antiparallel $sign(v_{\|}) = -1$ (backward motion of

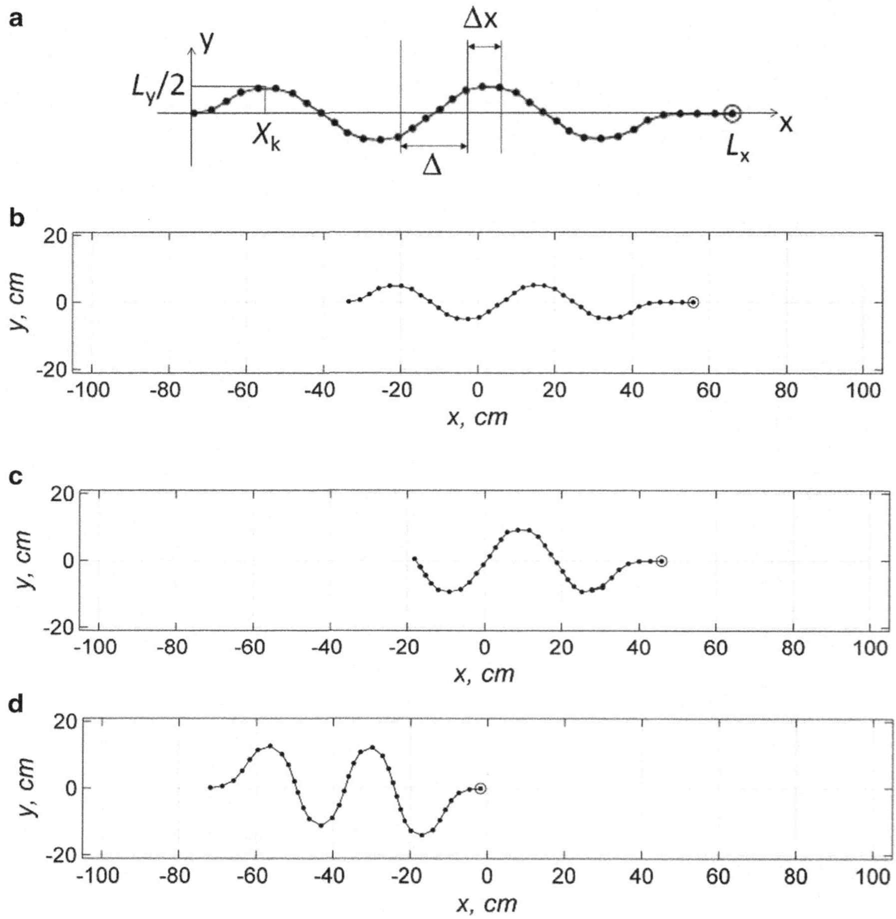

Fig. 5.14 Conceptual structure of the model. The snake segments are represented by black points. The "head" is indicated by the open circle. Some important shape parameters are shown in (**a**). Three different shapes of the snake corresponding to $\Psi = 8.9$, $\Psi = 3.7$, and $\Psi = 1.9$ are presented in (**b**)–(**d**), respectively. (From Filippov et al. 2018)

the stretched snake, craniad motion of the substrate) to the body. According to our observations, these forces must be connected by the inequality $F^{\parallel}_{v>0} \leq F^{\parallel}_{v<0}$. Within the frame of the model, it is convenient to normalize both forces onto a smaller one: $F^{\parallel}_{v>0}$. Below we will vary the relationship between the longitudinal forces in the relatively wide interval $F^{\parallel}_{v<0}/F^{\parallel}_{v>0} = [1, \ldots, 3.5]$ in order to explore and finally determine an optimal value. The same is correct for the force orthogonal to the body. Our estimation for this particular snake is $F^{\perp}_{friction}/F^{\parallel}_{v>0} = 1.75$. For generality of the model description, however, we will vary it in the interval $F^{\perp}_{friction}/F^{\parallel}_{v>0} = [0, \ldots, 5]$.

This paper supposes that the sum of the traveling waves caused by muscular forces $\sum_{k=1}^{4} F_k(x_j, t)$ that deform the snake's body in the $y$direction is the only source of locomotion. Thus, a complete system of the equations of motion can be written as follows:

$$\begin{cases} \gamma \dfrac{\partial x_j}{\partial t} = F_{j,x}^{eff} + F_{j,x}^{elastic} - F_{friction}^{\perp,x} - F_{friction}^{\|,x}; \\ \gamma \dfrac{\partial y_j}{\partial t} = F_{j,y}^{eff} + F_{j,y}^{elastic} + \sum\limits_{k=1}^{4} F_k(x_j, t) - F_{friction}^{\perp,y} - F_{friction}^{\|,y}, \end{cases} \tag{5.18}$$

where $F_{friction}^{\perp,x}$,$F_{friction}^{\|,x}$, $F_{friction}^{\perp,y}$ and $F_{friction}^{\|,x}$ denote x and y projections of the perpendicular and longitudinal friction forces, and γ is the damping constant that defines the time scale of the problem. In the system of Eq. (5.18), the motion is treated as an over-damped one, so all inertial terms are ignored in accordance with Hu et al. (2009).

The friction term here corresponds to so-called "dry dynamic friction". This means these forces are calculated as vectors of a constant length that do not depend on the velocity and are directed opposite to the instantaneous direction of motion of every segment. The nonzero time step Δt of the numerical procedure must also be controlled so that formally calculated components of the friction force do not exceed any values that completely stop the motion of the segment (because the friction force alone cannot cause any motion in other directions).

Figure 5.14 presents a conceptual structure of our model for three different configurations of the snake. The segments of the body are represented by black points. At a fixed total length of $L = 105$ cm, the shapes differ by the relation $\Psi = L_x/L_y$ between maximum and minimum in the x and y directions, $L_{x,\ y}$. As previously mentioned, the specific values of the form factor $|F_k| = F$ must be adjusted to match different Ψ values of the real snake. Three representative variants, corresponding to $\Psi = 8.9$, $\Psi = 3.7$, and $\Psi = 1.9$, are depicted in the subplots Fig. 5.14b–d, respectively.

The further procedure continued as follows. We fixed one of the form factors Ψ and varied two other parameters: longitudinal ($F_{friction}^{\|}$) and transversal ($F_{friction}^{\perp}$) friction forces. Each trial provides different dynamic scenarios of motion. Due to traveling waves sent in one direction along the body by the perpendicular force $|F_k|$ as well as the presence of anisotropic friction, the "head" of the snake permanently moves in a direction opposite to the motion of the waves. The kinetic behavior of the model (see supplementary movie 5.10) is quite similar to the retrograde wave propagation in a real snake as presented above.

The soliton-like waves used in our model differ considerably from the sinusoidal or triangular waves used in the model of, e.g., Wang et al. (2014). When the ratio $F_{friction}^{\perp}/F_{v>0}^{\|}$ is small, the traveling wave motion is also observed in the model of Alben (2013). The latter model, however, only coarsely addresses more complex motions such as when the ratio $F^{\perp}_{friction}/F^{\parallel}_{v>0}$ is neither large nor small (Alben 2013). Still, some of the locomotion patterns (e.g., bending/unbending or curling at the ends) are predicted by the Alben model. Both these processes minimize the energy expenses functional, but it is not related to the locomotion of real snakes.

Since the frictional coefficient ranges from 0.1 to 0.4, the energy expenses for overcoming friction are rather low. Therefore, we argue that locomotion speed is a more important biological parameter than the energy lost due to surface interactions. The speed of the "head" depends on Ψ, $F^{\parallel}_{friction}$ and $F^{\perp}_{friction}$. The simplest way to characterize the speed is to calculate the distance $x_1(t)$ which the "head" covers during a fixed time $t_{\max}$ of the run: $\Delta x = x_1\left(t_{\max}; F^{\perp}_{friction}, F^{\parallel}_{friction}\right) - x_1\left(0; F^{\perp}_{friction}, F^{\parallel}_{friction}\right)$.

If $\Psi = const$ and $F^{\perp}_{friction} = const$, this calculation leads to a family of trajectories $x_1(t)$. One example of such a family is presented in Fig. 5.15a (calculated with $\Psi = 8.9$).

It is possible to collect the final points of different runs, obtained at $t = t_{\max}$, and plot them as a function of $F^{\parallel}_{friction}$. If this calculation is repeated for various values of $F^{\perp}_{friction}$, it results in a family of final points depending on both parameters, $F^{\parallel}_{friction}$ and $F^{\perp}_{friction}$: $\Delta x = x_1\left(t_{\max}; F^{\perp}_{friction}, F^{\parallel}_{friction}\right) - x_1\left(0; F^{\perp}_{friction}, F^{\parallel}_{friction}\right)$ as shown in Fig. 5.15b.

A high density of lines at large values of transversal friction $F^{\perp}_{friction}$ is clearly visible in Fig. 5.15b. Thus, at given parameters, a further increase of $F^{\perp}_{friction}$ does not lead to an increase of speed, whereas in the snake model of Wang et al. (2014), the amplitude of triangular waves and the overall locomotion cost both monotonously decrease with increasing $F^{\perp}_{friction}$. To explain such behavior, we can present the same result as a gray-scale map of Δx within the parameter space $\{F^{\parallel}_{friction}, F^{\perp}_{friction}\}$ as shown in Fig. 5.16a. The interval of color gradations corresponds exactly to the limits of variation Δx clearly visible in Fig. 5.16b. Darker and lighter colors correspond to smaller and larger values of Δx, respectively. Contour lines are added to enhance the relief. As can be seen in the map, above some boundary (marked by a dashed, gray line), the transversal component of friction hardly influences the distance Δx covered by the snake up to $t = t_{\max}$.

We can also calculate the absolute value of the difference between the transversal friction force $F^{\perp}_{friction}$ and the transversal projection $\gamma v^{\perp}$ of the vector $\gamma\vec{v} = \left\{\gamma\frac{\partial x}{\partial t}, \gamma\frac{\partial y}{\partial t}\right\}$:

$$|\delta F^{\perp}| = \left|F^{\perp}_{friction} - \gamma v^{\perp}\right|. \tag{5.19}$$

In Fig. 5.16b, the gray-scale map of this combination is shown in the same coordinates $\{F^{\parallel}_{friction}, F^{\perp}_{friction}\}$ as Δx in Fig. 5.16a.

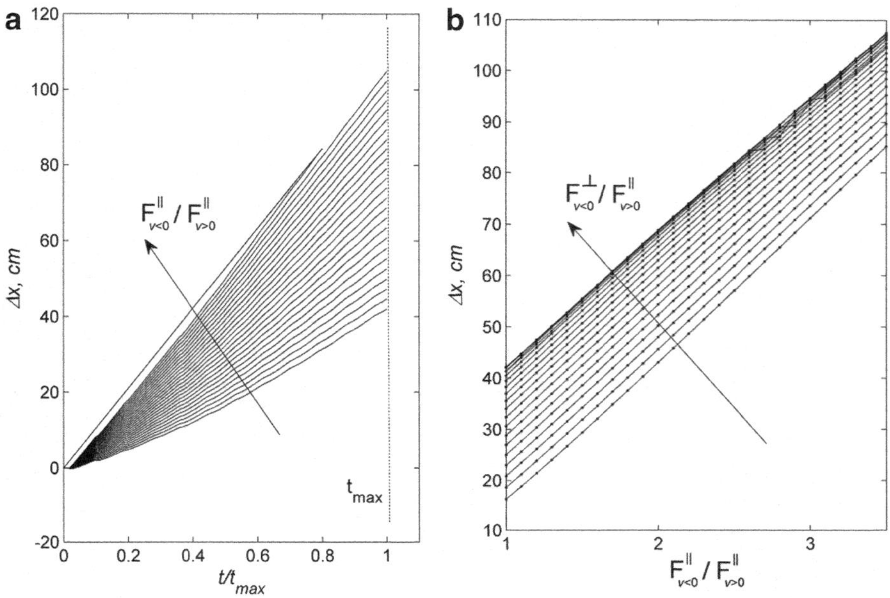

Fig. 5.15 (**a**) The temporal development of $x_1(t)$ and (**b**) the final position of the snake "head", $x_{final} = x_1\left(t = t_{\max}; F^{\perp}_{friction}/F^{\|}_{v>0}, F^{\|}_{v<0}/F^{\|}_{v>0}\right)$, shown as a function of the friction parameters $F^{\perp}_{friction}/F^{\|}_{v>0}$ and $F^{\|}_{v<0}/F^{\|}_{v>0}$: $\Delta x = x_{final} - x_1\left(0; F^{\perp}_{friction}/F^{\|}_{v>0}, F^{\|}_{v<0}/F^{\|}_{v>0}\right)$. Corresponding tendencies are indicated by the arrows. The vertical line represents the final time moment $t = t_{\max}$ which terminates the runs. (From Filippov et al. 2018)

It is evident that the equation $|\delta F^{\perp}| = \left|F^{\perp}_{friction} - \gamma v^{\perp}\right|$ approaches zero near the same line, above which $F^{\perp}_{friction}$ scarcely influences the speed any more. In turn, from the equations of motion set up above (Eq. 5.18) we see that when $\left|F^{\perp}_{friction} - \gamma v^{\perp}\right|$ is close to zero all other forces of the problem become mutually compensated for and the force $F^{\perp}_{friction}$ almost completely determines motion in the transversal direction. From a biological point of view, this means that above this line there is no advantage in spending more energy for bending the body because it no longer enhances propulsion during locomotion. This pattern is similar to the locomotion on a critical slope as described in the snake model by Wang et al. (2014).

We repeated the same calculations for different values of Ψ. To compare the results, it is useful to integrate both values, Δx and $|\delta F^{\perp}|$, over the vertical coordinate of Fig. 5.16, to normalize the obtained results on their maxima and to present them in one plot. Thus, Fig. 5.17 contains the results for $\Psi = 8.9$ as well as for $\Psi = 5.9$ and $\Psi = 3.7$. The curves for the integrated values $<\Delta x>$ and $<|\delta F^{\perp}|>$ are represented by the bold and the thin lines, respectively. Increasing the ratio $F^{\|}_{v<0}/F^{\|}_{v>0}$ above 2.5 does not lead to any further increase of $<\Delta x>$. Similarly, increasing the ratio

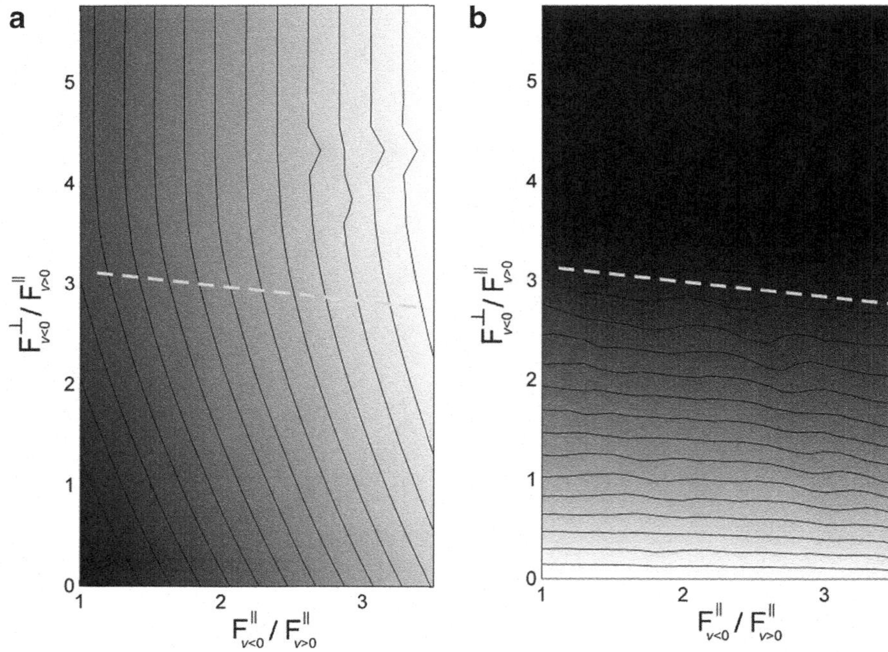

Fig. 5.16 Gray-scale maps (**a**) for the distance $\Delta x = x_{final} - x_1\left(0; F^{\perp}_{friction}/F^{\parallel}_{v>0}, F^{\parallel}_{v<0}/F^{\parallel}_{v>0}\right)$ and (**b**) for the difference $|\delta F^{\perp}| = \left|F^{\perp}_{friction} - \gamma v^{\perp}\right|$ between friction force and $F^{\perp}_{friction}$ and transversal projection $v^{\perp}$ of the vector $\gamma \vec{v}$. In both cases, the dashed, gray line marks a boundary above which any change of transversal friction does not enhance locomotion. (From Filippov et al. 2018)

$F^{\parallel}_{v<0}/F^{\parallel}_{v>0}$ above 3.5 does not lead to any further decay of <$|\delta F^{\perp}|$>. The tendencies of the values <Δx> and <$|\delta F^{\perp}|$> at increasing Ψ are indicated by the arrows in Fig. 5.17. The increase of the form factor Ψ attenuates both the speed (<Δx>) increase and the tangential friction (<$|\delta F^{\perp}|$>) decay when the $F^{\parallel}_{v<0}/F^{\parallel}_{v>0}$ ratio increases (Fig. 5.17). This effect is not observable in models that study typical locomotion on high-friction substrates (Alben 2013; Wang et al. 2014).

5.3.4 Biological Interpretation of the Numerical Results

It has been previously shown that, at high speeds, snakes lift the curved parts of their bodies off the ground as they travel in lateral undulation and in sidewinding (Gans 1984; Jayne 1986). In sidewinding locomotion, an animal pushes into the direction where the highest friction coefficient applies. Recently, theoretical modeling has also predicted that snakes might be able to redistribute their weight and thereby concentrate their weight on specific points of contact (Hu et al. 2009). These points of

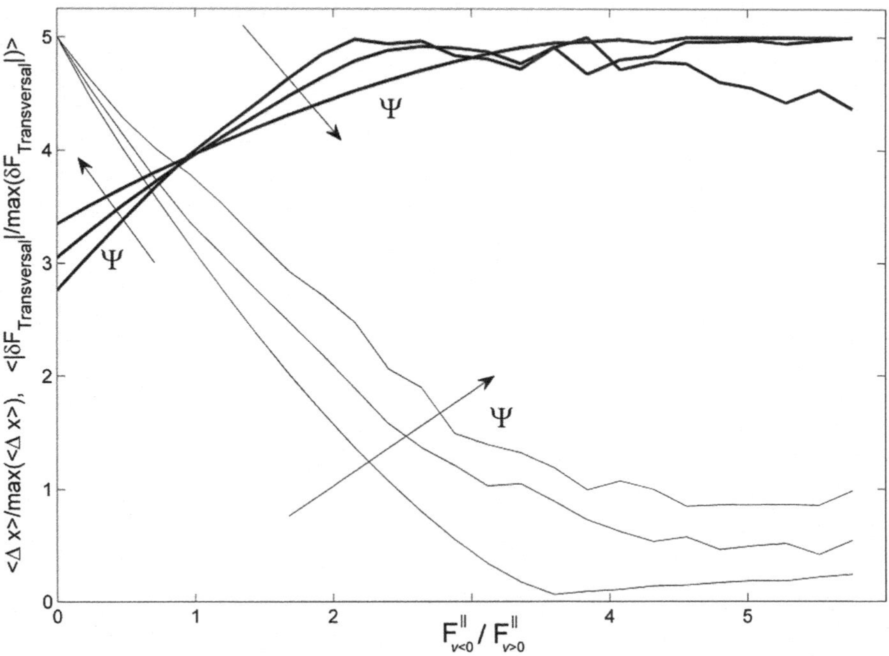

Fig. 5.17 Values of Δx and $|\delta F^{\perp}|$ integrated over a parallel friction coordinate $F^{||}_{friction}$ and normalized to their maxima, calculated for the parameters $\Psi = 8.9$, $\Psi = 5.9$ and $\Psi = 3.7$. The curves for $<\Delta x>$ and $<|\delta F^{\perp}|>$ are shown by the bold and thin lines, respectively. The tendencies for increasing values of Ψ are indicated by the arrows. (From Filippov et al. 2018)

contact approximately correspond to points of zero body curvature. Also, snakes are likely able to dynamically change their frictional interactions with a surface by adjusting the angle of their scales (Marvi and Hu 2012).

Here, we have modeled some aspects of the motion when snakes change their winding angles (1) when friction anisotropy is suppressed by a particular roughness of the substrate or (2) when the external force displacing the snake backwards overcomes its friction resistance during their locomotion on inclines. Adaptation of the winding curvature, as considered above, presumably has something to do with an enhancement of friction anisotropy under critical locomotory conditions, such as when encountering a low-friction substrate or moving up and down a slope.

The friction properties of the ventral skin have been previously tested in different snake species. Friction anisotropy was clearly demonstrated since the friction coefficient obtained in the cranial direction was always significantly lower than that in the lateral direction (Berthé et al. 2009). Our previous experiments comparing the cushioned skin (with a soft underlying layer) of the snake *Lampropeltis getula* to the uncushioned one (with a rigid underlying layer), revealed some enhancement of friction anisotropy when in contact with rough, rigid substrates (Baum et al. 2014b). Other friction experiments with anesthetized snakes on relatively smooth and rough surfaces demonstrated a friction anisotropy which almost completely disappeared on the smooth surface (Hu et al. 2009). Previous experimental and numerical data show that two hierarchical levels of the ventral skin structures (scales and microdenticles) are adapted to enhance friction anisotropy depending on the roughness of the substrate (Filippov and Gorb 2016). This is why snakes demonstrate significantly less locomotive ability on smooth substrates or substrates with fine roughness.

The undulating method of snake locomotion is modeled here by generating four solitary waves (two on each side), which correspond to the original action of lost extremities. These waves allow for both types of frictional anisotropy (longitudinal and transversal). Longitudinal friction anisotropy in snakes is limited due to the particular geometry of their skin microstructure, which allows for a maximum $F^{\|}_{v<0}/F^{\|}_{v>0} \leq 1.75$. This limitation requires an additional use of transversal anisotropy. In order to enhance transversal anisotropy, the snake changes its form factor (Alben 2013). According to our model, however, the growth of transversal friction, which intuitively must magnify the propulsion during any undulating locomotion, enhances propulsion to only a limited extent. We found that an increase of the ratio $F^{\perp}_{friction}/F^{\|}_{v>0}$ above around 3 leads to weaker advantages in locomotion and higher energetic costs. These costs are related to the fact that the snake must generate solitary waves in a caudal direction. Our model shows that the wave amplitude increase makes the snake's speed less dependent on the $F^{\|}_{v<0}/F^{\|}_{v>0}$ ratio, but entails higher energy costs (similar to the model of Wang et al. 2014), and the wave amplitude increase did not even enhance the overall propulsion.

This wave propagation pattern strongly differs from the motion of typical snake-inspired robots using wheels, because such robots do not create waves, but rather generate undulation by changing trajectory through turning their heads side-to-side.

References

Abdel-Aal HA (2018) Surface structure and tribology of legless squamate reptiles. J Mech Behav Biomed Mater 79:354–398

Abdel-Aal HA, Vargiolu R, Zahouani H, El Mansori M (2012) Preliminary investigation of the frictional response of reptilian shed skin. Wear 290–291:51–60

Alben S (2013) Optimizing snake locomotion in the plane. Proc R Soc A 469:20130236

Austin AD, Browning TO (1981) A mechanism for movement of eggs along insect ovipositors. Int J Insect Morphol Embryol 10:93–108

Autumn K, Liang YA, Hsieh ST, Zesch W, Chan WP, Kenny TW, Fearing R, Full RJ (2000) Adhesive force of a single gecko foot-hair. Nature 405:681–685

Bauer G, Klein MC, Gorb SN, Speck T, Voigt D, Gallenmüller F (2010) Always on the bright side: the climbing mechanism of *Galium aparine*. Proc R Soc B 278:2233–2239

Baum MJ, Heepe L, Gorb SN (2014a) Friction behavior of a microstructured polymer surface inspired by snake skin. Beilstein J Nanotechnol 5:83–97

Baum MJ, Kovalev AK, Michels J, Gorb SN (2014b) Anisotropic friction of the ventral scales in the snake *Lampropeltis getula californiae*. Tribol Lett 54:139–150

Benz MJ, Kovalev AE, Gorb SN (2012) Anisotropic frictional properties in snakes. In: Lakhtakia A, Martín-Palma RJ (eds) Bioinspiration, biomimetics, and bioreplication, Proc SPIE 8339, p 11

Berthé R, Westhoff G, Bleckmann H, Gorb SN (2009) Surface structure and frictional properties of the skin of the Amazon tree boa *Corallus hortulanus* (Squamata, Boidae). J Comp Physiol A 195:311–318

Bohn HF, Federle W (2004) Insect aquaplaning: *Nepenthes* pitcher plants capture prey with the peristome, a fully wettable water-lubricated anisotropic surface. Proc Natl Acad Sci U S A 101:14138–14143

Bowden FP, Tabor D (1986) The friction and lubrication of solids. Clarendon Press, Oxford

Chiasson RB, Lowe CH (1989) Ultrastructural scale patterns in *Nerodia* and *Thamnophis*. J Herpetol 23:109–118

Clemente CJ, Dirks J-H, Barbero DR, Steiner U, Federle W (2009) Friction ridges in cockroach climbing pads: anisotropy of shear stress measured on transparent, microstructured substrates. J Comp Physiol A 195:805–814

Conde-Boytel R, Erickson EH, Carlson SD (1989) Scanning electron microscopy of the honeybee, *Apis mellifera* L. (Hymenoptera: Apidae) pretarsus. Int J Insect Morphol Embryol 18:59–69

Dashman T (1953) The unguitractor plate as a taxonomic tool in the Hemiptera. Ann Entomol Soc Am 46:561–578

Elbaum R, Zaltzman L, Burgert I, Fratzl P (2007) The role of wheat awns in the seed dispersal unit. Science 316:884–886

Filippov AE, Gorb SN (2013) Frictional-anisotropy-based systems in biology: structural diversity and numerical model. Sci Rep 3:1240

Filippov AE, Gorb SN (2016) Modelling of the frictional behaviour of the snake skin covered by anisotropic surface nanostructures. Sci Rep 6:23539

Filippov AE, Popov V (2008) Directed molecular transport in an oscillating channel with randomness. Phys Rev E 77:N211114

Filippov AE, Westhoff G, Kovalev A, Gorb SN (2018) Numerical model of the slithering snake locomotion based on the friction anisotropy of the ventral skin. Tribol Lett 66:119

Fleishman D, Filippov AE, Urbakh M (2004) Directed molecular transport in an oscillating symmetric channel. Phys Rev E 69:011908

Gans C (1984) Slide-pushing: a transitional locomotor method of elongate squamates. Symp Zool Soc Lond 52:12–26

Goel SC (1972) Notes on the structure of the unguitractor plate in Heteroptera (Hemiptera). J Entomol 46:167–173

Gorb SN (1996) Design of insect unguitractor apparatus. J Morphol 230:219–230

Gorb SN (2001) Attachment devices of insect cuticle. Kluwer Academic Publishers

Gorb EV, Gorb SN (2002) Contact separation force of the fruit burrs in four plant species adapted to dispersal by mechanical interlocking. Plant Physiol Biochem 40:373–381

Gorb EV, Gorb SN (2009) Functional surfaces in the pitcher of the carnivorous plant *Nepenthes alata*: a cryo-SEM approach. In: Gorb SN (ed) Functional surfaces in biology: adhesion related systems, vol 2, pp 205–238

Gorb EV, Gorb SN (2011) The effect of surface anisotropy in the slippery zone of *Nepenthes alata* pitchers on beetle attachment. Beilstein J Nanotechnol 2:302–310

Gorb SN, Scherge M (2000) Biological microtribology: anisotropy in frictional forces of orthopteran attachment pads reflects the ultrastructure of a highly deformable material. Proc R Soc Lond B 267:1239–1244

Gorb SN, Sinha M, Peressadko A, Daltorio KA, Quinn RD (2007) Insects did it first: a micropatterned adhesive tape for robotic applications. Bioinspir Biomim 2:S117–S125

Gower DJ (2003) Scale microornamentation of uropeltid snakes. J Morphol 258:249–268

Greiner C, Schäfer M (2015) Bio-inspired scale-like surface textures and their tribological properties. Bioinspir Biomim 10:044001

Hazel J, Stone M, Grace MS, Tsukruk VV (1999) Nanoscale design of snake skin for reptation locomotions via friction anisotropy. J Biomech 32:477–484
Hoge AR, Santos PS (1953) Submicroscopic structure of "stratum corneum" of snakes. Science 118:410–411
Hu DL, Nirody J, Scott T, Shelley MJ (2009) The mechanics of slithering locomotion. Proc Natl Acad Sci U S A 106:10081–10085
Huber G, Gorb SN, Spolenak R, Arzt E (2005) Resolving the nanoscale adhesion of individual gecko spatulae by atomic force microscopy. Biol Lett 1:2–4
Irish FJ, Williams EE, Seling E (1988) Scanning electron microscopy of changes in epidermal structure occurring during the shedding cycle in squamate reptiles. J Morphol 197:105–126
Jayne BC (1986) Kinematics of terrestrial snake locomotion. Copeia 22:915–927
Klein M-CG, Deuschle JK, Gorb SN (2010) Material properties of the skin of the Kenyan sandboa *Gongylophis colubrinus* (Squamata, Boidae). J Comp Physiol A 196:659–668
Liley M (1998) Friction anisotropy and asymmetry of a compliant monolayer induced by a small molecular tilt. Science 280:273–275
Maderson PFA (1972) When? Why? And how? Some speculations on the evolution of vertebrate integument. Am Zool 12:159–171
Manoonpong P, Gorb S, Heepe, L (2017) Exploiting frictional anisotropy from a scale-like material for energy-efficient robot locomotion. ISBE Newsletter 6:9–10
Marvi H, Hu DL (2012) Friction enhancement in concertina locomotion of snakes. J R Soc Interface 9:3067–3080
Mickoleit G (1973) Über den Ovipositor der Neuropteroidea und Coleoptera und seine phylogenetische Bedeutung (Insecta, Holometabola). Z Morphol Tiere 74:37–64
Mühlberger M, Rohn M, Danzberger J, Sonntag E, Rank A, Schumm L, Kirchner R, Forsich C, Gorb SN, Einwögerer B, Trappl E, Heim D, Schift H, Bergmair I (2015) UV-NIL fabricated bio-inspired inlays for injection molding to influence the friction behavior of ceramic surfaces. Microelectron Eng 141:140–144
Müller HJ (1941) Über Bau und Funktion des Legeapparates der Zikaden (Homoptera Cicadina). Z Morphol Ökol Tiere 38:534–629
Murphy MP, Aksak B, Sitti M (2007) Adhesion and anisotropic friction enhancements of angled heterogeneous micro-fiber arrays with spherical and spatula tips. J Adhes Sci Technol 21:1281–1296
Nachtigall W (1974) Biological mechanisms of attachment. Springer, Berlin/Heidelberg/New York
Niederegger S, Gorb SN (2006) Friction and adhesion in the tarsal and metatarsal scopulae of spiders. J Comp Physiol A 192:1223–1232
Niitsuma K, Miyagawa S, Osaki S (2005) Mechanical anisotropy in cobra skin is related to body movement. Eur J Morphol 42:193–200
Picado C (1931) Epidermal microornaments of the crotalinae. Bull Antivenin Inst Am 4:104–105
Price RM (1982) Dorsal snake scale microdermatoglyphics: ecological indicator or taxonimical tool? J Herpetol 16:294–306
Price RM, Kelly P (1989) Microdermatoglyphics: basal patterns and transition zones. J Herpetol 23:244–261
Reif W-E, Dinkelacker A (1982) Hydrodynamics of the squamation in fast swimming sharks. Neues Jahrb Geol Paläontol 164:184–187
Renous S, Gasc JP, Diop A (1985) Microstructure of the tegumentary surface of the Squamata (Reptilia) in relation to their spatial position and their locomotion. Fortschr Zool 30:487–489
Roth-Nebelsick A, Ebner M, Miranda T, Gottschalk V, Voigt D, Gorb S, Stegmaier T, Sarsour J, Linke M, Konrad W (2012) Leaf surface structures enable the endemic Namib desert grass *Stipagrostis sabulicola* to irrigate itself with fog water. J R Soc Interface 9:1965–1974
Scherge M, Gorb SN (2001) Biological micro- and nanotribology. Springer, Berlin
Schmidt CV, Gorb SN (2012) Snake scale microstructure: phylogenetic significance and functional adaptations, Zoologica. Schweizerbart Science Publisher, Stuttgart

Schönitzer K (1986) Comparative morphology of the antenna cleaner in bees (Apoidea). Z Zool Syst Evolutionsforsch 24:35–51

Schönitzer K, Lawitzky G (1987) A phylogenetic study of the antenna cleaner in Formicidae, Mutillidae and Tiphiidae (Insecta, Hymenoptera). Zoomorphology 107:273–285

Schönitzer K, Penner M (1984) The function of the antenna cleaner of the honeybee (*Apis mellifica*). Apidologie 15:23–32

Seifert P, Heinzeller T (1989) Mechanical, sensory and glandular structures in the tarsal unguitractor apparatus of *Chironomus riparius* (Diptera, Chironomidae). Zoomorphology 109:71–78

Smith EL (1972) Biosystematics and morphology of symphyta. 3 External genitalia of *Euura*. Int J Insect Morphol Embryol 1:321–365

Tramsen HT, Gorb SN, Zhang H, Manoonpong P, Dai Z, Heepe L (2018) Inversion of friction anisotropy in a bio-inspired asymmetrically structured surface. J R Soc Interface 15:1–7

Wang X, Osborne MT, Alben S (2014) Optimizing snake locomotion on an inclined plane. Phys Rev E 89:012717

Zheng Y, Gao X, Jiang L (2007) Directional adhesion of superhydrophobic butterfly wings. Soft Matter 3:178–182

Chapter 6
Mechanical Interlocking of Biological Fasteners

Abstract Microstructures responsible for temporary arresting of contacting surfaces are widely distributed on surfaces in different organisms. They have different density of outgrowths and surprisingly not ideal distribution patterns. This is why they are often called probabilistic fasteners. Their size, shape and the density of their outgrowths do not correspond exactly to each other and interact by generating strong resistance force against acting force without precise positioning of both surfaces. For example, this kind of attachment is of importance for functioning of some biomechanical systems in insects. One can suggest that different structure of the interlocking devices is optimized by natural selection to get appropriate mechanical arrest. In this chapter, we simulate such a system numerically, both in the frames of continuous and discrete dynamical models. The feathers of modern birds are waterproof, breathable, lightweight constructions combining thermo-isolation, rigidity and flexibility due to the feather's ability to hold its parts together by a specific pattern of hooklets. The feather vane can be separated into two parts by pulling neighboring barbs apart, but original state can be re-established easily by lightly stroking through the feather. Hooklets responsible for holding vane barbs together are not damaged by multiple zipping and unzipping cycles. A model is developed which reproduces zipping and unzipping behavior in feathers similar to those observed in biomechanical experiments performed on real bird feathers.

6.1 Co-opted Contact Pairs in Arresting Systems of Insects

Specialized pairs of microstructure-covered surface fields of arthropod cuticles often allow for mechanical adhesion (attachment) between different body parts within an organism (Nachtigall 1974; Gorb 1998a, b; Scherge and Gorb 2001;

Electronic supplementary material The online version of this chapter (https://doi.org/10.1007/978-3-030-41528-0_6) contains supplementary material, which is available to authorized users.

A. E. Filippov, S. N. Gorb, *Combined Discrete and Continual Approaches in Biological Modeling*, Biologically-Inspired Systems 16,
https://doi.org/10.1007/978-3-030-41528-0_6

Gorb and Popov 2002). These microstructures responsible for a temporary arresting of contacting surfaces are widely distributed on surfaces in different organisms. Recent morphological studies show that these structures vary in terms of the density of their outgrowths and do not display an ideal distribution pattern on the two complementary parts of contact. Such systems are thus often called *probabilistic fasteners* because they are composed of cuticular micro-outgrowths of different origins (Richards and Richards 1979) and because the size, shape and density of these outgrowths do not exactly match on the two surfaces. Thus, mechanical interactions between the micro-outgrowths of both surfaces occur without precise positioning of both surfaces (Nachtigall 1974). When in contact, probabilistic fasteners prevent sliding of surfaces. Mechanical adhesion (attachment) in this case must depend on the size, shape, distribution density and material properties of these outgrowths (called *elements* in the following). This mechanical adhesion is fast, precise and reversible. The individual elements are not necessarily hook-shaped (Fig. 6.1). The mechanism of attachment in such systems definitely varies from the typical hook-and-loop Velcro principle but has only been marginally studied, as in the case of probabilistic fasteners with parabolic contact elements (Gorb and Popov 2002).

Since this kind of attachment structures occurs most widely among insects – e.g., in the unguitractor plate (Gorb 1996), in intersegmental fixators of leg joints (Gorb 2001), in synchronizing mechanisms of contralateral legs in cicada (Gorb 2004), in dragonfly head-arresting systems (Gorb 1999a) and especially in wing attachment devices (Schrott 1986; Gorb 1998a, b, 1999b) – it is of vital importance for the functioning of numerous biomechanical systems of insects.

The main similarity among these devices is that co-opted fields of cuticular outgrowths are present on two separate parts of the body. Sometimes, as in the case of the forewing fixators of beetles, more than just one pair (up to eight pairs in this case) of co-opted structures are involved in the proper functioning of the mechanical system. In this particular example, each pair favors movement of the closed wings in one preferred direction.

The different shapes of outgrowths in functionally corresponding fields have been previously described as: (1) hook-like, cone-shaped or parabolic elements (outgrowths) and mushroom-like elements from different sides, (2) cone-shaped or parabolic elements on both surfaces, (3) clavate elements on both surfaces, and (4) plate-like elements on both surfaces (Gorb 1998b; Gorb and Popov 2002). It may be assumed that the different structures of the interlocking devices are optimized by natural selection to achieve an appropriate mechanical arrest within the system. The goal is to prove this hypothesis and clarify other aspects of the optimization of such mechanical attachment systems.

In the following, we will simulate such a system numerically within the framework of continuous contact and as a discrete dynamical model.

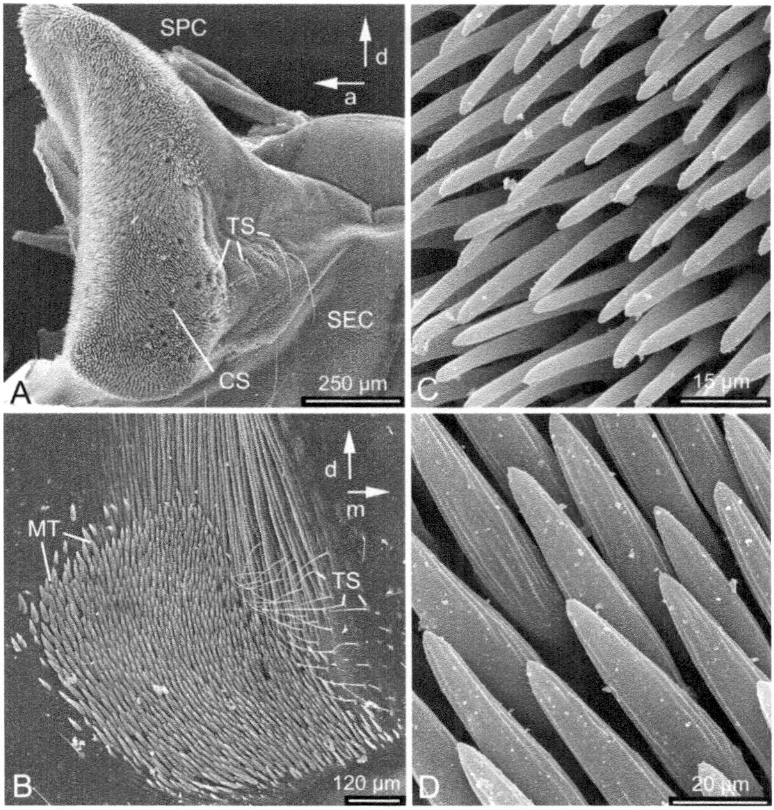

Fig. 6.1 Head-arresting system in the dragonfly *Aeshna mixta* (Anisoptera, Aeshnidae) (Gorb 2001). The postcervical sclerite of the neck (**a**, **c**) and the microtrichia field of the rear surface of the head (**b**, **d**) both show corresponding fields of cuticle outgrowths involved in the functioning of the arresting system. Abbreviations: *a* anterior direction, *d* dorsal direction, *m* medial direction, *CS* campaniform sensilla, *MT* microtrichia, *SEC* eucervical sclerite, *SPC* postcervical sclerite, *TS* trichoid sensilla. (From Filippov et al. 2015)

6.1.1 Some Arresting Structures Observed in Biological Systems

As mentioned above, surface outgrowths of probabilistic fasteners vary across biological systems and even between corresponding fields within the same system. In a pair of corresponding surfaces, however, some constraints in terms of size and distribution of co-opted areas of outgrowths may be expected. In the dragonfly head arrester, for example, the size of the outgrowths is usually larger in large species, and their density is lower (Gorb 1999a), with a higher density on the neck sclerite than on the rear surface of the head (Gorb 2001). Surprisingly, however, the density of outgrowths on co-opted fields varies even within a species. This was observed, for example, in the co-opted microtrichia fields of the forewing fixators of beetles (Gorb 1998a). And in the head arrester of adult dragonflies (*Aeshna mixta*), the density of microtrichia is higher on the neck sclerite than on the rear surface of the head (Fig. 6.1c, d) (Gorb 2001). The difference in microtrichia density on the corresponding structures is more distinct in larger species (Gorb 2001). An increase in the size of single microtrichia in larger species may be explained by a possible correlation with increasing cell size in larger species. It is more difficult, however, to explain the difference in outgrowth density on the corresponding structures. Also, the distribution of outgrowths roughly resembles a hexagonal pattern, which is generally accepted as representing the highest packing density of structures. This pattern, though, is far from an ideal one. It remains to be seen whether these features have any functional significance when it comes to the function of co-opted fields.

Understanding the role of the dimensions and distribution of outgrowths of co-opted fields is crucial for understanding the mechanical behavior of such systems in contact. To the best of our knowledge, the only model that exists is one of the frictional behavior between two single parabolic pins in sliding contact (Gorb and Popov 2002), but there are no models predicting the behavior of arrays of outgrowths in such systems. Thus, here we aim to establish simple models that realistically describe the behavior of co-opted fields of probabilistic fasteners in contact. We are especially interested in understanding the role of the distribution of outgrowths on corresponding fields. The main question raised in this chapter is: how do the differences in densities and in the type of distribution of single elements influence the contact forces generated by the system of co-opted fields?

6.1.2 Continuous Model of an Arresting System

First, we analyze the contact properties of a continuous numerical model (Filippov et al. 2015). A conceptual structure of the model is shown in Fig. 6.2. We numerically constructed two flexible contacting surfaces (upper and lower) with either regular or partially randomized arrays of outgrowths of different lengths, periods of distribution, symmetry and widths. For precision and numerical simplicity, we chose to simulate the outgrowths by arrays of Gaussians:

$$G^{kj}(x,y;\{x_k,y_j\}) = G^{kj}\exp\left\{-\left[\left(x-x^k\right)^2+\left(y-y^j\right)^2\right]/\Delta_{kj}^2\right\}; \tag{6.1}$$
$$k=1,2,\ldots,N_x,\ \ j=1,2,\ldots,N_y,$$

with numerically regulated height $G^{k,j}$, width Δ_{kj}^2 and individual positions of their centers $\{x^k, y^j\}$. By manipulating these parameters and by varying the distances

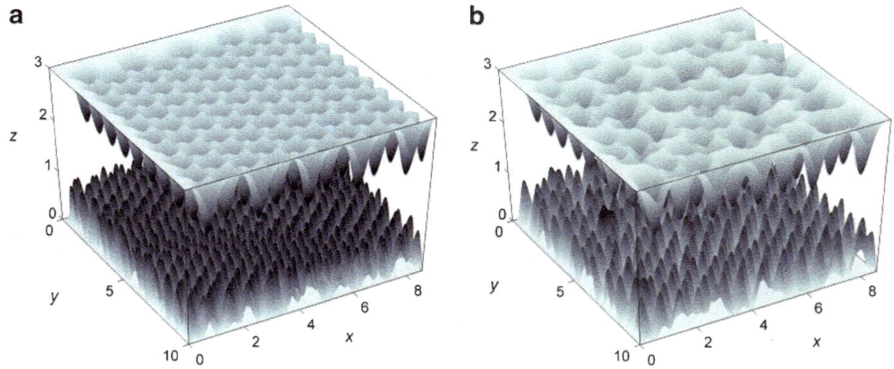

Fig. 6.2 Two typical examples of (**a**) ordered and (**b**) partially randomized contacting surfaces with different periods of asperities. For the lower surface, the dark color on the color map corresponds to maxima and the white color to minima. A reverse color map is used for the upper surface. (From Filippov et al. 2015)

between the outgrowths on the upper and lower surfaces, we were able to simulate different combinations of forms of outgrowths on both contacting surfaces.

It should be noted that the majority of structures observed in biological systems present a randomized hexagonal ordering (Gorb 1999a, b; Gorb and Popov 2002). Such structures are to be expected in the case of this self-organized, densely packed array of relatively rigid outgrowths repulsing each other by their cores. In many specific cases in real biological systems, however, these structures have a lower degree of symmetry and a rhombic or striped quasi-periodic ordering instead of an ideal hexagonal one. In any case, it can be suggested that this ordering results from natural selection as a compromise between the self-organized dense packing of outgrowths and the specific functional requirements of the particular system – for example, its arresting efficiency, generated friction forces in contact, or other functional properties – which might demand a specific symmetry for a required function. This hypothesis has not been proven experimentally, although a few experimental studies show that this type of structure is generally very promising for the purpose of biomimetics (Gorb and Popov 2002; Lilienthal et al. 2010; Pang et al. 2012a, b, c).

Unfortunately, it is difficult to approach this problem experimentally because designing and comparing structures of low- and high-grade symmetry is challenging. We therefore decided to build and analyze a numerical model that allows for simulating how a particular ordering of the outgrowths satisfies different prescribed functional demands. We started with an analysis of the contact properties of two surfaces with an ideal hexagonal distribution of their outgrowths:

$$Z^{u,d}(x,y) = \sum_{k=1,j=1}^{Nx,Ny} G^{u,d,kj}\left(x,y;\left\{x^{u,d}{}_{k}, y^{u,d}{}_{j}, G^{u,d}{}_{kj}, \Delta^{u,d}{}_{kj}\right\}\right), \tag{6.2}$$

where arrays of the coordinates $\{x_k, y_j\}$ belong to the regular hexagonal lattice $\{x_{0k}, y_{0j}\}$ with constant Gaussian parameters for each of the two (upper and lower) surfaces: $G^u = G^{0u} = const.$, $G^d = G^{0d}$, $\Delta^u = \Delta^{0u}$, $\Delta^d = \Delta^{0d}$. Despite the perfect symmetry of each surface, we allow for different periods of outgrowths on the upper and lower surfaces λ^{0u}, λ^{0d}. Other fixed parameters of the surfaces can also differ: $\lambda^{0u} \neq \lambda^{0d}$, $G^{0u} \neq G^{0d}$, $\Delta^{0u} \neq \Delta^{0d}$. This corresponds to our biological observations (Gorb 1998a, 2001).

Next, after studying the case of perfect symmetry, we will modify the surfaces by randomizing some – or all – of the mentioned parameters:

$$\begin{cases} x^{u,d}{}_k = x^{u,d}{}_{0k}\left(1 + \zeta^{u,d}{}_{x,k}\right) \\ y^{u,d}{}_j = y^{u,d}{}_{0j}\left(1 + \zeta^{u,d}{}_{y,j}\right) \end{cases}. \tag{6.3}$$

Here, $\zeta_{x,\ k}$ and $\zeta_{y,\ j}$ are δ-correlated random numbers, which are independent for every surface, position and coordinate$\{x, y\}$:

$$< \zeta^u{}_{x,k} >=< \zeta^d{}_{x,k} >=< \zeta^u{}_{y,k} >=< \zeta^d{}_{y,k} >= 0, \tag{6.4}$$
$$< \zeta^u{}_{x,k}\zeta^d{}_{x,k} >= 0, \quad < \zeta^u{}_{x,k}\zeta^u{}_{y,k} >= 0,$$
$$< \zeta^u{}_{x,k}\zeta^u{}_{x,k'} >= \Delta^u{}_x\delta_{kk'}, \ \ldots \{x = y; u = d\}.$$

The intensity of the randomization can be regulated by choosing varying amplitudes $\Delta^{u,\ d}{}_{x,\ y}$ of the random noise ζ. It is obvious that at the limit $\Delta^{u,\ d}{}_{x,\ y} \to 0$, the corresponding surfaces will be perfectly ordered ones, and at the opposite limit $\Delta^{u,\ d}{}_{x,\ y} > > 1$, the system becomes completely disordered. Thus, free choice of the constants $\Delta^{u,\ d}{}_{x,\ y}$ allows for a modeling of different combinations of weakly and strongly disordered surfaces. It also allows for a study of various contacts between completely ordered and differently randomized surfaces. Typical renderings of ordered and intermediately randomized surfaces are shown in Fig. 6.2 (with subplot (a) less randomized, and (b) more randomized).

When the surfaces come in contact, they deform proportionally to their elasticity $k^{u,\ d}$. If both surfaces are made of the same material, their elasticities coincide ($k^u = k^d \equiv k$) and they deform equally. In a numerical simulation, this means when converging surfaces $Z^u(x, y)$ and $Z^d(x, y)$ formally intersect with one another, they form some equilibrium contact surface in the regions of intersection. This equilibrium contact surface coincides with the mean value $Z^{contact}(x, y) = (Z^u(x, y) + Z^d(x, y))/2$. The deformation of both surfaces also generates local elastic forces, which in this case are equal to: $f^{elastic}(x, y) = 2k(Z^u(x, y) - Z^d(x, y))/2 \equiv k\delta Z(x, y)$. These forces correspond to the load acting on both contacting surfaces. Even at an unknown coefficient k, the load can be characterized by the deviation of the surfaces $\delta Z(x, y)$ from their equilibrium positions. From observation of biological structures, it can be estimated that typical deformations $\delta Z(x, y)$ for dragonfly species, such as those depicted in Fig. 6.1, occur in the interval 10–20 μm and create corresponding elastic forces $f^{elastic}(x, y) = k\delta Z(x, y)$ of about 10–20 μN at the typical spring constants of the insect micro-outgrowths of 1 N/m (Niederegger et al. 2002). This estimation will be useful to validate the proposed model in further experimental studies.

If both surfaces represent ideal hexagonal lattices, it can be expected that $\delta Z(x, y)$ depends on the relation between their periods $\lambda^{u,\ d}$ and the mutual orientation between the surfaces (angle φ between their axes of symmetry). It is clear that the minima of $\delta Z(x, y)$ should appear at some specific orientations of φ, where asperities of the upper and lower surfaces fit between one another. This is especially true when the periods $\lambda^{u,\ d}$ are commensurate or simply equal: $\lambda^u = \lambda^d$. The opposite situation is expected for the *strongest incommensurate* relation $\lambda^u/\lambda^d = \tau$, where $\tau = \left(1 + \sqrt{5}\right)/2 = 1.618\ldots$, i.e., when it represents the so-called *golden ratio*. For example, in the dragonfly head arrester, this relationship ranges from 1.5 to 6.0.

Thus, it is interesting to note that this relation $\lambda^u/\lambda^d = \tau$ between the periods of outgrowths on both surfaces is quite close to some relations previously discovered in biological systems (Gorb 2001). As a result, the hypothesis seems quite plausible that some incommensurateness between the periods $\lambda^{u,\ d}$ is biologically important. To prove it, we performed a calculation of the angle-depending configurations, $\delta Z(x, y; \varphi)$, for different $\lambda^{u,\ d}$. As expected, at some angles φ, regular patterns of $\delta Z(x, y; \varphi)$ were observed for the commensurate cases (especially for $\lambda^u = \lambda^d$). For the majority of arbitrary angles φ, however, projections of both lattices do not coincide, but instead create rather complex contact patterns.

From a biological point of view, such a mean absolute value of $\delta Z(x, y; \varphi)$ averaged over the spatial coordinates $<|\delta Z(x, y; \varphi)| >|_{x,\ y} \equiv <|\delta Z|>$ should be important. We calculated this value for different values of $\lambda^{u,\ d}$. The typical angle dependence of $<|\delta Z|>$ normalized to the maximum values for the two most important and representative cases, $\lambda^u = \lambda^d$ and $\lambda^u/\lambda^d = \tau$ is shown in Fig. 6.3a. The figure also clearly shows that, in contrast to initial expectations, the curves in both commensurate and incommensurate cases have their own fine structure of minima corresponding to some special orientations of φ.

According to the general definition by Popov (2010), the absolute value of the inclination angle (or mathematical gradient) of the contact surface, calculated as $|\nabla Z(x, y)| = \sqrt{(\partial Z(x, y)/\partial x)^2 + (\partial Z(x, y)/\partial y)^2}$, is necessary to estimate the local impact on friction between two rough surfaces. When integrated over the entire system at different mutual orientations φ and normalized per unit area, this results in an angle-dependent mean gradient $<|\nabla Z(x, y; \varphi)| >|_{x,\ y} \equiv <|\nabla Z|>$. The results of the integration for two representative cases, $\lambda^u = \lambda^d$ and $\lambda^u/\lambda^d = \tau$, are shown in Fig. 6.3b.

The local impact on friction is proportional to the product of the local gradient $|\nabla Z(x, y)|$ and the deviation $|\delta Z(x, y)|$, determining local load. In this case, the local load is

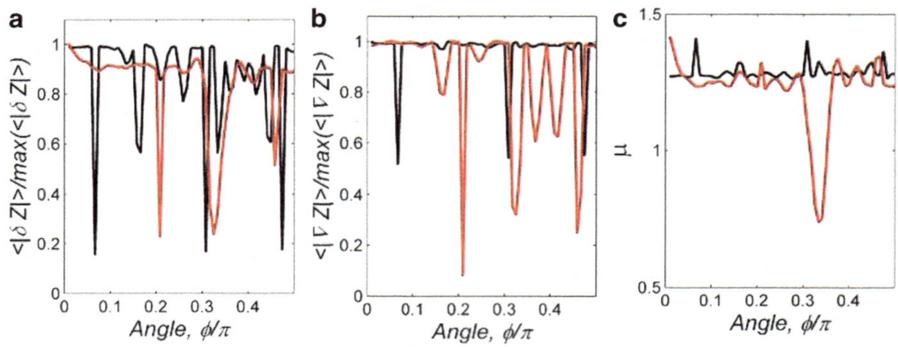

Fig. 6.3 The angle-dependent mean values of (**a**) deviation <|δZ|>, (**b**) gradient <|∇Z|>, and (**c**) friction coefficient $\mu = \ <|\nabla Z||\delta Z|> / <|\delta Z|>$ of regular hexagonal lattices are shown as red and black lines for commensurate ($\lambda^u = \lambda^d$) and incommensurate ($\lambda^u/\lambda^d = \tau$) cases, respectively. (From Filippov et al. 2015)

proportional to the elastic force $f^{elastic}(x, y) = k\delta Z(x, y)$. The mean friction coefficient μ is given by a sum of these products, $|\nabla Z(x, y)||\delta Z(x, y)|$, normalized to the total load:

$$\mu =< k|\nabla Z(x, y)||\delta Z(x, y)| > / < k|\delta Z(x, y)| > . \tag{6.5}$$

At constant k, it does not depend on the elasticity: $\mu = \ <|\nabla Z||\delta Z|> / <|\delta Z|>$. As we found for all relations λ^u/λ^d, this practically important combination is much less dependent on the orientation φ than on both solitary values <|∇Z|> and <|δZ|>. Furthermore, there is a large disparity between commensurate and incommensurate cases. As seen clearly in Fig. 6.3c, the red line representing case $\lambda^u = \lambda^d$ has a well-pronounced deep minimum, but the black curve corresponding to $\lambda^u/\lambda^d = \tau$ is practically flat (or has only small maxima at some angles).

As mentioned above, the distribution of outgrowths on biological surfaces is not perfectly hexagonal, but partially randomized. Thus, the same calculations as for the cases of regular distribution were repeated for the randomized hexagonal lattices $x^{u,\ d}{}_k = x^{u,\ d}{}_{0k}(1 + \zeta^{u,\ d}{}_{x,\ k})$, $y^{u,\ d}{}_j = y^{u,\ d}{}_{0j}(1 + \zeta^{u,\ d}{}_{y,\ j})$ with conserved equal amplitudes $\Delta^u{}_{x,\ y} = \Delta^d{}_{x,\ y}$ of the random noises $<\zeta^u{}_{x,\ k}\zeta^u{}_{x,\ k'}> \ = \Delta^u{}_x\delta_{kk'}, \ldots \{x \rightarrow y;\ u \rightarrow d\}$. It was found that even with intensive randomization $\Delta^{u,\ d}{}_{x,\ y} = 0.15$, the curves for $|\nabla Z(x, y)|$ and $|\delta Z(x, y)|$ still result in well-pronounced minima. Physically meaningful friction coefficients for both commensurate $\lambda^u = \lambda^d$ and incommensurate $\lambda^u/\lambda^d = \tau$ cases, however, demonstrate relatively small fluctuations. These fluctuations are only related to the limited size of the numerically simulated systems, so each individually calculated line varies from one realization to another.

All results obtained for <|∇Z|>, <|δZ|> and the friction coefficient μ of the randomized system are summarized in Fig. 6.4. Figure 6.4c clearly shows that in the incommensurate case ($\lambda^u/\lambda^d = \tau$) the friction coefficient is definitely higher than in the commensurate case ($\lambda^u = \lambda^d$).

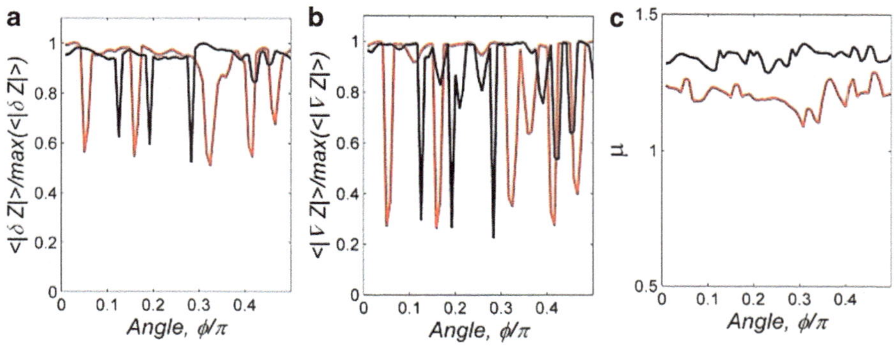

Fig. 6.4 The angle-dependent mean values of (**a**) deviation <|δZ|>, (**b**) gradient <|∇Z|>, and (**c**) friction coefficient μ of randomized hexagonal lattices are shown as red and black lines for commensurate ($\lambda^u = \lambda^d$) and incommensurate ($\lambda^u/\lambda^d = \tau$) cases, respectively. (From Filippov et al. 2015)

6.1.3 Discrete Model of an Arresting System and Dynamic Simulations

Unfortunately, our continuous model only partially reflects any real situation. Besides, in many cases it is too time-consuming to calculate dependencies of the system properties based on varied parameters, especially in the case of dynamic simulations. Therefore, an alternative model that allows for studying these systems by means of numerical simulation of Brownian dynamics of simple, mutually interacting subsystems would be useful. Below, we will develop and study such a discrete model.

A conceptual structure of the discrete model is presented in Fig. 6.5. We simulated contact between two surfaces of probabilistic fasteners by combining arrays of outgrowths which are attached by one of their ends to the upper or to the lower flat surfaces. The outgrowths were allowed to rotate around the attachment roots with some flexibility and to interact with each other according to the simple rules defined by corresponding interaction potentials. These properties of the model are presented in Fig. 6.5, where two arrays of outgrowths are depicted as black and red lines with attachment roots shown as black and red circles, respectively.

Due to their interactions with each other and due to their flexibility, the outgrowths appear instantly bent, as is shown in Fig. 6.5. As in the continuous model above, we can distribute the roots of these outgrowths differently. In particular, they can be ordered into a perfect or partially randomized hexagonal "densely packed" lattice, or they can be placed in some other order.

By default, the outgrowths are vertically oriented at the equilibrium $\alpha_0 = \pi/2$. The simplest way to simulate the deviations of the outgrowths from the perfectly vertical positions is to use an effective elastic interaction, $f^{p} = B(\pi/2 - \alpha)$, where α is the instant inclination angle. When the deviations are relatively small, the difference between α and $\pi/2$ is proportional to the distance between the projections of the

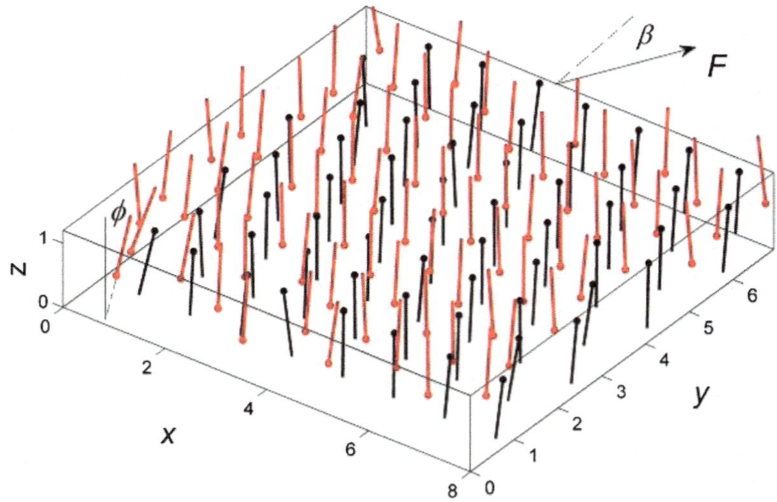

Fig. 6.5 Conceptual view of the discrete model of an arresting system. Fixed bases of the outgrowths of the upper and lower surfaces are shown as black and red circles, respectively. Movable fibers are shown as lines of the same colors. Angle β is calculated starting from one of the coordinates coinciding with any of the symmetry axes (y-axis for definiteness). For a dynamic representation of this picture, see the Supplementary movies 6.1, 6.2, 6.3 and 6.4 and the accompanying descriptions. (From Filippov et al. 2015)

specific positions of the outgrowth ends $\left\{\vec{r}_j\right\}$ and their root coordinates $\left\{\vec{r}_j^{\,0}\right\}$. Here, the discrete index $j = 1, 2, \ldots N$ runs across all N outgrowths, and $\left\{\vec{r}_j^{\,0}\right\}$ is the two-dimensional vector $\left\{\vec{r}_j^{\,0}\right\} = \left\{x_j^0, y_j^0\right\}$. Thus, below we simulate the elasticity of the outgrowths by a simple elastic relation:

$$f_j^{elastic} = k_{eff}\left(\vec{r}_j^{\,0} - \vec{r}_j\right) \tag{6.6}$$

with an effective elastic constant k_{eff}.

Real biological outgrowths consist of rather stiff polymers and, in approximation, can be modeled by rigid outgrowths which can rotate around their root positions but do not bend in their middle segments. In the course of their interactions, they are repulsed by each other's hard cores and cannot get any closer than a characteristic distance, which corresponds to their widths. The width of the outgrowths, w, can vary from one system to another, and it can be different for upper and down surfaces (w^u and w^d, respectively). In real biological systems, these widths, $w^{u,\,d}$, normally correlate with the density of the outgrowths – in other words, with the periodicity $\lambda^{u,\,d}$ of the structure $w^{u,\,d} \approx \lambda^{u,\,d}$.

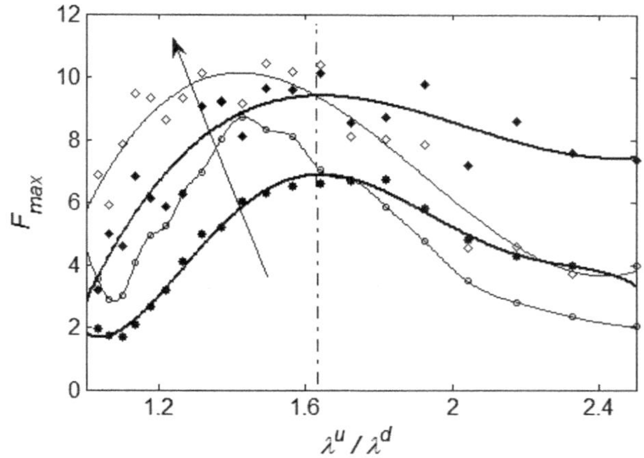

Fig. 6.6 Dependence of the friction force on the relationship between the periods λ^u and λ^d. The vertical dash-dotted line marks the golden ratio $\lambda^u/\lambda^d = \tau$. Black and open symbols, as well as bold and thin fitting lines, correspond to values of the elastic constants $k^a = k^b = 10$ and $k^a = k^b = 1$, respectively. Circles and diamonds correspond to regular and randomized ($\Delta^{u,\ d}_{x,\ y} = 0.15$) lattices, respectively. The general trend of the friction force with randomization of the lattices is indicated by the arrow. (From Filippov et al. 2015)

The simplest way to numerically simulate the repulsion between the outgrowths is to use Gaussian interaction potentials:

$$U_{kj}^{a,b} = U_0^{a,b} \exp\left\{-\left[\left(\vec{r}^{a,b} - \vec{r}_k^{a,b}\right)/S_{ab}\right]^2\right\}; \tag{6.7}$$

where $k = 1, 2, \ldots, N_x$, $j = 1, 2, \ldots, N_y$; $U_0^{a,b}$, $S_{ab}^2 = (w_a + w_b)/2$ represent the intensities and decays of the interactions, and the indices $\{a, b\}$ can independently take the values $a = u$, $a = d$, $b = u$ and $b = d$ corresponding to the upper and lower surface lattices. To perform this numerical experiment, the center of mass $\vec{R}$ of the upper lattice was pulled by an external elastic force $\vec{F}_{external} = K\left(\vec{V}t - \vec{R}\right)$ with the velocity $\vec{V} = \{V\sin(\beta), V\cos(\beta)\}$. If the system has hexagonal (or any other) symmetry, then the angle β is calculated by starting from one of the y coordinates that coincides with any of the symmetry axes.

This model was used for a number of experimental runs at different angles of pull β, for different symmetries and mutual orientations φ of the upper and lower lattices, as well as different amplitudes of the randomization of the outgrowth distributions $\Delta^{u,\ d}_{x,\ y}$ and their rotational stiffness $k^{a,\ b}$. These simulations extend and generally confirm the results obtained with our continuous model.

Figure 6.6 summarizes the most important results for the general orientation $\varphi = \pi/5$ and the direction of pull $\beta = 3\pi/10$ obtained for two contrasting elastic constants for both surfaces, $k^a = k^b = 10$ and $k^a = k^b = 1$. The simulations for all mentioned sets of parameters were completed at varied relations λ^u/λ^d between the periods of perfect or randomized lattices in the interval $1 < \lambda^u/\lambda^d < 2.5$.

The numerical experiment was organized as follows. At different λ^u/λ^d, the external force $\vec{F}_{external}$ pulls the upper plate from its initial static position, and after a brief process, it begins to move. We recorded the maximum friction force, $F_{\max} = \max\left(\left|\vec{F}_{external}\right|\right)$, which according to the definition determines a static friction threshold, and determined its dependence on the relation λ^u/λ^d in Fig. 6.6.

Figure 6.6 clearly shows that for relatively stiff outgrowths the results for friction behavior of the discrete dynamic model are in a very good agreement with those of the continuous model. Function $F_{\max}(\lambda^u/\lambda^d)$ has a maximum around $\lambda^u/\lambda^d \approx \tau$. The vertical dash-dotted line in the plot indicates the golden ratio $\lambda^u/\lambda^d = \tau$. With an increasing degree of randomization, the curve $F_{\max}(\lambda^u/\lambda^d)$ generally shifts toward higher values. In particular, the function $F_{\max}(\lambda^u/\lambda^d)$, calculated at $\Delta^{u,\,d}_{x,\,y} = 0.15$, is shown in Fig. 6.6 (curve marked by black diamonds).

It is probable that these results concur with our continuous model as a result of the proper selection of the outgrowth rigidity $k^{a,\,b} = 10$, which allows the outgrowths to remain almost vertical in the course of motion. This was to be expected, as our continuous model ignores the horizontal motion of any outgrowths.

It is interesting to note that in the system with softer outgrowths, friction behavior is relatively unaffected. For $k^{a,\,b} = 1$, both curves of $F_{\max}(\lambda^u/\lambda^d)$, obtained for the cases of perfect and randomized distributions (Fig. 6.6), slightly shift to higher values of $F_{\max}$ at $\lambda^u/\lambda^d \leq 1.5$ and to lower values of $F_{\max}$ for $\lambda^u/\lambda^d \geq 2$. Thus, the optimal relationship between the periods of outgrowths of co-opted surfaces, corresponding to $F_{\max}(\lambda^u/\lambda^d)$, is achieved at smaller values of $\lambda^u/\lambda^d \approx 1.45$. Perhaps the extreme flexibility of these outgrowths allows the fibers from one surface to quickly adapt to positions between the outgrowths of the other surface. Due to this adaptation there is little difference between the curves for perfect and randomized cases, as it is easier for outgrowths with almost freely rotating ends to leave their ideal lattice positions.

6.1.4 Biological and Biomimetic Significance of the Obtained Results

A previous analytical model of a pair of sliding parabolic outgrowths in contact showed that the pull-off force which can be generated by this kind of probabilistic fasteners is of the same order of magnitude as the loading force needed to bring such arrays of individual outgrowths in contact (Gorb and Popov 2002). This model, however, is only valid for the small deformations at the initial phase of contact formation between sliding outgrowths. A full analysis of outgrowth deformation for just a single pair of outgrowths can only be done numerically. Also, in reality, all the forces rely on the collective behavior of numerous outgrowths, and therefore depend on the spatial distribution of the outgrowths on their co-opted lattices. This hypothesis can only be tested numerically.

It has been previously shown that in biological systems the density of outgrowths correlates with their length, width, and especially with the distance between them, so that the longest and widest outgrowths are usually sparsely distributed across the field (Gorb 1998a). There is only a weak correlation, however, between the length and width of outgrowths – longer outgrowths are not always wider. The distances between the tips of longer and wider outgrowths are larger. The interplay of these parameters may result in different behavior of contact element assemblages. The present chapter studies two numerical models that predict the behavior of probabilistic fasteners with different densities of single outgrowths and different distribution periodicities in co-opted lattices. The main result of our virtual modeling experiments is that there is an optimal relationship of densities on co-opted lattices that generates the strongest contact force. Another important result is that the optimal distribution of outgrowths is not a perfect (grid-like) formation, but should instead be slightly randomized. Both of these features are observed in various biological probabilistic fastener systems capable of fixating body parts due to friction enhancement.

Our finding that co-opted lattices within the same organism have differently distributed individual outgrowths was initially rather surprising (Gorb 2001), as an organism would be able to achieve a similar density of outgrowths on both surfaces due to the similar size of cells responsible for the development of such outgrowths. By maintaining the same cell distribution on both surfaces, a highly ordered type of outgrowth distribution could be easily generated, as in the case of the tarsal adhesive hairs of the fly. The numerical simulations presented here provide a strong argument for the functional significance of both (1) a certain difference in outgrowth density on co-opted lattices and (2) a certain imperfection in outgrowth distribution on each lattice in biological probabilistic fasteners.

The Velcro principle has been transferred from biological observations to industrial applications. Recent patent databases contain a multitude of ideas dealing with various applications of existing hook-and-loop fasteners. These fasteners are widely used in industries such as textiles, medicine, sports, transportation, etc. Most of these applications, however, use the same type of hook-like tapes. Indeed, these systems generate very high attachment forces when compared with the applied load. Their function is long-term attachment, whereas the probabilistic systems discussed in this paper may be important for designing short-term, quickly attaching and detaching devices, as previously described for a macroscopic case (Gorb and Popov 2002) and for microscopic systems (Lilienthal et al. 2010; Pang et al. 2012a, b, c). For the purpose of optimizing such technical devices, further research on the optimal distributions of single outgrowths on co-opted lattices in biological systems, combined with the numerical study presented here, may be of great importance.

6.2 Mechanical Interlocking and Unzipping in Bird Feathers

6.2.1 General Properties of Bird Feathers

The feathers of recent birds are complex epidermal outgrowths. Their structures have evolved since the Late Jurassic period (Clarke 2013). Birds likely inherited feathers from a common ancestor of birds and crocodilians (Alibardi et al. 2006). The main function of feathers in birds and their ancestors is to support flight. The discovery of non-flying Late Cretaceous feathered dinosaurs, however, suggests that a display function played a key role in the early evolution of pinnate feathers (Xu et al. 2012). In smaller dinosaurs, thermoregulation and/or waterproofing could also have been the original primary function of feathers (Chen et al. 1998; Sumida and Brochu 2000). This hypothesis is supported by the fact that feather density in smaller birds is higher than in larger birds, since smaller birds lose more heat due to their larger surface area in proportion to their body weight (Pettingill Jr. 1970).

The feathers of modern birds are thus waterproof, breathable and of a lightweight construction combining thermo-isolation, rigidity and flexibility (Carboneras 1992; Dimond et al. 2011). These properties are partly related to an elaborate pattern of microhooks (hooklets) that is responsible for the cohesion of the vane. Such feather interlocking devices have already been portrayed in zoology textbooks, e.g. by Miller and Harley (1996).

The vane can be separated by pulling the barbs apart. The original state can then be easily reestablished by lightly stroking the feather. Hooklets responsible for holding vane barbs together are not damaged by multiple "zipping" and "unzipping" cycles. Since numerous microhooks maintain the integrity of the feather, their properties are of great interest for understanding the mechanics of the entire feather structure.

The vane of a typical contour feather consists of numerous side branches, called barbs, aligned in parallel and at an angle to the shaft. On either side, the barbs carry a similar array of second-level side branches called barbules. In most birds, the upper row of barbules (oriented toward the tip of the feather) bears very fine backward-facing hooklets, as is evident in the scanning electron microscopy (SEM) images in Fig. 6.7a, b. These barbules are called hook barbules. The barbules on the other side of the barb are sharply curved and have only very short spine-like outgrowths (called bow barbules). Hook barbules automatically anchor on bow barbules due to the preferred orientation of the hooklets and micro-irregularities on both types of barbules. These hooklets usually separate synchronously (with an average of 20 hooklets at a time), with the highest observed separation force amounting to 1.74 mN (and an average force of 0.27 mN), whereas the single hooklet separation force was 14 μN.

Separation of the barbules can be caused by some external force. This prevents the feathers from heavy damage by allowing for easy repair of a micro-rupture in the feather coverage. Although the structure and function of these systems are generally understood, the rupture forces have not yet been rigorously evaluated. Since

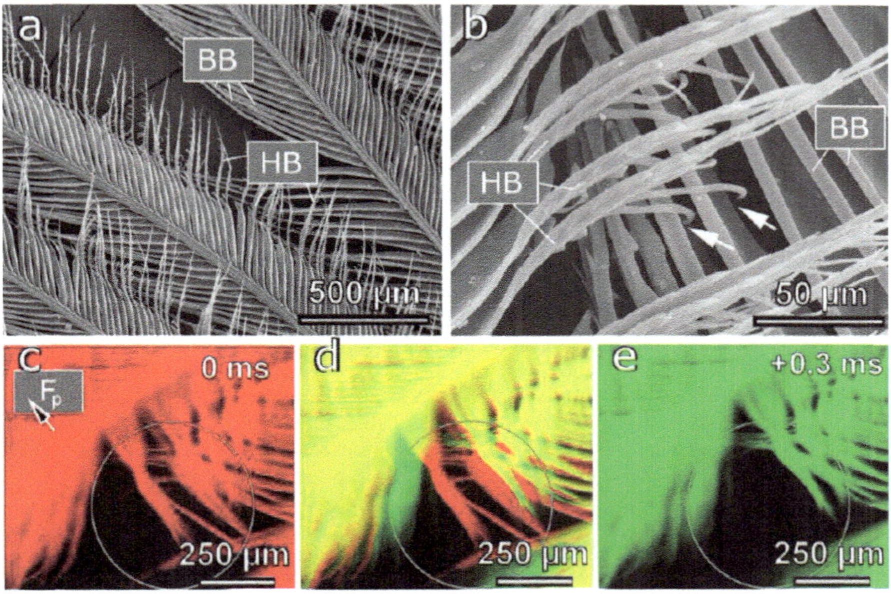

Fig. 6.7 (**a, b**) Structure of a feather; SEM micrographs of hook barbules (HB) with microhooks (arrows) attached to a row of bow barbules (BB). (**c–e**) Rupture area (inside grey circles) (**c**) before the rupture takes place: the pulling force F_p is applied to one of the barbs (direction indicated by the arrow), whereas the other barb is fixed; (**e**) after the rupture (0.3 ms after the onset of the pulling force). Shutter time = 0.1 ms. (**d**) Superposition of images (**c**) and (**e**). The yellow color indicates the immobile feather elements, red is for the initial position, and green represents the final position of different feather elements (barbules). (From Kovalev et al. 2014)

numerous feather microhooks maintain the integrity of the feather, their collective properties are of great interest for understanding the rupture/recovery mechanics of the entire feather. Behavior similar to the collective behavior of microhooks in feather rupture is observed in many different processes, such as, for example, in the motion of plasticity carriers (Lebyodkin et al. 2002) or in Barkhausen noise in ferromagnetic materials (Durin and Zapperi 2000).

A simple numerical model is suggested below for a better understanding of the zipping and unzipping process in feathers. This modeling was undertaken to estimate the forces necessary for separating single hooklets and their arrays using force measurements in an unzipping feather vane. The model demonstrates features similar to those observed in experiments with real feathers which will be reported first.

6.2.2 Basic Experimental Results

Pennaceous feathers of the swan (*Cygnus olor*) were used for force measurements, as shown in Fig. 6.8. The swan feather was clamped at its vane in a position that oriented barbs parallel to the ground. A force transducer (FORT 10; World Precision

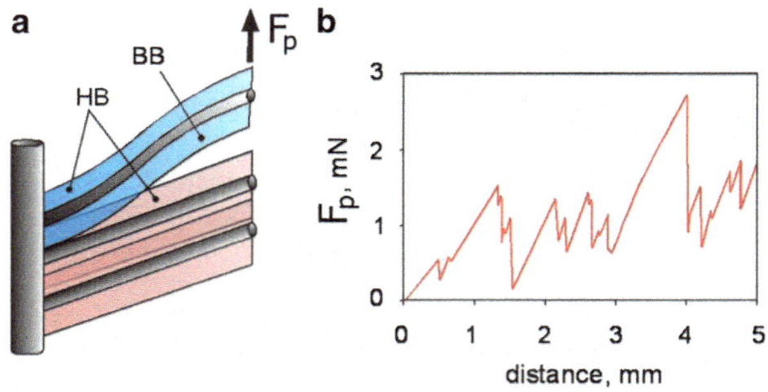

Fig. 6.8 (**a**) Scheme of the experimental setup and (**b**) the force–distance diagram of the force measurement. The feather was separated at the interface between two neighboring barbs using a micromanipulator equipped with a force transducer attached to the vane. This separation caused the detachment of hook barbules (HB) from bow barbules (BB). The arrow in (**a**) indicates the direction of pulling. After some final degree of displacement was reached, the force transducer was returned to its initial position (healing). (From Kovalev et al. 2014)

Instruments, Sarasota, FL, USA) was clamped to the tip of the next free barb. This force sensor was attached to a motorized micromanipulator (DC3001R; World Precision Instruments, Sarasota, FL, USA), similar to the procedure described by Gorb and Popov (2002). Vertical displacement of the transducer by the micromanipulator caused the separation of two neighboring barbs (this is seen in Fig. 6.8a). Separation was performed at a speed of 200 μm/s. Afterward, the force transducer was moved back to its initial position. Ten such measurements were performed on 5 individual feathers at a sampling frequency of 1 kHz.

A video of the barbule separation process in pennaceous feathers of swans and a rectrice of a pigeon (*Columba palumbus*) was recorded at 10,000 frames per second (50 μs exposition time) using a high-speed camera (Photron Fastcam SA1) mounted on an inverted microscope (Axio Observer.A1; Carl Zeiss MicroImaging GmbH, Jena, Germany). The high-speed video recordings of the separation process demonstrated that – since barbules are oriented at some angle to barbs – the barbules were bent near their bases when the barbs were separated. For example, compare the left upper (free barb) and the right lower (fixed barb) corners in Fig. 6.7c. Normally, during this separation process, bending of the linked barbules increases, and at a certain point the linked barbules uncouple from one another (Fig. 6.7c–e). This uncoupling process is very quick and proceeds within <0.1–0.3 ms. Previously deformed barbules then return to their initial positions.

When the barbs are separated by pulling, as seen in Fig. 6.8a, the force decreases each time some individual barbule(s) uncouple(s) (Fig. 6.8b). When the pulling force decreases to zero, the barbs return to their initial positions guided by elastic forces (Fig. 6.8b) and by the anchoring between hook and bow barbules as they recover (healing). Hysteresis (cf. the distance curve at pulling/healing in Fig. 6.8b) occurs because barbule bending takes place at pulling but not at healing.

Our observations show that instead of separating one by one, many barbules separate at once. The average force increments resulting in the separation of multiple

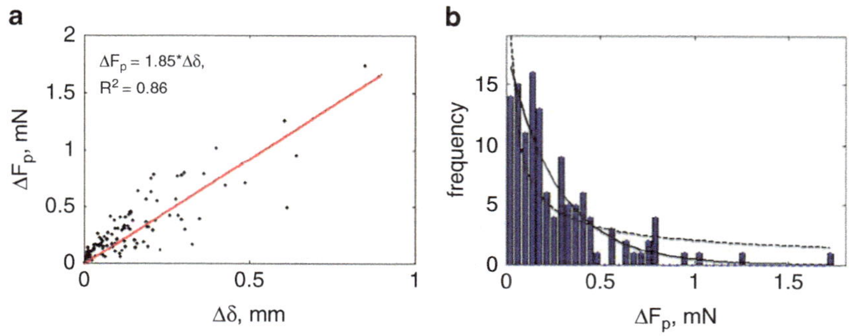

Fig. 6.9 (**a**) Dependence of the force required for a single separation event (ΔF_p) on the distance ($\Delta\delta$) traveled by the force sensor. (**b**) Frequency histogram of ΔF_p. 126 force peaks were measured on five feathers. The histogram was fitted with an exponential function, $17.9e^{-3.56\Delta F_p}$ ($\chi^2 = 105$) (solid line), and with a power function, $2.05\Delta F_p^{-0.6}$ ($\chi^2=326$) (dashed line). (From Kovalev et al. 2014)

hooklets was 0.27 mN (sd = 0.27, n = 126). The highest force increment was 1.74 mN. Assuming that the force increments are proportional to the number of hooklets holding the barbs together before the next uncoupling event takes place, the number of simultaneously separating barbules can be estimated from the measured pulling force and from the force necessary for the separation of a single hooklet. Thus, while pulling the row of barbules, 1–40 hooklets (mean = 19.4, sd = 19.4, n = 126) were separated at once.

At 100 μm of separation distance, up to 3.53 single hooklets (mean = 0.79, sd = 0.73, n = 158) were separated at once. The histogram of force increments was fitted with a power function and an exponential function. The exponential fit has a lower value of χ^2 (Fig. 6.9b).

6.2.3 Modeling of Feather Unzipping

To analyze the dynamics of the number of hooklets uncoupling during the feather rupturing process, a mathematical model of a feather was used (Kovalev et al. 2014). The discrete numerical model describes the dynamics of two initially parallel elastic barbs. The conceptual structure of the model is shown in Fig. 6.10. Each barb was constructed of 100 elastic segments, each having a length of $dR = 60$ μm, as seen in Fig. 6.10b, according to the measurements on SEM images (Fig. 6.7a, b). The barbs were provided with longitudinal ($K^{\|}$) and transverse ($K^{\perp}$) stiffness, $K^{\|} = K^{\perp}$. Transverse stiffness between neighboring segments tends to maintain an angle near 180°.

As in a real feather, the barbs in the model carry barbules. We constructed the barbules in a manner similar to that of the real barbs: each of the barbules was elastically attached to a barb. In the model, each barbule was constructed of

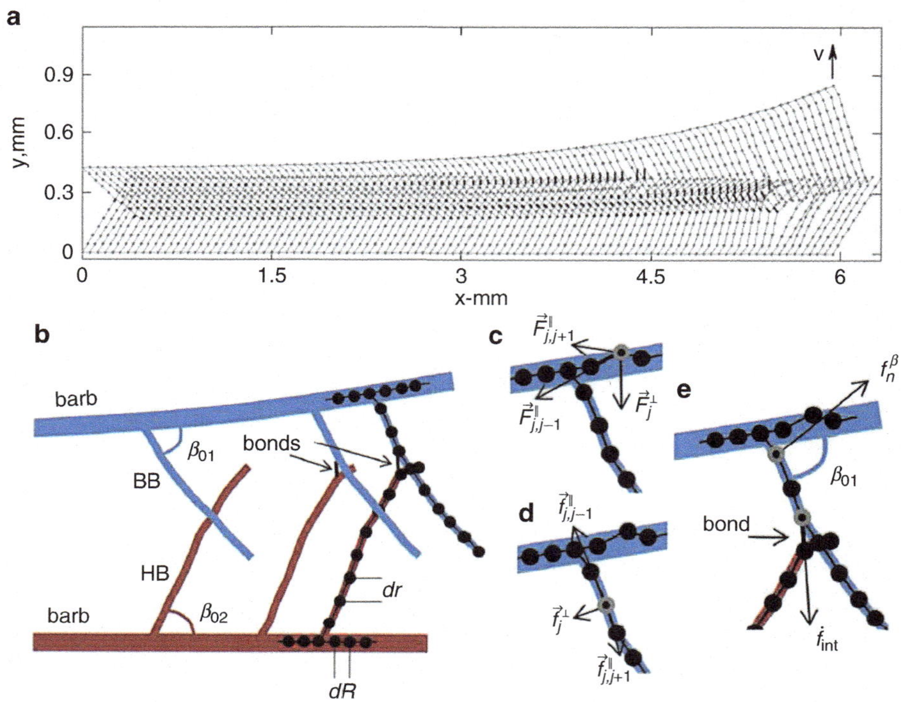

Fig. 6.10 Conceptual structure of the feather interlocking model. (**a**) Two connected elastic subsystems of fibers, simulating two barbs with barbules, are shown as gray lines. The nodes and elastically connected segments are depicted as gray circles and gray lines, respectively. The pairs of instantly interacting nodes from two different subsystems (which are close enough to build an interaction bond by meshing the hooks) are connected by bold black lines. The arrow indicates that the right terminal point of the upper subsystem is attached to an external cantilever which is moving with constant velocity *V*. (**b**) Schematic of the structural parameters of the mathematical model and (**c–e**) the forces presented in the model. The forces are shown only for some selected nodes (small black circles inside gray circles). (From Kovalev et al. 2014)

10 segments. According to the measurements on SEM images, the segment length (*dr*) was identical to that of the barbs, $dr = dR$.Þ

In real feathers, the barbules are oriented at some angles to the barbs. The barbules in the model were arranged at $\beta_{01} = 30^\circ$ (bow barbules, upper side) and $\beta_{02} = 45^\circ$ (hook barbules, lower side), respectively (Fig. 6.10b). The barbules and barbs have different thicknesses and therefore different elastic constants $k^{\|}$ and $k^{\perp}$. Based on SEM images, we assume that the barbules' transverse elastic constant is approximately ten times lower than that of the barbs, and their longitudinal elastic constants are the same: $k^{\|} = K^{\|}$, $k^{\perp} = 0.1K^{\perp}$.

As in the experiment, the tip of the barb (Fig. 6.10a, top right corner) was moved with constant velocity, $V = 50$ μm/s, until the tip reached the final displacement of 2.1 mm. Then, different to the experiment, the barb was released and returned spontaneously to its initial position. In the model, the deformation of a barb produced an elastic force proportional to the barb's stiffness.

The forces are described by the following equations:

$$\vec{F}_{jk}^{\parallel} = K^{\parallel}\left(\vec{R}_j - \vec{R}_k\right)\left[1 - \left(\frac{\vec{R}_j - \vec{R}_k}{dR}\right)^2\right], \text{ and } \vec{F}_j^{\perp} = K^{\perp}\left(2\vec{R}_j - \vec{R}_{j+1} - \vec{R}_{j-1}\right), \quad (6.8)$$

where $\vec{R}_j$ is a position vector at the middles of each segment (numerated as the nodes) j; $k = j \pm 1$. The longitudinal force, $\vec{F}_{jk}^{\parallel}$, is described with a double-well potential, which tends to maintain the distance between the nodes $\vec{R}_j$ and $\vec{R}_{j\pm1}$ close to the equilibrium length of the segment dR. This current form of the equation for the longitudinal force was chosen as it is linear at small displacements and increases non-linearly at large displacements. The transverse force $\vec{F}_j^{\perp}$ is directly proportional to the lateral deflection and tends to keep $\vec{R}_j$ close to the mean value between its nearest neighbors $\left(\vec{R}_{j+1} + \vec{R}_{j-1}\right)/2$. The transverse force in the present form is easy to realize but is not purely a bending force, as it may also include a longitudinal component.

The segments of the barbules are connected by forces in the same way as the segments of the barbs (Eq. 6.8):

$$\vec{f}_{jk}^{\parallel} = k^{\parallel}\left(\vec{r}_j - \vec{r}_k\right)\left[1 - \left(\frac{\vec{r}_j - \vec{r}_k}{dr}\right)^2\right], \text{ and } \vec{f}_j^{\perp} = k^{\perp}\left(2\vec{r}_j - \vec{r}_{j+1} - \vec{r}_{j-1}\right), \quad (6.9)$$

where $\vec{r}_j$ is a position vector of the barbule's node j; $k = j \pm 1$.

Additionally, a barbule–barb joint rotational stiffness B was introduced, which serves to maintain constant joint angles. The rotational force acting on adjacent segments of a barb and its barbules is linearly proportional to the difference $\beta_{0n} - \beta_n$ (Fig. 6.10b):

$$f_n^{\beta} = B(\beta_{0n} - \beta_n), \quad (6.10)$$

where index $n = 1, 2$ numerates the barbs. At $B > 0$, the force in Eq. 6.10 tends to keep the angle as close to the equilibrium value as possible. As a result, the system dynamically maintains a shape similar to the one depicted in Fig. 6.10.

The main feature of the feather is that bow and hook barbules stay connected during pulling until some critical force/displacement with regard to the connection between hooklets and bows is attained. To simulate this in the model, we assume that if a node of a hook barbule is close enough to the nodes of the bow barbules, it can form a "bond" (a connection between hooklet and bow, Fig. 6.10b). For simplicity, in an initial state, all possible bonds appear at a set distance between the barbules' segments.

The bonds can be characterized by the vectors $\vec{r}_{0nm} = \vec{r}_{0n} - \vec{r}_{0m}$, where indices n and m numerate the contacting nodes on both subsystems (hooklets and bows), $\vec{r}_{0n}$ represents an equilibrium, and $\vec{r}_n$ is an actual position vector. When one end of a barb starts to move vertically, it pulls barbule segments from their initial positions to new ones: $\vec{r}_{nm} = \vec{r}_n - \vec{r}_m$. This deformation causes an interaction force $\vec{f}_{\text{int}}$, which tends to return the distance between bound barbules to an equilibrium. For small deviations, $\vec{f}_{\text{int}}$ linearly increases with the distortion:

$$\vec{f}_{\text{int}} = k_{\text{int}}\left(\vec{r}_{0nm} - \vec{r}_{nm}\right), \tag{6.11}$$

where k_{int} is the stiffness of the bond. At large deformations, the interaction force $\vec{f}_{\text{int}}$ becomes too strong, and the hooklets detach. In the framework of this model, this means that if the deformation exceeds some threshold value r_{th}:

$$\left|\vec{r}_{0nm} - \vec{r}_{nm}\right| \geq r_{th}, \tag{6.12}$$

the bond breaks. In a real system, the threshold depends on the spatial realization of the particular contact and differs for different interconnections. In the model, this can be realized by a random distribution of the thresholds:

$$r_{th} = \overline{r}_{th}\xi, \tag{6.13}$$

where $\overline{r}_{th}$ is a constant and $0 < \xi < 1$ is a uniformly distributed, random number. For further simulations, we use $k^{\|} = 10k_{\text{int}}$; $\overline{r}_{th} = dR$. Figure 6.10 reproduces the described model system with some intermediate configuration.

A single hooklet can form only one bond, therefore all the bonds can be numerated. Additionally, we can define the total interaction force between two barbs as the sum of these forces:

$$F = \sum_{i=1}^{N} \vec{f}_{\text{int}}(i), \tag{6.14}$$

where i is an index and N is the number of bonds. This is the force that has to be applied to the pulled edge of the upper barb to allow for movement at a constant velocity.

The motion of barb/barbule segments may be calculated according to a standard equation of particle motion in a viscous medium:

$$m\ddot{r} = -\gamma\dot{r} + \sum f, \tag{6.15}$$

where m is the mass of the segment ($m = 1$), γ is a coefficient damping oscillations in the system ($\gamma = 1.1$), and $\sum f$ is the sum of all the forces acting on the segment. The integration time step was taken as 0.02 s.

The further the upper barb moves from its initial position, the higher the number of broken bonds. Due to the barbules' elasticity, the system can, at various intervals, gradually deform without breaking any bond (see Fig. 6.10 and Supplementary movie 6.5). The longer such intervals last, the larger the force F, and the higher the probability of a simultaneous separation of many hooklets or the generation of an avalanche of subsequent bond breakage (Fig. 6.7c–e). Due to such a collective effect the feather unzipping process is a non-trivial problem.

To quantitatively describe the collective behavior of the bonds, we introduce a specific measurement: the difference $\Delta N(t) = N(t + \tau) - N(t)$ between the total number of connected bonds at the beginning $N(t)$ and at the end $N(t + \tau)$ of a time interval $[t, t + \tau]$. When normalized to the length of the interval τ, this value coincides with a time derivative of the function $N(t)$ averaged over the corresponding time interval:

$$\left\langle \frac{dN}{dt} \right\rangle \Bigg|_{\tau} \equiv \frac{N(t+\tau) - N(t)}{\tau}. \tag{6.16}$$

This value is a convenient measurement to characterize the rate of bond breakage and restoration. The time interval was set at $\tau = 0.5$ s. During this time, a typical avalanche of bond breaking develops.

The typical results of our numerical simulations are summarized in Fig. 6.11. This figure presents the dependences of N, of the averaged derivative $\left\langle \frac{dN}{dt} \right\rangle \big|_{\tau}$ and the force $F = \sum_{n=1}^{N} \vec{f}_{\text{int}}(n)$ on the displacement of the pulled end of the barb, as shown in panels a–c of Fig. 6.11, respectively. The dashed line crossing these subplots separates an initial transient process – during which no bond breaking occurs – from the subsequent stationary regime of bond rupture. Figure 6.11d–f shows details of the model behavior in smaller time intervals. Here, the dash-dotted lines mark pronounced correlations between force drops after long periods of continual growth and avalanches of bond breakage, which are clearly recognizable by minima of $\left\langle \frac{dN}{dt} \right\rangle \big|_{\tau}$.

6.2.4 *Recovery of Ruptured Feathers*

The upper barb gradually returns to its equilibrium position when the external pulling force disappears. At some stage of the recovery (healing) process, the broken bonds begin to restore themselves. A bond forms again when the distance between a hooklet and its correct contact position becomes less than r_{th}.

The thin line in Fig. 6.12 represents the evolution of N during a typical unzipping–zipping cycle. First, the unzipping takes place until the displacement reaches a terminal distance, as indicated by the vertical black line. Then, the healing takes place until the displacement reaches the initial position.

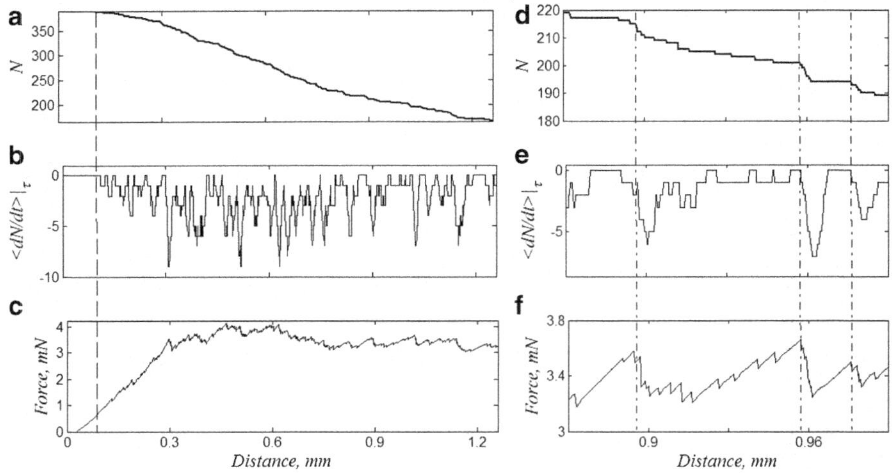

Fig. 6.11 Typical evolution of the model system at the parameters specified in the text. The dependences of the specific number of unbroken bonds N at each instant, the averaged time derivative of the number of bonds $\left\langle \frac{dN}{dt} \right\rangle \Big|_{\tau}$ and the total force of interaction, F_p, as functions of distance are shown in panels **a–c**, respectively. The dashed line separates an initial transient process, during which the upper subsystem moves without breaking the contacts. Panels **d–f** show the same values during a small interval of stationary evolution in more detail. The dash-dotted lines in these subplots mark correlations between long intervals of continued increase of the force without bond breaking and subsequent avalanches of bond breakage. See also Supplementary movie 6.6. (From Kovalev et al. 2014)

It is evident that the barb heals with a smaller number of contacts than it had in its initial state, as indicated by the distance between the bold and the thin line in the inset in Fig. 6.12. The recovery is, therefore, not complete, since not all hooklets find the correct contact position to restore the bonds, even if the barbs and barbules have returned to their original configuration. The plot clearly shows that the thin line deviates from the original line, which starts at the moment when the periodic force is applied. When forced oscillations are applied, the curve returns to a higher point than without them and ends up with a number of bonds that practically coincides with the initial number for the original unperturbed system.

Figure 6.13 shows an obvious correlation between lateral oscillations of the force and the rate of the bonds' recovery. Dash-dotted lines mark the relationship between the force maxima (associated with maximal displacement of the elastic fibers) and stimulated avalanches of bond recoveries, as quantitatively represented by the positive bursts of the derivative $\left\langle \frac{dN}{dt} \right\rangle \Big|_{\tau}$.

The zigzag shape of the distance–force dependence during the separation of interlocking hooklets – situated on hook barbules from the spine-like outgrowths, which in turn are situated on bow barbules – is important for the mechanical stability of covering feathers under external forces caused by air microturbulences in flight. The variation of the rupture force from 0.014 to 1.7 mN allows a feather to withstand a higher pulling force (up to 1.7 mN, in our case) than the average (0.27 mN). A

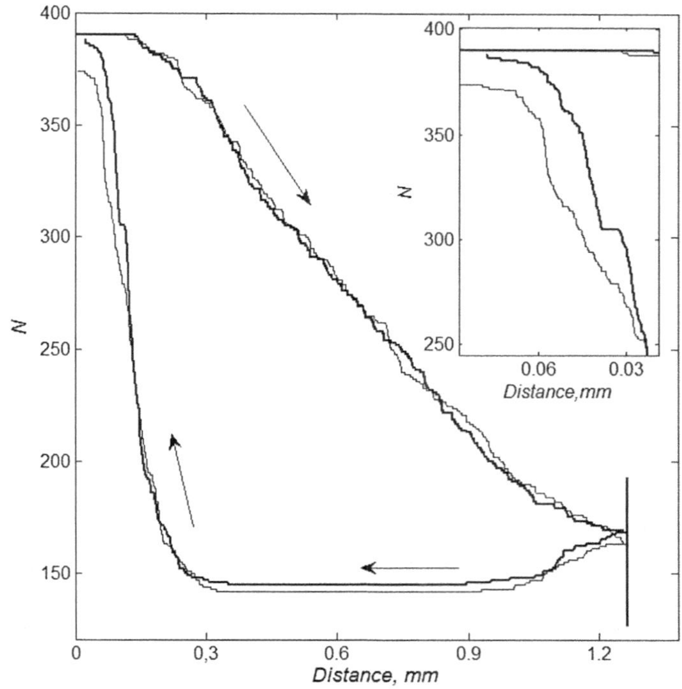

Fig. 6.12 Typical unzipping–zipping cycle. The thin line represents the number of bonds, *N*, remaining intact during a cycle of unzipping (pulling the upper barb to some final displacement indicated by the vertical black line) and subsequent zipping (healing), which returns the barb to its initial position. The same loop but with additional lateral oscillations (the bird's preening activity) – that encourage more bonding – applied at the end of the process is shown by the bold line. In the inset, the region of the graph at small distances is magnified. (From Kovalev et al. 2014)

strong correlation between the force increment (ΔF_p) and the pulling distance ($\Delta\delta$) required for a single separation event (Fig. 6.9a) indicates that during the stationary phase of pulling, the feather possesses a stationary stiffness. The numerical model demonstrated the same correlation (Fig. 6.14a).

The histogram of jump (avalanche) sizes in the motion of plasticity carriers in alloys (Lebyodkin et al. 2002) or in Barkhausen noise (Durin and Zapperi 2000) follows a power law distribution. In our model and in experiments with real feathers, however, the best fit for the histograms of the force increments' distribution ΔF_p is an exponential function (cf. Figs. 6.9b and 6.14b). This implies that a separation event is independent from a previous one. This conclusion is supported by a low correlation coefficient for consecutive force increments ($R^2 = -0.087$).

An interesting feature of such an interlocking mechanism is that microstructures are not damaged by the separation process but can resist multiple zipping–unzipping cycles. The emergence of hooklets permitted the development of compacted, closed and aerodynamically efficient vanes in large asymmetric contour feathers (Alibardi 2005, 2007). The hooklets, which are built from keratin with a Young's modulus of

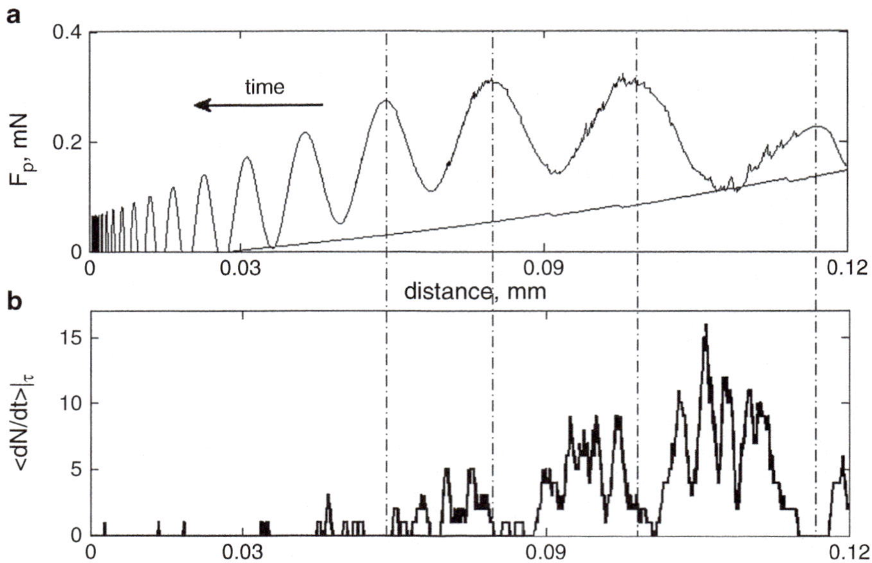

Fig. 6.13 Last phase of the healing process, when lateral force oscillations are applied. Distance decreases with time (as indicated by the arrow). (**a**) Oscillations of the pulling force induced by lateral force oscillations. (**b**) The avalanches of the contact recoveries represented by positive bursts of the derivative $\left\langle \frac{dN}{dt} \right\rangle \Big|_{\tau}$. Dash-dotted lines mark the positions of pulling force maxima (corresponding to maximal shifts of the elastic fibers in the left-hand direction). (From Kovalev et al. 2014)

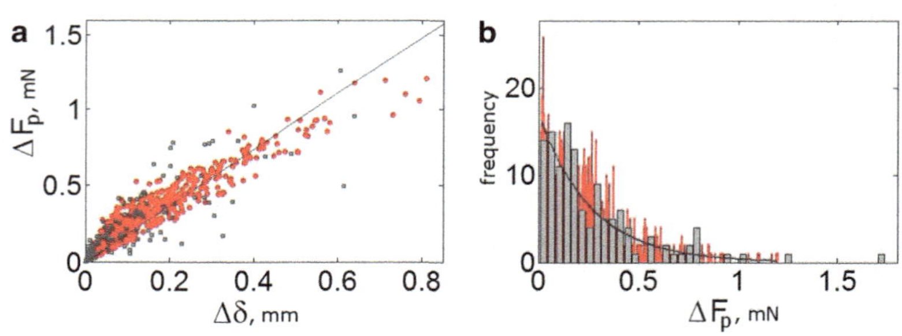

Fig. 6.14 (**a**) Dependence of the force required for a single separation event (ΔF_p) on the pulling distance ($\Delta\delta$) traveled by the force sensor in the model (closed red circles) and in the experiment (open gray circles). Time in the model was scaled to have the same fit parameters for the regression function as in the experiment. (**b**) Frequency histogram of ΔF_p for the model (red bars) and for the experiment (gray bars). Force constants in the model were scaled to have the same exponent in the exponential fit as in the experiment. 931 model separation events were analyzed to prepare the graphs. (From Kovalev et al. 2014)

about 2.5 GPa (Bonser and Purslow 1995), are grasping devices holding together barbules and barbs. Microhooks responsible for maintaining the integrity and mechanical stability of the wing structure during flight are also found in insects, such as the wing interlocking mechanisms in representatives of Heteroptera and Hymenoptera. In insects, such hooks provide functional diptery. Forewing microhook arrays holding the hind wings of the bug *Coreus marginatus* in flight can be separated from the hind wing at a force of 0.8 mN. The array consists of numerous (more than 20), 40–60 μm long hooks (Perez-Goodwyn and Gorb 2004). The number of wing hooks holding fore and hind wings together correlates with the possible flight distance in bees (Abrol 1986).

Since, in the plumage of birds, feathers have different morphologies, it would be interesting to compare how the rupture force differs between types of feathers, how it is related to the stiffness of the vane, and how it corresponds to the particular function of the feather. Also, interspecific comparisons might provide some insight into the microstructural adaptations to feather specializations.

The numerical model presented here differs from similar models of artificial systems, such as Velcro® models in geometry (Pugno 2007), especially due to the introduction of bonding strength distribution (compare to Maddalena et al. 2009). Our model demonstrates the features of real feathers as observed in mechanical experiments. It demonstrates that even when an external force disappears, the force–distance curves and system healing retain their same shape. Moreover, the model demonstrated a more effective system healing when a periodic force, such as the modeled bird's preening, was applied to the feather's vane.

References

Abrol DP (1986) Flight range and significance of wing hooks in *Megachile femorata* Smith (Hymenoptera, Megachilidae). J Anim Morphol Physiol 3:107–112

Alibardi L (2005) Fine structure of juvenile feathers of the zebrafinch in relation to the evolution and diversification of pennaceous feathers. J Submicrosc Cytol Pathol 37:323–343

Alibardi L (2007) Cell organisation of barb ridges in regenerating feathers of the quail: implications of the elongation of barb ridges for the evolution and diversification of feathers. Acta Zool 88:101–117

Alibardi L, Knapp LW, Sawyer RH (2006) Beta-keratin localization in developing alligator scales and feathers in relation to the development and evolution of feathers. J Submicrosc Cytol Pathol 38:175–192

Bonser RHC, Purslow PP (1995) The Young's modulus of feather keratin. J Exp Biol 198:1029–1033

Carboneras C (1992) Family Anatidae (ducks, geese and swans). In: Handbook of birds of the world, V. 1: Ostrich to ducks. Lynx Edicions, Barcelona

Chen PJ, Dong ZM, Zhen SN (1998) An exceptionally well-preserved theropod dinosaur from the Yixian formation of China. Nature 391:147–152

Clarke J (2013) Feathers before flight. Science 340:690–692

Dimond CC, Cabin RJ, Brooks JS (2011) Feathers, dinosaurs, and behavioral cues: defining the visual display hypothesis for the adaptive function of feathers in non-avian theropods. Bios 82:58–63

Durin G, Zapperi S (2000) Scaling exponents for Barkhausen avalanches in polycrystalline and amorphous ferromagnets. Phys Rev Lett 84:4705–4708

Filippov AE, Popov VL, Gorb SN (2015) The functional significance of density and distribution of outgrowths on co-opted contact pairs in biological arresting systems. Phil Trans Roy Soc B 370:20140032
Gorb SN (1996) Design of insect unguitractor apparatus. J Morphol 230:219–230
Gorb SN (1998a) Frictional surfaces of the elytra to body arresting mechanism in tenebrionid beetles (Coleoptera: Tenebrionidae): design of co-opted fields of microtrichia and cuticle ultrastructure. Int J Insect Morphol Embryol 27:205–225
Gorb SN (1998b) Functional morphology of the head-arrester system in Odonata. Zoologica 148:1–132
Gorb SN (1999a) Evolution of the dragonfly head-arresting system. Proc Roy Soc Lond B 266:525–535
Gorb SN (1999b) Ultrastructure of the thoracic dorso-medial field (TDM) in the elytra-to-body arresting mechanism in tenebrionid beetles (Coleoptera: Tenebrionidae). J Morphol 240:101–113
Gorb SN (2001) Attachment devices of insect cuticle. Kluwer, Dordrecht
Gorb SN (2004) The jumping mechanism of cicada *Cercopis vulnerata* (Auchenorrhyncha, Cercopidae): skeleton-muscle organisation, frictional surfaces, and inverse-kinematic model of leg movements. Arthropod Struct Dev 33:201–220
Gorb SN, Popov VL (2002) Probabilistic fasteners with parabolic elements: biological system, artificial model and theoretical considerations. Phil Trans Roy Soc Lond A 360:211–225
Kovalev A, Filippov AE, Gorb SN (2014) Unzipping bird feathers. J Roy Soc Interface 11:20130988
Lebyodkin MA, Dunin-Barkovskij LR, Lebedkina TA (2002) Universality and scaling of unstable plastic flow. JETP Letters 76(10):612–615. Translated from *Pis'ma v Zhurnal Éksperimental'noj i Teoreticheskoj Fiziki* 76(10): 714–718
Lilienthal K, Stubenrauch M, Fischer M, Schober A (2010) Fused silica '*glass grass*': fabrication and utilization. J Micromech Microeng 20:025017
Maddalena F, Percivale D, Puglisi G, Truskinovsky L (2009) Mechanics of reversible unzipping. Cont Mech Therm 21:251–268
Miller SA, Harley JP (1996) Zoology. Wm. C. Brown Publishers, Dubuque
Nachtigall W (1974) Biological mechanisms of attachment. Springer, Berlin
Niederegger N, Gorb SN, Jiao Y (2002) Contact behaviour of tenent setae in attachment pads of the blowfly *Calliphora vicina* (Diptera, Calliphoridae). J Comp Physiol A 187:961–970
Pang C, Kang D, Kim T, Suh K-Y (2012a) Analysis of preload-dependent reversible mechanical interlocking using beetle-inspired wing locking device. Langmuir 28:2181–2186
Pang C, Kim SM, Rahmawan Y, Suh K-Y (2012b) Beetle-inspired bi-directional, asymmetric interlocking using geometry-tunable nanohairs. ACS Appl Mater Interfaces 4:4225–4230
Pang C, Kwak MK, Lee C, Jeong HE, Bae W-G, Suh K-Y (2012c) Nano meets beetles from wing to tiptoe: versatile tools for smart and reversible adhesions. Nano Today 7:496–513
Perez-Goodwyn PJ, Gorb SN (2004) Frictional properties of contacting surfaces in the hemelytra-hindwing locking mechanism in the bug *Coreus marginatus* (Heteroptera, Coreidae). J Comp Physiol A 190:575–580
Pettingill OS Jr (1970) Ornithology in laboratory and field, 4th edn. Burgess, Minneapolis
Popov VL (2010) Contact mechanics and friction. Physical principles and applications. Springer, Berlin
Pugno NM (2007) Velcro® nonlinear mechanics. Appl Phys Lett 90:121918
Richards AG, Richards PA (1979) The cuticular protuberances of insects. Int J Insect Morphol Embryol 8:143–157
Scherge M, Gorb SN (2001) Biological micro- and nanotribology. Springer, Berlin

Schrott A (1986) Vergleichende Morphologie und Ultrastruktur des Cenchrus-Dornenfeldapparates bei Pflanzenwespen (Insecta: Hymenoptra, Symphyta). Ber Nat Med Verein Innsbruck 73: 159–168
Sumida S, Brochu CA (2000) Phylogenetic context for the origin of feathers. Am Zool 40:486–503
Xu X, Wang K, Zhang K, Ma Q, Xing L, Sullivan C, Hu D, Cheng S, Wang S (2012) A gigantic feathered dinosaur from the lower cretaceous of China. Nature 484:92–95

Chapter 7
Biomechanics at the Microscale

Abstract Male and female genitalia usually evolve rather quickly and yield remarkable diversity in comparison with other characters. Since natural selection influences shapes indirectly through corresponding properties, links between genital features and their functional mechanisms are critically important, in order to understand evolutionary mechanisms of morphological diversity. In this chapter, we numerically modeled hyper-elongated male and female genitalia especially focusing on their physical properties previously observed in cassidine beetles. These male beetles bear an elongated flagellum and female genitalia have a helically coiled spermathecal duct including some reversal turns called knots. The flagellum was demonstrated to have stiffness gradient, which results in a specific mechanical behavior of the entire copulatory system, because the flagellum stiffness properties might influence its motion within the female spermathecal duct. These properties might be partially due to rubber-like protein resilin that is assumed to be an almost perfect elastomer in insects. Resilin is an ideal material for elastic joints that are subjected to repeated cyclical stress. During the lifetime of an insect, resilin shows neither tearing nor fatigue, when stressed within its natural limits. The resilin, almost unique among biological materials returns back to its original position, when the stress is relieved. In this chapter, the simple sphere-on-the-flat geometry was used in a microindentation experiment as one with the most precise determined boundary conditions. Then the mechanical response of resilin was simulated using a generalized Maxwell model with two characteristic time constants and alternatively using a 1D model with just one characteristic time constant. We also applied newly-developed method of dimension reduction allowing us utilization of functional model along with the generalized Maxwell model.

Electronic supplementary material The online version of this chapter (https://doi.org/10.1007/978-3-030-41528-0_7) contains supplementary material, which is available to authorized users.

A. E. Filippov, S. N. Gorb, *Combined Discrete and Continual Approaches in Biological Modeling*, Biologically-Inspired Systems 16,
https://doi.org/10.1007/978-3-030-41528-0_7

7.1 Model of Penile Propulsion in a Chrysomelid Beetle

It is well known that sexual selection is the main driving force of the substantial diversity of genitalia found in animals (Eberhard 1985; Arnqvist 1998; Hosken and Stockley 2004; Simmons 2014). However, how it facilitates the diversity is still largely unknown, because the mechanics of genital structures in sexual intercourse are not well understood for the vast majority of organisms.

Male and female genitalia evolve more quickly and demonstrate a remarkable diversity in comparison with other characters (Eberhard 1985; Hosken and Stockley 2004; Macagno et al. 2011; Rowe and Arnqvist 2012; Simmons 2014). This phenomenon has previously attracted the attention of evolutionary biologists, and its mechanisms have been intensively investigated by various sophisticated experiments (Rowe and Arnqvist 2012; Arnqvist and Rowe 1995; Rodriguez 1995; Arnqvist and Danielsson 1999; House and Simmons 2003; Cordoba-Aguilar 2005; Holwell et al. 2009; Simmons et al. 2009; van Lieshout and Elgar 2010; Sakurai et al. 2012; Dougherty et al. 2015).

Most researchers in this field have focused on male genitalia (see review by Ah-King et al. 2014). It is clear, however, that females also play an active role in sexual selection by choosing males (Villavaso 1975; Rodriguez 1994) and that the interaction between the sexes is a very important factor for the genital evolution of both sexes (Miller and Pitnck 2002; Tanabe and Sota 2013).

Knowledge of the way in which male and female genital structures and properties affect the physical interaction during sexual intercourse is essential for the evolutionary study of genital structures. Since natural selection does not act directly on shapes, but rather influences shapes indirectly through corresponding functions, links between genital morpho-physical features and their functional mechanisms are critically important for understanding the evolutionary mechanisms of morphological diversity. Yet, the biomechanical aspects which are necessary for a comprehensive understanding of the kinematics and dynamics of movements in relation to structural features have been largely ignored. Only recently has this field started to contribute more to our knowledge, due to modern experimental techniques and an increasing number of interdisciplinary studies (Matushkina and Gorb 2007; Matsumura et al. 2014; Dougherty et al. 2015; Filippov et al. 2015; Michels et al. 2015; Sentenská et al. 2015; Willkommen et al. 2015).

Here, we numerically model hyper-elongated male and female genitalia, especially focusing on the physical properties observed in cassidine beetles, to gain insights about how genital features affect penile propulsion, which is an initial process of male and female physical interactions during copulation. In these beetles, the male penis is an elongated cuticular flagellum and the female genital tract, into which the flagellum is inserted, is a helical spermathecal duct that includes some reversals called knots (Rodriguez et al. 2004). Recently, the impact of the shape of the female spermathecal duct on male penile propulsion in cassidine beetles was numerically evaluated by Filippov et al. (2015).

7.1.1 CLSM Examination of the Genitalia of **Cassida rubiginosa**

Since stiffness gradients play an important role in many cuticular and fibrous biological structures (see Peisker et al. 2013; Gorb and Filippov 2014), the stiffness properties of the male flagellum and the female spermathecal duct were hypothesized to be important parameters which might influence flagellum motion within the female spermathecal duct. In order to test this hypothesis, we analyzed the autofluorescence composition of the genitalia of *Cassida rubiginosa* (Chrysomelidae) using confocal laser scanning microscopy (CLSM). This method provides indications of the material composition of the arthropod cuticle structures (Michels and Gorb 2012).

CLSM analysis revealed a gradient in the autofluorescence composition along the flagellum (Fig. 7.1). Except for the apical regions, the flagellum exhibited strong green and red autofluorescence. The tip of the flagellum (nearly 1% of the total length) (Filippov et al. 2016) showed mainly blue autofluorescence on the convex side and green autofluorescence on the other side (Fig. 7.1d, f). In the material of the spermathecal ducts of the females, we found no longitudinal but a transverse gradient in the autofluorescence composition (Fig. 7.1g, h) (Filippov et al. 2015). The internal wall showed mainly orange to green autofluorescence, and the external layer, blue autofluorescence (Fig. 7.1g, h). Generally, the females showed a stronger intensity of green and red autofluorescence in comparison to the males.

These CLSM results indicate a longitudinal gradient in the material composition of the flagellum and the absence of such a longitudinal gradient in the female spermathecal duct. According to Michels and Gorb (2012), the uniform orange to green autofluorescence observed in the inner wall of the female spermathecal duct suggests that the wall is composed of sclerotized cuticle. In contrast, the observed autofluorescence composition in the flagellum suggests that while large parts of the flagellum are mainly composed of sclerotized cuticle, the flagellum tip contains mainly non-sclerotized cuticle and relatively high proportions of the rubber-like protein resilin, whose concentration increases in the distal direction.

For other cuticular outgrowths, such as hairy attachment pads found in many insects (Gorb and Beutel 2001; Gorb et al. 2002), correlations between material and stiffness gradients were recently demonstrated (Peisker et al. 2013; Michels et al. 2015). Moreover, the female duct, which has been analyzed simultaneously with the male flagellum, had generally stronger intensity of green to red autofluorescence, which suggests larger thickness of the cuticular wall in females (Filippov et al. 2016).

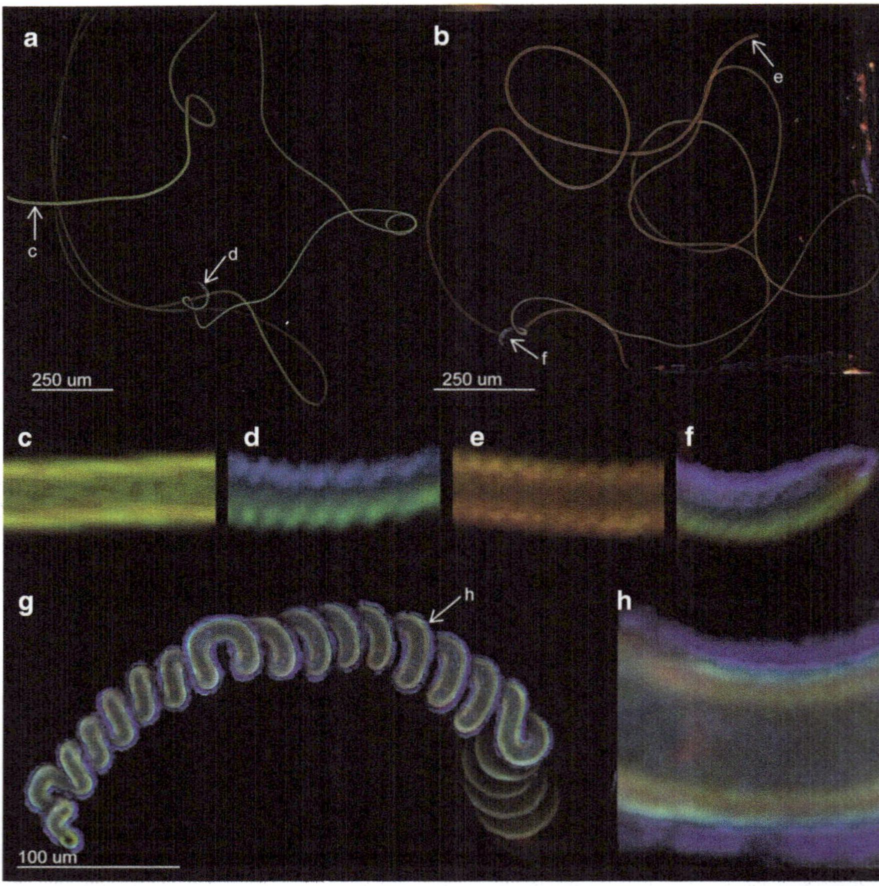

Fig. 7.1 (**a, b**) Confocal laser scanning micrographs of flagella and a spermathecal duct of different individuals of *Cassida rubiginosa*. (**c–f**) Details of the base (**c**, **e**) and the tips (**d**, **f**) of flagella enlarged from images **a** and **b**, respectively. The stripes do not represent any structures but are artifacts of image processing. (**g**) Spermathecal duct. The left end corresponds to the beginning of the spermatheca and the right one to that of the vagina. (**h**) Details of the duct. (From Filippov et al. 2015)

7.1.2 Simplified Model of the Flagellum and the Helical Spermathecal Duct

To simulate the system of flagellum and spermathecal duct of *Cassida rubiginosa*, we apply a numerical model which consists of the following two conceptual components:

1. an elastically movable, incompressible fiber with regulated transversal rigidity, which is moving inside an elastic channel being pressed by an external force along the direction x in 3-dimensional space $\{x, y, z\}$ from one of its ends;

2. an elastic channel, inside which the above-mentioned fiber moves. The channel has an adjustable flexibility, but is generally assumed to be stiffer than the fiber to provide sufficient confinement of the fiber inside the channel.

The fiber can be mathematically constructed using an approach which has been successfully applied in the simulation of a number of different systems. In the framework of this approach, we construct it using a thousand ($N = 1000$) elastically connected discrete segments. Each segment has a length of $dr = 1$ μm, adding up to the total length of the system ($L = 1000$ μm) as measured in microscopic images. The fiber is provided with longitudinal ($k^{\|}$) and transverse ($k^{\perp}$) stiffness. Transverse stiffness tends to keep the fiber straight by maintaining the angle between the neighboring segments close to 180°. Here, we assume that the transversal stiffness is much weaker than the longitudinal stiffness ($k^{\|} >> k^{\perp}$), which prevents extension and compression of the practically rigid segments.

A deformation of the fiber produces elastic forces $\vec{f}^{elastic} = \{f^{\|}, f^{\perp}\}$ proportional to its stiffness. The forces are described by the following equations:

$$\vec{f}^{\|}_{jk} = k^{\|}\left(\vec{r}_j - \vec{r}_k\right)\left[1 - \left(\frac{\vec{r}_j - \vec{r}_k}{dr}\right)^2\right], \text{ and } \vec{f}^{\perp}_j = k^{\perp}\left(2\vec{r}_j - \vec{r}_{j+1} - \vec{r}_{j-1}\right), \quad (7.1)$$

where $\vec{r}_j$ is a position vector of each node connecting neighboring segments j; $k = j \pm 1$.

The longitudinal force, $\vec{f}^{\|}_{jk}$, is described here by a double-well potential which tends to maintain the distance between the nodes $\vec{r}_j$ and $\vec{r}_{j\pm1}$ close to the equilibrium length of the segment dr. The current form of the equation for the longitudinal force was chosen since it is linear at small displacements and increases non-linearly for large displacements. The transverse force, $\vec{f}^{\perp}_j$, is directly proportional to the lateral deflection and tends to keep $\vec{r}_j$ close to the mean value between its nearest neighbors, $\left(\vec{r}_{j+1} + \vec{r}_{j-1}\right)/2$. The transverse force in the present form is easy to realize, but it is not purely a bending force since it may include a longitudinal component.

The spiral channel is organized as follows. We first define the core of the spiral which according to the images of the real system can essentially vary along the channel. For definiteness, we choose the X-coordinate of the channel along its longest dimension $X \in [0, L_x]$ and vary the core of the spiral in 3-dimensional space along this coordinate in the Y-direction at fixed Z: $Y_0 = A_0 \cos(2\pi X/\lambda_0)$; $Z = const.$, with an amplitude A and wavelength λ. In the real system, as is evident from the observations, the following applies: $0.1L_x < A_0 < 0.5L_x$ and $0.2L_x < \lambda_0 < 0.4L_x$.

The spiral turning around this core with radius R and period λ can now be defined as follows:

$$X \in [0, L_x]; \;\; Y = Y_0(X) + A\cos(2\pi X/\lambda); \;\; Z = Z_0(X) + A\cos(2\pi X/\lambda), \qquad (7.2)$$

The amplitude and radius of this spiral are equal to the following values: $R = L_x/30$; $\lambda = L_x/40$.

Besides the basic structure with simple rotations, defined by Eq. (7.2), the real spiral also has some number N_{knots} of "knots" where the direction of the spiral rotation reverses. These rotation reversals may be implemented into the model by introducing the variable phase $\varphi(X)$ which is practically constant inside a certain interval that corresponds to the fixed direction of spiral rotation and quickly changes to $\varphi(X) \to \varphi(X) + \delta\varphi(X)$ at some rotation point X_n to $\delta\varphi(X) = \pm\pi$. The typical interval of these phase changes is equivalent to one period of the spiral.

The simplest way of describing the phase rotation is by applying a step-like function with regulated width: $\delta\varphi(X) = \pm\left[\pi/2 + \arctan\left(\frac{X}{2\lambda}\right)\right]$. Each pair of rotations in opposite directions corresponds to one of the signs "+" or "–" in this expression. For example, two pairs of such rotations will be caused by the function

$$\delta\varphi(X) = \sum_{n=0}^{1}\left[\arctan\left(\frac{X_{2n+1}}{2\lambda}\right) - \arctan\left(\frac{X_{2n+2}}{2\lambda}\right)\right]. \qquad (7.3)$$

Such a spiral with four knots, $N_{knots} = 4$, and $R = L_x/30$; $\lambda = L_x/40$, is shown in Fig. 7.2a. The corresponding function defined by Eq. (7.3) is shown in Fig. 7.3.

One of our main hypotheses here is that the biological role of these knots is to reduce (or maybe regulate) the speed of the propulsion of the penile tip along the channel. To separate the effect of the knots from that of the other peculiarities of the channel, below we will first neglect the variations of the spiral core $Y_0 = A_0\cos(2\pi X/\lambda_0) = 0$ and deal with the straight spiral only, as shown in Fig. 7.2b.

In both panels of Fig. 7.2, the bold red line indicates a momentary position of the fiber inside the spiral channels and its first point (tip) is marked by the black circle at the beginning. The channel itself may be numerically defined as an elastic confinement with corresponding returning forces acting on every point, with the coordinates $\{z_j(x_j), y_j(x_j)\}$ connecting each segment $f^{confinement}{}_y(X = x_j) = K(Y(X) - y_j)$; $f^{confinement}{}_z(X = x_j) = K(Z(X) - z_j)$ with the isotropic elastic constant K. For a sufficiently rigid channel (corresponding to the real biological situation) it is expected that $K > k^{||}$ or even $K >> k^{||}$.

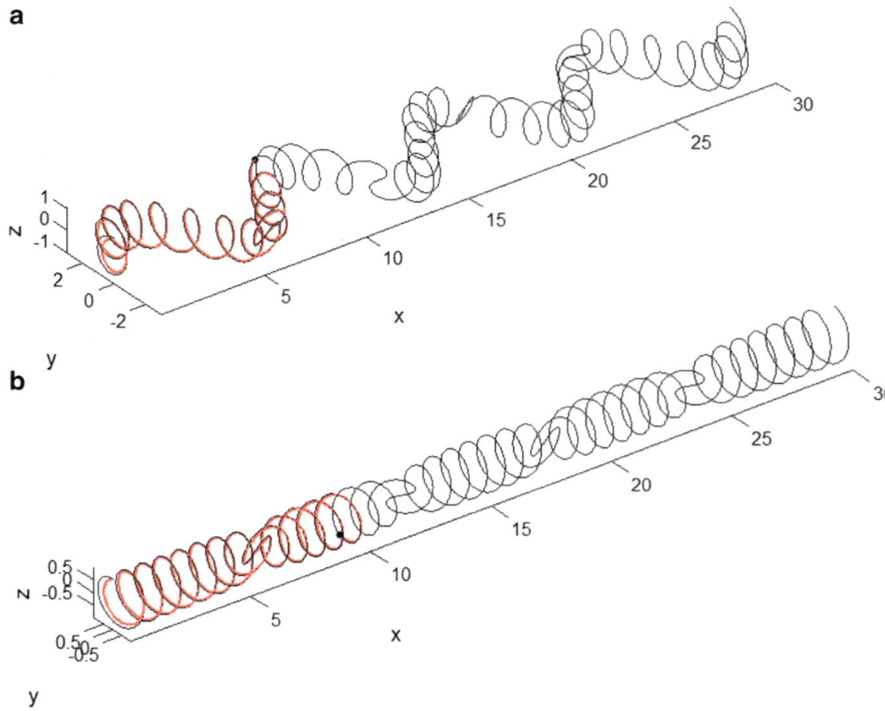

Fig. 7.2 Configurations of model spirals: (**a**) general case with secondary phase rotations, (**b**) simplified spiral with a straight axis. In both cases, the center of the spiral channel is shown by a black line, and an example of the configuration of the fiber inside the spiral is represented by the red line with the position of the tip marked by a black dot. The motion of the fiber inside the spirals in cases (**a**) and (**b**) is illustrated in Supplementary movies 7.1 and 7.2, respectively. (From Filippov et al. 2015).

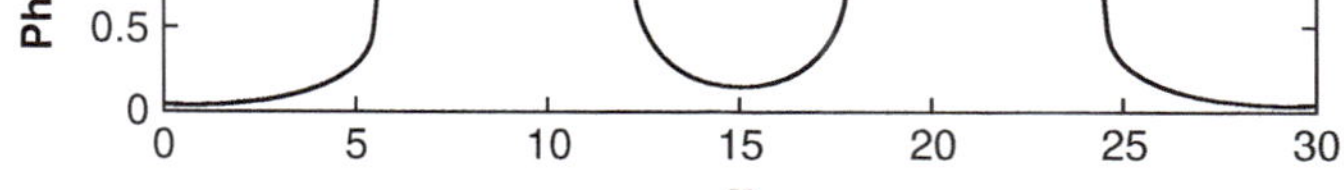

Fig. 7.3 Smooth step-function defining the rotation of the spiral phase which corresponds to the particular realization of the spiral channel (with 4 knots) shown in Fig. 7.2b. (From Filippov et al. 2015)

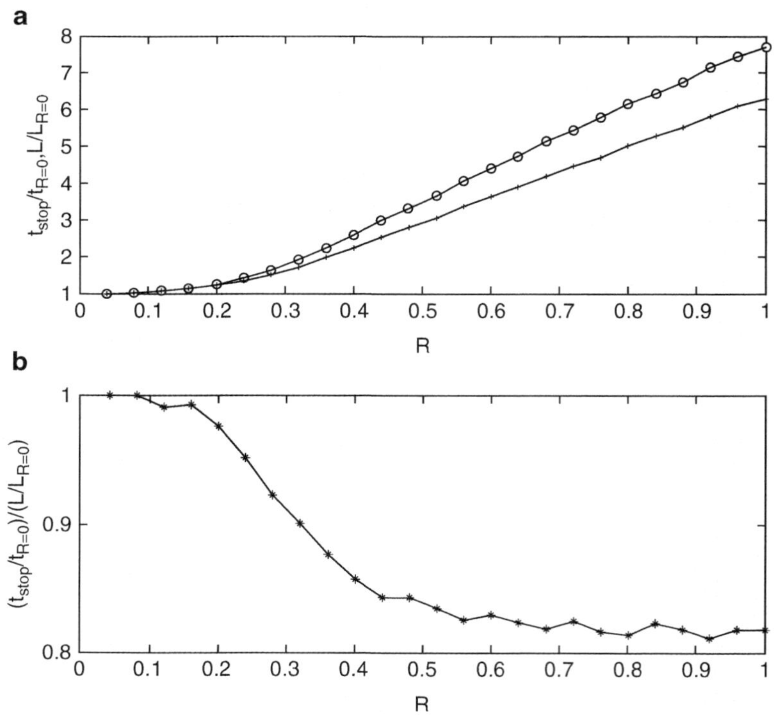

Fig. 7.4 (**a**) Dependencies of the time the tip of the fiber needs to reach the end of the spiral channel (black vertical dashes) and of the channel length (open circles) on the radius R of the spiral, normalized to these values for the straight tube $R = 0$; (**b**) relationship between these normalized values (stars). (From Filippov et al. 2015)

Liquid inside the channel causes a damping force $f_j^{damping} = \gamma \vec{v}_j$ proportional to the velocity $\vec{v}_j$. This force also has to be added to the equation of motion. Thus, the complete equation of motion for 3D vectors has the following form:

$$m\frac{\partial \vec{v}_j}{\partial t} = f_j^{elastic} + f_j^{confinement} f_j^{damping}. \tag{7.4}$$

Results of the simulations using Eq. (7.4) are summarized in Figs. 7.4, 7.5 and 7.6.

First of all, we studied the dependence of the time necessary for the tip to pass the whole length of the ideal channel to a predefined stop time t_{stop} without knots ($N_{knots} = 0$) on the radius R of the spiral. This passage time, normalized to the same time needed to pass the straight channel $R = 0$ is presented in Fig. 7.4. It is directly evident in Fig. 7.4a that the passage time increases with the radius R of the spiral. The increase is mainly related to an increase of the channel length L, especially. However, the behavior of the passage time is nonlinear and the coefficient of the proportionality between t_{stop} and L varies with the radius R of the spiral.

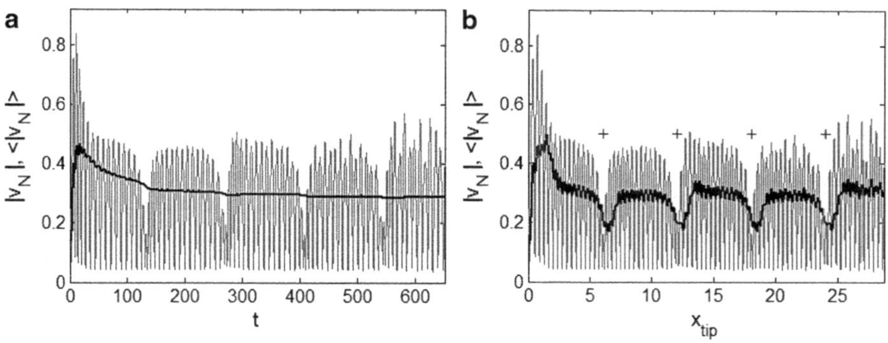

Fig. 7.5 Correlation between the minima of the absolute velocity of the tip along the channel and in the locations where the rotation of the spiral changes. (**a**) Time dependence of the velocity (gray curve) and the mean velocity (black curve, averaged over the interval that starts with the beginning of motion). (**b**) The same velocity (gray curve) shown as a function of the tip position and compared with the velocity averaged over the moving time window (with a width corresponding to the time of minimum intervals) (black line). The exact positions of the centers where the rotation changes are marked in the plot by crosses. (From Filippov et al. 2015)

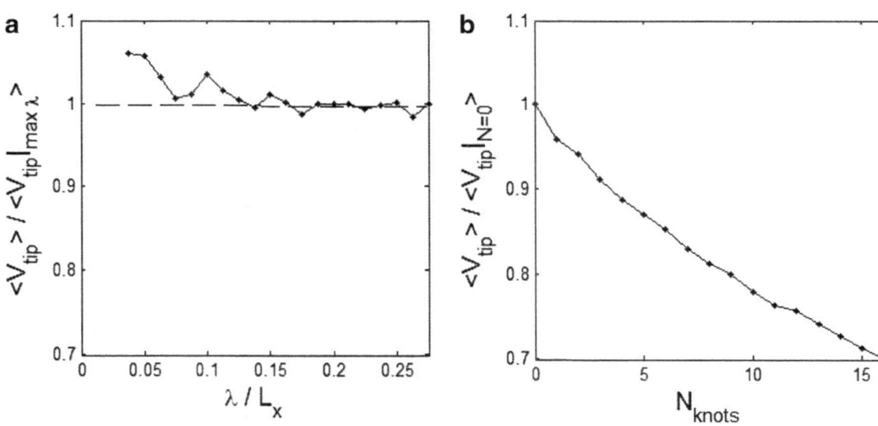

Fig. 7.6 Dependencies of the mean tip velocity along the channel on (**a**) the period and (**b**) the number N_{knots} of the changes of the rotation direction. The horizontal dashed line in (**a**) indicates the tip velocity in a straight channel. (From Filippov et al. 2015)

It is interesting to note that this dependence almost has the form of a step-like function. The value drops from the unit 1.0 to approximately 0.8 in the interval between $R = 0.2$ and $R = 0.4$.

Starting from $R > 0.5$, the behavior of the system practically does not change any further. Thus, we will now use $R = 1$ for definiteness and turn to the systems with a nonzero number of knots N_{knots}. As mentioned above, our main goal here is to study how the rotation changes influence the momentary velocity $|v^{||}(j = N)| = v_N(t; x)$ of the tip at a given N_{knots} along the channel and the mean velocity $<v_N>$ at different values of N_{knots}.

Typical time and space dependencies of the velocity $v_N(t;x)$ for a particular system with four knots ($N_{knots} = 4$) are illustrated in Fig. 7.5a, b, respectively. The correlation between the minima of the absolute velocity of the tip along the channel $|v^{||}(j = N)| = v_N(t;x)$ and the spatial locations of the spiral rotation changes ("knots") is directly evident in the plots. For clarification, the exact positions of the knots are indicated in the plot by crosses. To correlate them with decreases in the integral mean velocity $<v_N>$ averaged over the time interval starting from the beginning of motion, Fig. 7.5a presents the time dependence of the velocity (gray curve) along with the mean velocity (black curve). Besides, the momentary velocity (gray curve) is shown in Fig. 7.5b as a function of the tip position. It is compared with the velocity averaged over the moving time window (with a width corresponding to the duration of the minimum intervals) (black line).

Now we will perform the same numerical experiment at different numbers of knots N_{knots} and plot the mean velocity $<v_N>$ as a function of N_{knots}. First of all, we check whether the mean velocity depends on the period of the spiral. The result is shown in Fig. 7.6a. Next, we accumulate the velocity data at fixed $\lambda = L_x/40$ in intervals $N_{knots} = 0, 1, \ldots, 16$. The result of this accumulation is presented in Fig. 7.6b. It is evident from the figure that with an increasing number of knots ($N_{knots} \rightarrow 16$) the mean velocity gradually decreases to 0.7 of the velocity in the channel free of knots ($N_{knots} = 0$). At this limit, the numerically generated channel practically degenerates into a sequence of the knots. At the same time, it is clearly evident in Fig. 7.5 that the minima of the velocity, caused by the knots, do not tend to zero but reach at least half the original mean velocity. Thus, the total mean velocity cannot fall below them either. These observations suggest that the biologically reasonable channel structure has a relatively small number of knots ($4 < N_{knots} < 7$).

7.1.3 The Stiffness Gradient of the Beetle Penis Facilitates Propulsion in the Female Spermathecal Duct

The main question here is how the male–female interactions during copulation are affected by the material properties of the genitalia. Above we mentioned the presence of a material gradient; here, we will numerically study the effect of the stiffness gradient of the beetle penis during its propulsion through the female duct. Microscopic investigation suggests that the tip of the hyper-elongated penis of *Cassida rubiginosa* is softer than the rest of it, and a numerical model confirms that this kind of stiffness gradient aids in faster propulsion (Filippov et al. 2016). This result indicates that previously ignored physical properties of genital materials are of crucial importance in evolutionary studies of genitalia.

As described in Sect. 7.1.1, the autofluorescence composition of the male flagellum and the female spermathecal duct of the beetle *Cassida rubiginosa* was investigated with confocal laser scanning microscopy (Zeiss LSM 700; Zeiss, Germany). Of the treated samples, one male and one female were analyzed simultaneously to

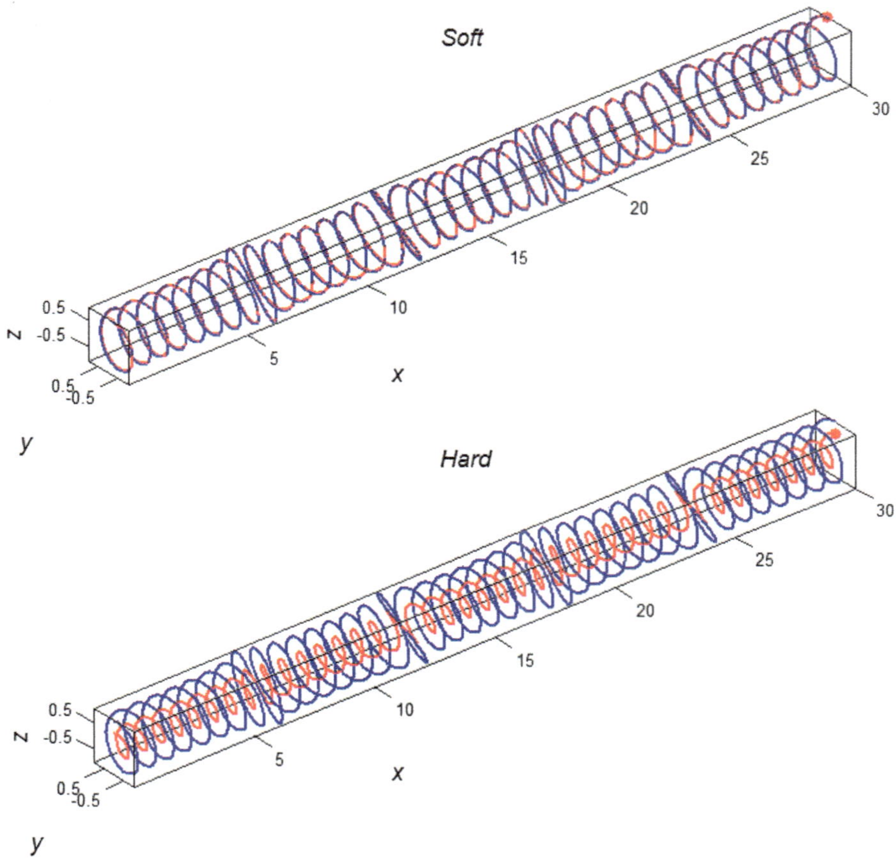

Fig. 7.7 3D view of the female duct, represented by the blue line, and the male flagellum, represented by the red line, for two extreme variants of flagellum stiffness studied with our numerical model: uniformly soft (upper panel) and uniformly hard (lower panel). In both cases, the entire flagellum is inserted in the female spermathecal duct. (From Filippov et al. 2016)

compare the intensity of autofluorescences, as described by Michels and Gorb (2012).

Below, we will apply the same model as in Sect. 7.1.2, in which we mathematically created the male flagellum as an elastic incompressible fiber and the helical spermathecal duct as an elastic channel (Figs. 7.7 and 7.8). For simplicity's sake and in contrast to reality we created the female model as a helical duct along a linear axis, without the convoluted shape observed in nature (Figs. 7.7 and 7.8), but including the "knots" as above (Sect. 7.1.2). The transversal rigidity of the fiber (flagellum) and the elasticity of the channel are adjustable. As suggested by our analysis of the autofluorescence composition, the female duct is generally stiffer than the flagellum. Flagellum motion is generated by using an external force to push its base so as to imitate the genital motion during penile penetration.

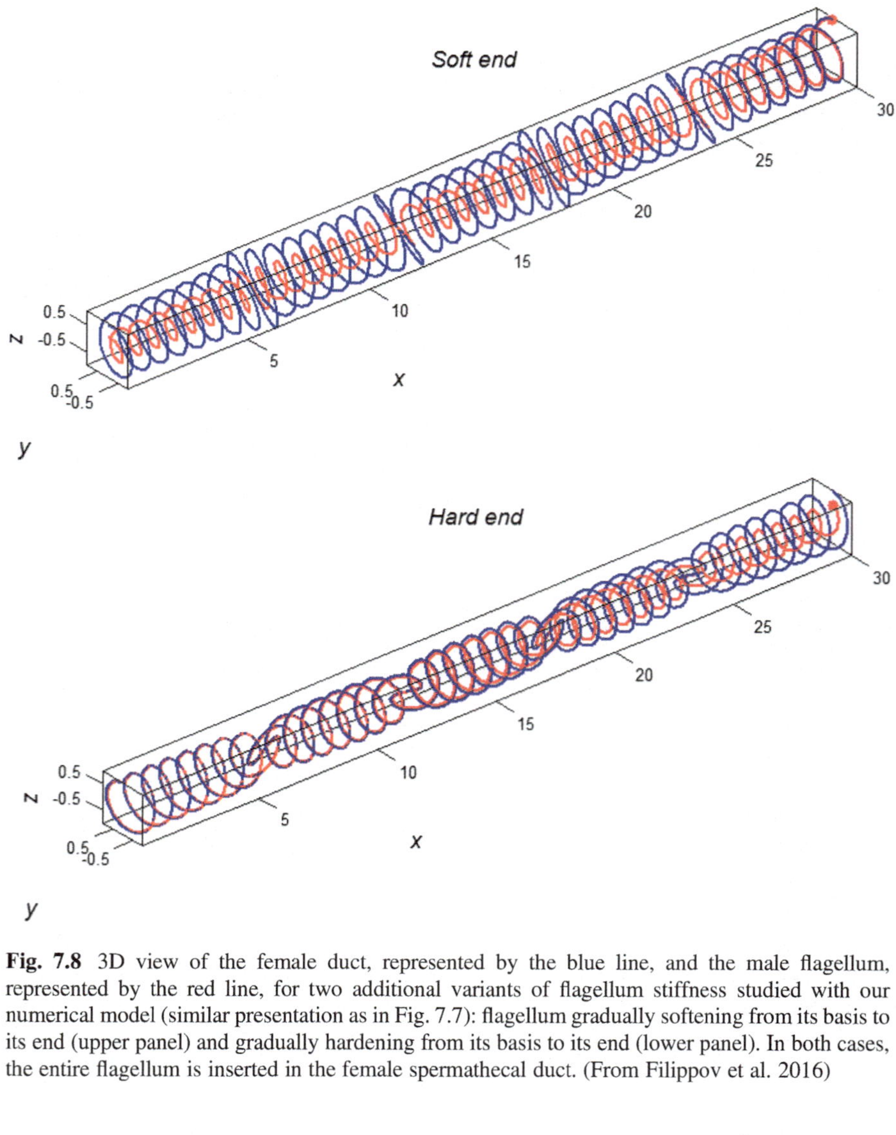

Fig. 7.8 3D view of the female duct, represented by the blue line, and the male flagellum, represented by the red line, for two additional variants of flagellum stiffness studied with our numerical model (similar presentation as in Fig. 7.7): flagellum gradually softening from its basis to its end (upper panel) and gradually hardening from its basis to its end (lower panel). In both cases, the entire flagellum is inserted in the female spermathecal duct. (From Filippov et al. 2016)

The flagellum is again constructed with a large number (up to $N = 10^5$) of elastically connected discrete segments. Here, we need a more accurate calculation, so we created more segments than for the model above (Sect. 7.1.2). Each segment has a length of $dr \leq 1$ μm, defined by the total length of the flagellum ($L = 10$ mm) measured on microscopic images of our model beetle, *Cassida rubiginosa*. The flagellum model is provided with a strong longitudinal stiffness $k^{||}$, which prevents extension and compression of the practically rigid segments, and two components of elastic forces, $\vec{f}^{elastic} = \{f^{\parallel}, f^{\perp}\}$: $\vec{f}^{\parallel}_{jj\pm1} = k^{\parallel}\left(\vec{r}_j - \vec{r}_{j\pm1}\right)\left[1 - \left(\left(\vec{r}_j - \vec{r}_{j\pm1}\right)/dr\right)^2\right]$ and $\vec{f}^{\perp}_j = k^{\perp}\left(2\vec{r}_j - \vec{r}_{j+1} - \vec{r}_{j-1}\right)$, acting between the neighboring nodes $\vec{r}_j$. The longitudinal force, $\vec{f}^{\parallel}_{jj\pm1}$, tends to maintain the distance between the nodes $\vec{r}_j$ and $\vec{r}_{j\pm1}$ close to the equilibrium *dr*.

To account for the probably uniform stiffness along the main length of the flagellum (without the tip), as suggested by CLSM images, we used a general flagellum rigidity, $k_0^{\perp} = K$ for a more rigid flagellum and $k_0^{\perp} = 0.01K$ for a uniformly soft flagellum ($g = 0$), respectively. The parameters $g = 0.99$ and $k_0^{\perp} = K$ were chosen for a rigid flagellum with a soft end, $g = -99$ and $k_0^{\perp} = 0.01K$ for a soft flagellum with a stiff end. As above, we assume that the transverse stiffness $k^{\perp} = const$ is much weaker than the longitudinal one: $k^{\parallel} >> k^{\perp}$. In the real system, the stiffness of the flagellum seems to vary along its length (see above), which was taken into account in the current analysis. This means that $k^{\perp}$ varies from segment to segment. From preliminary experiments, we expect that $k^{\perp}$ should be a nonlinear function which is almost constant at the beginning and quickly decreases near the end. For definiteness, we apply:

$$k^{\perp} = k_0^{\perp}\left[1 - g\left(\frac{r_j - r_{j=1}}{r_N - r_{j=1}}\right)^{\alpha}\right]. \tag{7.5}$$

Here, the parameters $k_0^{\perp}$ and g determine a general rigidity plus a rigidity gradient along the flagellum. In particular, at $g = 0$, a large or a small constant $k_0^{\perp}$ corresponds to a uniformly hard or soft flagellum, respectively. If $g = 0.99$, the tip of the flagellum is 100 times softer than its base. Using different combinations of the parameters, we analyzed four different variants of the system: uniformly soft, uniformly hard, almost uniformly soft with a hard tip, and almost uniformly hard with a soft tip. To reproduce the latter two cases, we used the exponent $\alpha = 5$.

As above, the female duct along the coordinate $X \in [0, L_x]$ was defined by its spiral core: $Y = A_0 \sin(2\pi X/\lambda_0) + R \sin(2\pi X/\lambda)$; $Z = Z_0(X) + R \cos(2\pi X/\lambda + \phi(X))$ with the parameters $R = L_x/30$; $\lambda = L_x/40$, $0.1L_x < A_0 < 0.5L_x$ and $0.2L_x < \lambda_0 < 0.4L_x$. It has a secondary structure with some number N_{knots} of knots at the positions X_n where the direction of the spiral rotation changes to a reverse one. The knots of phase rotations are caused by the phase function $\varphi(X)$, defined as a sum of step-like functions, $\varphi(X) = \sum_{n=1}^{N_{knots}} (-1)^n[\pi/2 + \arctan((X - X_n)/2\tau)]$, with the factors $(-1)^n = \pm 1$ defining opposite phase rotations around the knots.

Although the female spermathecal duct is not only helical, but the entire helical duct is also convoluted (Filippov et al. 2015), we neglected all the variations of the spiral core, $Y_0 = A_0 \cos(2\pi X/\lambda_0) = 0$, and assumed a straight helical duct in order to

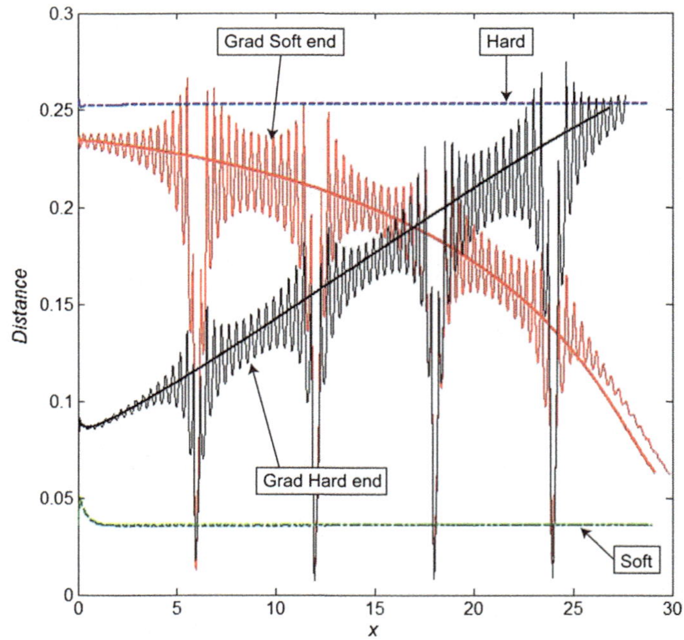

Fig. 7.9 Local deviation of the flagellum from the female duct axis along the x-coordinate of the channel for the four main variants presented in Figs. 7.7 and 7.8 as indicated by the labels. The dashed curves represent the variants with a uniformly soft/stiff flagellum. The oscillating curves show the distances for the variants in which the flagellum had a gradient of material properties and the channels had four knots. Additionally, the smooth curves represent the variants with channels without any knots. (From Filippov et al. 2016)

concentrate only on the most relevant effects. Such spirals with four knots ($N_{knots} = 4$) and different combinations of flagellum rigidities are shown in Figs. 7.7 and 7.8. In each figure, the female elastic channel is represented by a blue line. The flagellum inside the channel is indicated by a red curve, and the flagellum tip by a red dot.

Numerically, the female duct may be defined as an elastic confinement acting on every segment $\{z_j(x_j), y_j(x_j)\}$ of the male flagellum with the forces $f_y^{confinement}(X = x_j) = K(Y(X) - y_j)$; $f_z^{confinement}(X = x_j) = K(Z(X) - z_j)$ which are regulated by the elastic constant K. The presence of liquid inside the channel was taken into account by introducing a damping force proportional to the velocity $\vec{v}_j$: $f_j^{damping} = \gamma \vec{v}_j$. As above, this force was also added to the equations of motion of all the segments: $\partial \vec{v}_j / \partial t = f_j^{elastic} + f_j^{confinement} + f_j^{damping}$.

The results of our simulations for a flagellum with a uniform and a gradient distribution of rigidity are summarized in Figs. 7.9 and 7.10. Because a longitudinal material gradient was found only in the male flagellum, whereas the inner wall of the female spermathecal duct showed uniformly green and red autofluorescence, we focused on the analysis of the material gradient of the male flagellum. First of all, we analyzed the local deviation of the flagellum from the intact female duct axis along

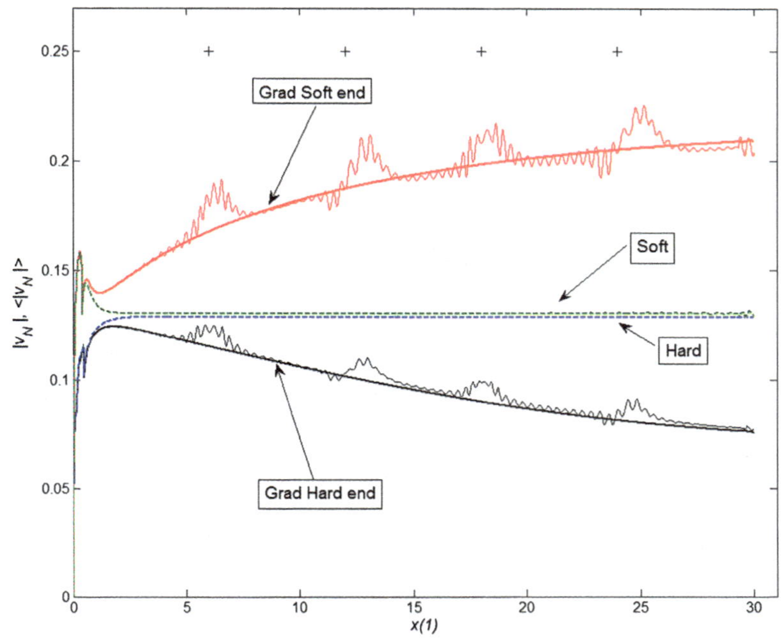

Fig. 7.10 Correlation between maxima of the tip velocity along the female channel coordinate *x* (1) and the positions of the knots. The locations of the knots are indicated by cross symbols (+). The curves corresponding to flagella with stiffness gradients and female channels with four knots are shown as oscillating lines. The curves corresponding to flagella with stiffness gradients and channels without knots are shown as thick lines. Finally, the curves corresponding to flagella without stiffness gradients and female channels without knots are shown as dashed lines. The tip velocity curves are averaged over a moving time window with the width corresponding to the time interval between neighboring minima. (From Filippov et al. 2016)

the x-coordinate for the following four cases of flagellum stiffness: (1) uniformly soft, (2) uniformly hard, (3) soft with a hard tip, (4) and hard with a soft tip (Fig. 7.9).

There is a clear difference between the curves obtained for the female system with knots and without knots. This difference is mainly due to the different degrees of oscillation caused by the phase rotations at the knots.

A correlation between the maxima of the absolute tip velocity along the female duct and the positions of the knots is presented in Fig. 7.10. The velocity of the flagellum tip was calculated for each case as a function of time. However, for better comparison, this velocity is plotted here as a function of the tip coordinate $x(1)$. Every curve in Fig. 7.10 presents the velocity averaged over a moving time window with the width corresponding to the time interval between neighboring minima. The averaging process renders the velocity maxima more pronounced. The thick lines in Fig. 7.10 show the velocities calculated for the systems without knots. As is evident in the figure, the propulsion velocity strongly depends on the stiffness of the flagellum. The velocity of the rigid flagellum with a soft tip is much higher than that of all other cases. An accelerated flagellum propulsion was observed at the sites of the knots.

7.1.4 Comparison of the Model Results and Microscopical Observations

The presented model of flagellum motion clearly demonstrated that the stiffness gradient along the flagellum impacts its penetration into the relatively stiff, helical female duct.

For a flagellum that is uniformly stiff along its length, passage through the female duct results in a strong deformation according to our model. This means the flagellum simply "ignores" the duct's mechanical disturbances as demonstrated by Filippov et al. (2015). In contrast to these results, a hard flagellum with a soft tip, which possibly reflects the actual system (Fig. 7.7), shows quite different results (Figs. 7.9 and 7.10), especially regarding velocity. It seems that the soft end of the flagellum is flexible enough to quickly adjust to small curvatures in the duct, and at the same time, it is strongly pushed by the rigid basal part. In reality, the apical region of the flagellum not only demonstrates a longitudinal stiffness gradient, but also a transverse stiffness gradient, i.e. a relatively soft inner curve and a rigid outer curve; this might aid the flagellum tip to adjust to the small curvatures of the female duct. Interestingly, a uniformly soft flagellum and an almost uniformly soft flagellum with a stiff tip show worse performances in comparison with the rigid flagellum with a soft end, i.e. either a strong deformation of the female duct and/or a much lower velocity of the male flagellum (Figs. 7.9 and 7.10). Thus, the results obtained from our simulations strongly support the hypothesis that the stiffness gradient in the beetle flagellum facilitates penile propulsion.

In Sect. 7.1.2, we analyzed the effects of various shapes of the female duct on penile propulsion. It was found that certain parameters of the female duct, such as the presence of knots and a slight curvature, result in a velocity decrease of penile propulsion or an increase of the energy expenditure for propulsion. Because sexual intercourse in the field may be disrupted by other males or by natural enemies, it is highly likely that males which can quickly insert the penis and ejaculate will be preferred. Therefore, the stiffness gradient of the flagellum, which was demonstrated to be able to penetrate the female duct quickly, might have been selected as a favorable structure for both males and females, even though it may increase the energy expenditure for male penile propulsion (Filippov et al. 2015).

Several sophisticated experimental studies have investigated the driving and stabilizing mechanisms of hyper-elongation (Rodriguez 1995; Arnqvist 1998; Tadler 1999; Rodriguez et al. 2004; van Lieshout and Elgar 2010; Kamimura 2013; Dougherty et al. 2015). None of these authors took the material properties of the genitalia into account. Recently, however, it has been shown that cutting off the tip of the flagellum resulted in lower paternity rates of these males (Dougherty et al. 2015). This result may be partly explained by the results of our simulation. It is likely that the absence of the softer tip causes trouble in sexual behavior due to less effective propulsion, although the question why slower insertion causes lower paternity under laboratory conditions will have to be considered in further studies. As pointed out by Dougherty et al. (2015), functional morphology has been largely ignored in studies of genital evolution. Hyper-elongated genitalia are ubiquitous in animals, occurring not only in insects but also in vertebrates such as ducks (Brennan et al. 2007), and it would be interesting to study more taxa for a broader comparison among animals.

7.2 Slow Viscoelastic Response of Resilin

7.2.1 General Properties and Biological Importance of Resilin

The importance of the rubber-like protein resilin in invertebrate biomechanics has recently been rediscovered. Resilin is a multifunctional rubber-like cuticular protein found in many arthropods (Lyons et al. 2009; Michels et al. 2016). In locusts and in many other insects, the flight apparatus includes highly elastic resilin-rich structures, which are responsible for storing elastic energy during the wing upstroke and releasing it back during the succeeding downstroke (Weis-Fogh 1960). Resilin was also found to act as a spring powering the high-speed catapult mechanism in the jumping flea *Spilopsyllus cuniculi* (Bennet-Clark and Lucey 1967). Furthermore, it was found in an elastic spring structure acting as an antagonist to the muscle in the feeding pump of the bug *Rhodnius prolixus* (Bennet-Clark 2007), and in the amazingly complex wing-folding mechanism of the earwig *Forficula auricularia* (Dermaptera), where it aids in packing quite large and fully functional hindwings under tiny, hard forewings (Haas et al. 2000a). Resilin was also recently discovered at the tips of the tarsal setae in the ladybird beetle *Coccinella septempunctata* (Peisker et al. 2013), where it facilitates setal adhesion due to its strong adaptability to the substrate roughness at a low degree of setal clusterization (Gorb and Filippov 2014). In the joints between wing veins in dragonflies resilin is responsible for providing a very specific wing flexibility and adaptability to the aerodynamic situation during flight (Appel and Gorb 2011).

Resilin is an ideal material for elastic joints, such as wing hinges, which are subjected to repeated cyclical stress. Indeed, in the course of its adult life, a locust may fly for 8 h per day for about 30 days with wingbeat frequencies of about 25 Hz, requiring over 20 million wing beats (Bennet-Clark 2007). During the lifetime of an insect, resilin shows neither tearing nor fatigue when stressed within its natural limits. Weis-Fogh (1960) had originally claimed that resilin, almost uniquely among biological materials, shows perfect elasticity: he observed that even when strained to over twice its original length for 2 weeks, a dragonfly's resilin tendon returns back to its original length when the stress is relieved.

The elastic force in resilin arises from the decrease in conformational entropy caused by deformation as is the case in true rubbers (Weis-Fogh 1961). Resilin does not have any regular structure. The high content of glycine and proline is responsible for the folding of the protein in random coils. Resilin lacks sulphur-containing amino acids and tryptophan (Bailey and Weis-Fogh 1961; Lombardi and Kaplan 1993) and it is cross-linked at regular intervals by stable covalent di- and tri-tyrosine links

(Elliott et al. 1965). The amino acid composition of resilin is unique and different from that of elastin, collagen, and silk fibroin. Although resilin is highly hygroscopic and represents a kind of hydrogel, once it has been cross-linked it is completely insoluble in water (Andersen 1963).

The basic mechanical properties of resilin were studied in the early sixties (Weis-Fogh 1960). It was shown that resilin behaves like a true physical rubber with an elastic efficiency of 97% (Weis-Fogh 1960). Elastic moduli found for resilin range from 0.6 MPa in elastic tendons of dragonflies to 10 MPa in the rubber-like cuticle of locusts (Jensen and Weis-Fogh 1962). In the latter work, the authors explored the mechanical properties of resilin, showing that the energy loss, even at 200 Hz, was below 5%. It was mentioned that the loss factor does not appear to increase linearly with frequency, suggesting that the losses are not due to viscous damping.

While the mechanical properties of resilin have so far mainly been determined at stretching, we determined them at compression. Microindentation of the wing hinge of *Locusta migratoria* revealed the viscoelasticity of resilin (Kovalev et al. 2018). In our experiments, about a quarter of the mechanical response could be assigned to a viscous component. Here the mechanical response is characterized using a generalized Maxwell model with two characteristic time constants and alternatively using a 1D model with just one characteristic time constant. Slow viscous responses with characteristic times of 1.7 s and 16.0 s were observed during indentation (Kovalev et al. 2018). These results demonstrate that the locust flight system is adapted to both fast and slow mechanical processes. The fast, highly elastic process is related to the flight system, and the slow, viscoelastic process may be related to wing folding. Our measurement of the dynamic mechanical properties strongly indicates that for fast processes (within the time range of milliseconds) the resilience of resilin is actually small (Gosline et al. 2002). In the modeling presented in the following, an attempt will be made to obtain quantitative data about the actual energy loss in resilin at compression.

Due to its precisely determined boundary conditions, a simple sphere-on-flat geometry was used in the microindentation experiment. Since the energy dissipation in resilin at long time scales has not been studied before, the force dynamics within seconds of indentation were observed and are reported here. For data processing, we reduced the dimensions, using a functional 1D model with just one characteristic time constant along with the generalized Maxwell model with two characteristic time constants for the description of the mechanical properties of resilin. Dimension reduction is also best suitable for the visualization of stress evolution in such a viscoelastic sample.

7.2.2 Physical Properties of Resilin and Experimental Methods

The wing hinge ligament of the locust *Locusta migratoria* (L.) was used in the experiments. The ligament is 0.1 mm thick and 1 mm long, homogeneous, and one of the most important resilin-containing structures in locusts (Andersen and Weis-Fogh 1964). Moreover, the resilin of the wing hinge is produced by the epidermis in the form of thick continuous layers in a relatively pure form less mixed with other proteins and chitin (Weis-Fogh 1960; Bailey and Weis-Fogh 1961).

For mechanical testing, a relatively large and homogeneous native resilin patch from the wing hinge ligament of *L. migratoria*, about 0.15 mm thick and about 1 mm long, was mechanically fixed to a piece of plastic (Fig. 7.11b). For relaxation, the specimen was placed in a phosphate buffer solution (pH 7.0) 2 h before the force measurements, which were carried out with a load cell force transducer (FORT100; World Precision Instruments, Sarasota, FL, USA). Compressive load was applied by a sapphire sphere 3 mm in diameter firmly fixed to the force transducer that was moved by the motorized micromanipulator. The compressive type of the experiment was selected because under natural conditions the wing hinge bends, experiencing not only tensile, but also compressive forces. Load was applied by moving the force transducer down onto the resilin sample. After 10 to 80 s of load application, the transducer was pulled back. Different forces were applied in consecutive loading cycles. The contact force was measured during multiple loading and unloading cycles and recorded with AcqKnowledge 3.7 software. The measurements were performed in phosphate buffer solution. The obtained slow viscoelastic response of resilin (Fig. 7.12) was characterized with a generalized Maxwell model by fitting experimental force–time curves. Nonlinear regression analysis was used to fit the data and to estimate the 95% confidence intervals of the fit parameters (Seber and Wild 1989).

The wing hinge sections were stained with toluidine blue, which stains the rubber-like cuticle, but not the solid cuticle, bright blue. Within the rubberlike cuticle, chitin-bearing lamellae are visible (Fig. 7.11a), running nearly parallel to each other. In the proximal part of the rubber-like cuticle of *L. migratoria*, chitin lamellae are missing, indicating a patch of probably pure resilin as previously observed (Weis-Fogh 1960).

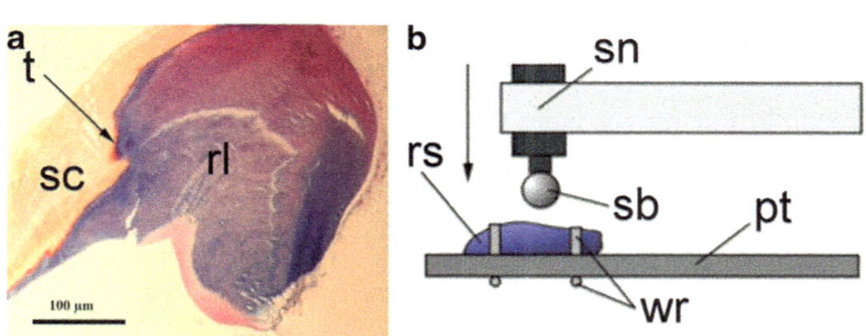

Fig. 7.11 (**a**) Section through the wing hinge of *L. migratoria*. After the standard fixation and dehydration procedure (Gorb 2004), the wing hinges were embedded in Technovit 7100 glycol methacrylate; 3.5 μm thick sections were prepared and stained with toluidine blue. (**b**) Experimental setup for force measurements on a native resilin patch from the wing hinge ligament with the arrow indicating the loading direction. Abbreviations: *pt* plate, *rl* resilin with chitin lamellae, *rs* resilin sample, *sb* sapphire sphere, *sc* solid cuticle, *sn* force transducer, *t* transition zone, *wr* stainless steel wire. (From Kovalev et al. 2018)

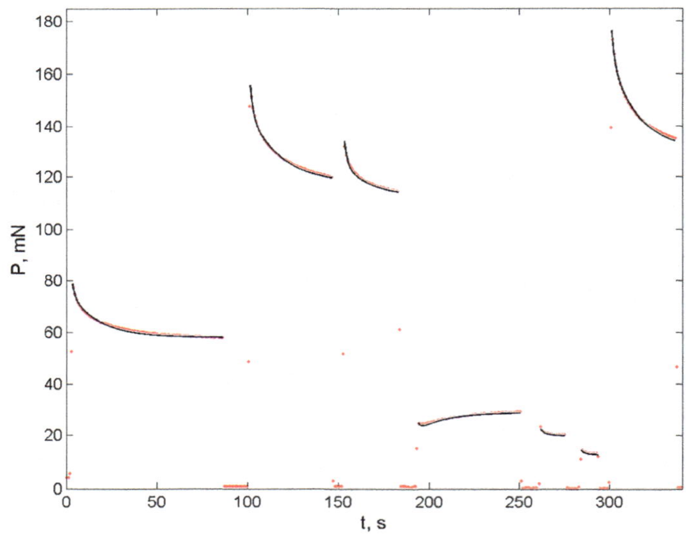

Fig. 7.12 Consecutive loadings with a spherical indenter on the wing hinge of *L. migratoria* (embedded in phosphate buffer solution). The temporal evolution of the loading force at different constant displacements demonstrates the viscous stress relaxation. The force in the experiment is shown by the thick red line. The generalized Maxwell model fit is shown by the thin black line. (From Kovalev et al. 2018)

7.2.3 Two Procedures for Modeling the Experimental Results

Standard Approach As previously shown for viscoelastic materials (Cheng et al. 2005), the time-dependent response may be considered independent of the contact geometry. That is why a generalized Maxwell model (Christensen 1982) was used as a standard approach to characterize the viscoelastic properties of resilin. Three elements were insufficient to properly fit the experimental data, but a model with five elements fit the main features of the experimental force curves well (Fig. 7.12).

The experimental force curves were fitted to the generalized Maxwell model using a formalism similar to that presented by Cheng et al. (2005) according to the following equation:

$$P(t) = \frac{4\sqrt{R}}{3(1-\nu^2)}\left(E_\infty \delta_k^{1.5} + \sum_{i=1}^{k} \delta_i^{1.5} \sum_{j=1}^{2} E_j e^{-\left(t-t_i^l\right)E_j/\eta_j} - \sum_{i=1}^{k-1} \delta_i^{1.5} \sum_{j=1}^{2} E_j e^{-\left(t-t_i^u\right)E_j/\eta_j}\right), \tag{7.6}$$

where t is a time within a certain range ($t_k^l < t < t_k^u$); t_i^l and t_i^u are the times of the ith loading application and release, correspondingly; k is the number of loading/unloading cycles (1–7); E_∞, E_1, and E_2 are the effective Young's moduli of the

Table 7.1 Parameters of the generalized Maxwell model

E (kPa)		η (kPa·s)		τ (s)	
E_∞	1927.0 (± 44.0)				
E_1	536.0 (± 12.0)	η_1	8680.0 (± 190.0)	τ_1	16.0
E_2	175.6 (± 8.9)	η_2	305.6 (± 1.5)	τ_2	1.7
E_0	2638.0 (± 59.0)				

Elasticity moduli (E), viscosities (η) and relaxation times (τ) for different components are labeled with different indexes
The values in parentheses are standard errors

static and the two dynamic components; η_1 and η_2 are the viscosities of the two dynamic components; ν is the Poisson ratio (assumed to be equal to 0.49); δ is the indentation depth; R is the indentation sphere radius (1.5 mm).

The pure elastic modulus (E_∞) and the two elastic moduli related to the two viscous components are equal to 1.927 ± 0.044 MPa (mean ± standard error), 536 ± 12 kPa, and 175.6 ± 8.9 kPa, respectively (Table 7.1). Young's modulus, which is observed immediately after the loading application (E_0), was 2.638 ± 0.059 MPa. This value is in good agreement with the value of 2 MPa previously determined for the resilin in the prealar arm of *Schistocerca gregaria* in static bending experiments (Weis-Fogh 1960).

The stress relaxation is composed of two components with effective relaxation times of 16.0 s and 1.7 s, respectively. The fact that the higher effective elasticity modulus corresponds to the longer relaxation time component indicates a strong contribution of viscous response in resilin. Indeed, the contribution of the viscous response was rather strong, namely 27% (Fig. 7.12). The two viscous components have the following viscosity values: 8.68 ± 0.19 MPa·s and 305.6 ± 1.5 kPa·s.

The fit presented in Fig. 7.12 is still not ideal, and knowing the various viscosities or decay times does not help to understand the microscopic behavior of resilin. Besides, at different loading values the indentation sphere forms different contact areas and the time-dependent response may become dependent on the contact geometry. Unfortunately, using the approach presented by Cheng et al. (2005) does not allow us to take into account the relaxation process at different locations. That is why, in order to visualize the resilin surface profile and stress distribution in the present experiment, we additionally used a 1D model.

One-Dimensional (1D) Model Dimension reduction, an approach used here, may for some applications be used as an alternative to the time-consuming finite element method. It can be shown that contact with an elastic foundation (an array of independent springs) provides correct force–displacement relationships for a large number of simple surface profiles (Filippov and Popov 2007; Geike and Popov 2007; 2010; Popov and Filippov 2012; Pohrt et al. 2012). The surface profile in this case, however, has to be modified according to some specially formulated rules. For spherical profiles, the rule was provided by Geike and Popov (2007). They showed that the relationship between force, indentation depth and the contact radius of a spherical indenter with radius R pressed into a half-space (Fig. 7.13a) may be exactly reproduced by a contact with a one-dimensional elastic foundation (Fig. 7.13b) by

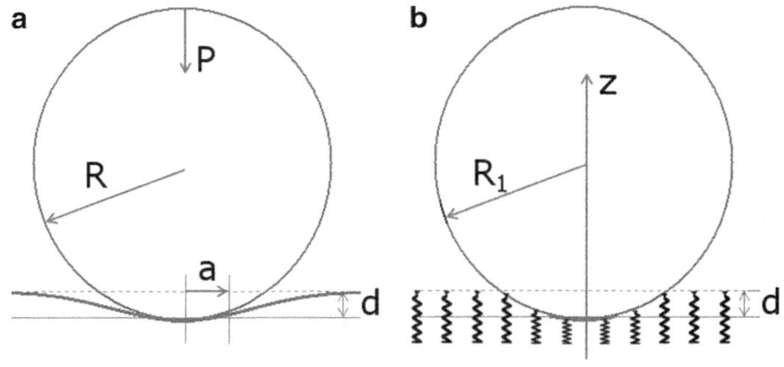

Fig. 7.13 Conceptual structure of the 1D model. (**a**) Sphere with the radius R indenting a flat surface with the loading force P, the indentation depth d and the contact radius a. (**b**) One-dimensional representation of the sphere with radius R_1. The sphere indents the flat surface represented by an array of springs. The coordinate axis is shown as z. (From Kovalev et al. 2018)

formally changing the radius R to R_1. If a sphere with the radius R is brought into contact with an elastic foundation, in this model the contact radius is equal to $a = \sqrt{2R_1 d}$, d is the penetration depth, and the normal force is equal to $P = (4/3)E\sqrt{2R_1 d^3}$ (where E is Young's modulus). If we choose a radius $2R_1 = R$, the results exactly coincide with the Hertz theory.

In our case, this means that we can replace a 3-dimensional problem with a much simpler 1-dimensional model consisting of an indenting circle and a set of non-interacting springs having a specific spatial distribution of equilibrium stiffness, $K_{\mathrm{eq}}(y)$ which depends on the indentation depth d: $K_{eq}(y) = K_{\min} + (K_{\max} - K_{\min}) \exp\{-|y|/d\}$, where $K_{\min}$ and $K_{\max}$ are the final and initial stiffness of a spring, and y is the vertical coordinate of the substrate. The springs relax to the equilibrium stiffness $K(y; t) \rightarrow K_{eq}(y)$ with a characteristic rate γ that models the strain softening. In turn, being disturbed, the surface must relax to the complete contact with the circle between the two points where the circle crosses the line $y = 0$. This process is characterized by the rate η.

Observable properties of the system are strongly determined by the relationship between two characteristic times, $1/\eta$ and $1/\gamma$. To obtain a qualitatively correct behavior, we generally assume that the relaxation of the substrate surface is faster than the relaxation of the elasticity, $1/\gamma > 1/\eta$, but it is important that the time span of surface adaptation to the circle does not equal zero, $1/\eta \neq 0$. This substrate surface adaptation to the indenting circle could be clearly demonstrated in an experiment with repeated indentations (Fig. 7.14), where the contact area continuously advances.

Initially, the springs possess a high stiffness ($K_{\max}$), and during the first indentation, the springs' stiffness distribution approaches the equilibrium distribution from above (as indicated by the arrow in Fig. 7.15a). When, after the first indentation, a second indentation to a lower depth is performed, the equilibrium distribution will

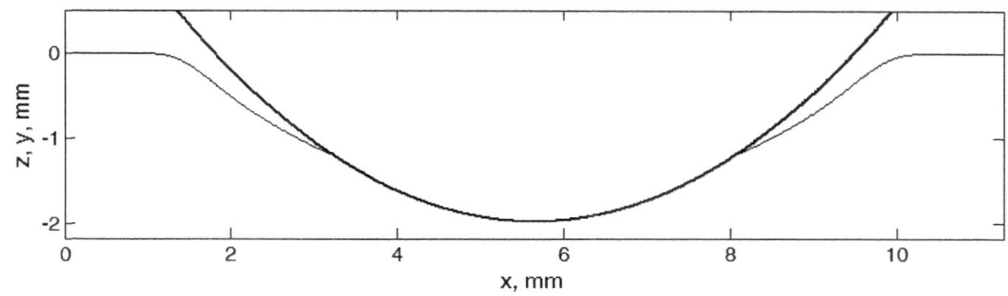

Fig. 7.14 Configuration where the sphere (thick line) is immersed for a second time into a depth smaller than during the first indentation. The figure shows an intermediate stage at which the substrate (represented by a 1D array of springs, thin line) is moving in a positive (vertical) direction, slowly adapting to the surface of the sphere. The vertical coordinates z and y are the coordinates of the sphere surface and of the upper substrate surface. (From Kovalev et al. 2018)

shift to a higher stiffness since the distribution depends on the indentation depth, and the springs' stiffness distribution approaches the equilibrium distribution from below (as indicated by the arrow in Fig. 7.15b).

This example emphasizes the importance of the springs' stiffness distribution for the visualization of the relaxation processes in a viscoelastic substrate. The springs' stiffness distribution may be derived as follows: $P(K) \sim dy/dK$, $y = z_0 + R\sqrt{(x - x_0)}$, where x and y are the horizontal and vertical coordinates of the substrate, respectively, z is the vertical coordinate of the indenter with radius R, and x_0 and z_0 are the vertical coordinates of the deepest position of the substrate and of the indenter, respectively (Fig. 7.13). After differentiation the following equation could be obtained:

$$P\{K_{eq}(y)\} = \frac{d(R - y + \min(y))\exp(-|y|/d)}{(K_{\max} - K_{\min})\sqrt{R^2 - (R - y + \min(y))^2}}. \tag{7.7}$$

Total indentation force may be presented as a sum: $P = \sum\limits_{x} K(y)$. During the process described above, i.e. the two indentations with lower depth during the second indentation, the force–time curves demonstrate nonstandard behavior. Immediately after the second indentation, the force does not decrease, but rather increases. This phenomenon is called inverse stress relaxation (Kothari et al. 2001; Misak et al. 2013) (Fig. 7.16). Here we should remark that viscosity (controlled by the damping factor η in the model) plays a crucial role in the formation of this specific behavior.

Since the 1D model well reproduces the inverse stress relaxation observed in our experiments, the model may also be used to describe the data derived from a real experiment. Indeed, the simulation using the 1D model corresponds well to a fragment of experimental data (the same as shown in Fig. 7.12) demonstrating inverse stress relaxation(cf. Fig. 7.17). Generally, the force curve is well reproduced in the simulation. The temporal development of the elasticity distribution $P\{K(y);t\}$ is presented by the waterfall map in Fig. 7.18. It is directly evident that every cycle of

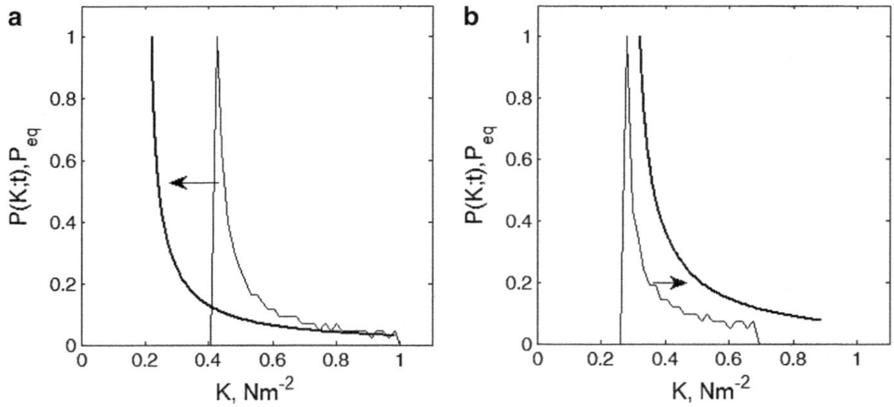

Fig. 7.15 Hypothetical equilibrium distribution (probability) of the effective elasticity (Eq. 7.7) $P\{K_{eq}(y)\} = d(R - y + \min(y))\exp(-|y|/d)/(K_{\max} - K_{\min})\sqrt{R^2 - (R - y + \min(y))^2}$. (**a**) and (**b**) represent two possible variants of how the pressure may be attracted to the equilibrium distribution depending on the prior history of the system (two curves moving to $P\{K_{eq}(y)\}$ from the right and the left side, respectively). (From Kovalev et al. 2018)

Fig. 7.16 Modeling results for the typical time-dependent total pressure *P* caused by two cycles of the circle's motion: indentation during the first cycle is deeper than during the second one. (From Kovalev et al. 2018)

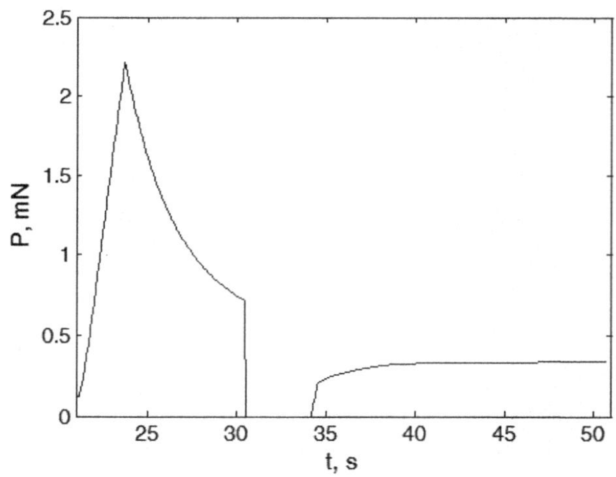

the circle's motion is accompanied by the evolution of $P\{K(y);t\}$ to $P\{K_{eq}(y)\}$, which each time starts from the new initial condition determined by the prior history of the process.

In conclusion, we can note that the slow viscoelastic response of resilin revealed in stress relaxation experiments is similar to the reversible strain softening in crosslinked natural rubber (Santangelo and Roland 1992). This behavior is presumably due to the friction present in low-ordered resilin molecular chains as well as water flow within the resilin matrix due to changes in hydrostatic pressure (Andersen and Weis-Fogh 1964).

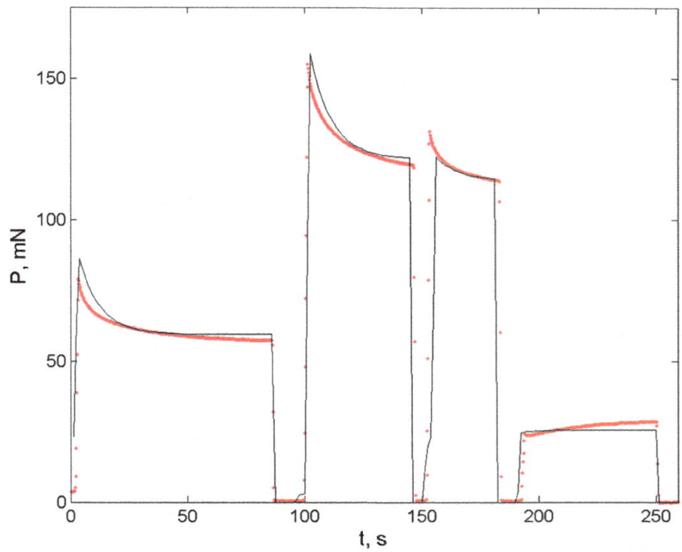

Fig. 7.17 Comparison between the 1D modeling results and the experimentally found time-dependent behavior of *P*. $\gamma = 1.5\ \mathrm{s}^{-1}$, $\eta = 0.5\ \mathrm{s}^{-1}$, $K_{\min} = 0.6$, $K_{\max} = 1$, $d = 0.5$ mm. The experimental results are shown as thick red lines, the numerically obtained curve is shown as a thin black line. (From Kovalev et al. 2018)

Previously, resilin was described as a resilient material with 97% efficiency in storing mechanical energy (e.g., Weis-Fogh 1960). Surprisingly, in our compression experiments a wing hinge with high resilin content lost 27% of energy due to viscous losses on a time scale of seconds. Additionally to the mechanisms mentioned above, the interaction of the resilin matrix with water molecules or ions in the buffer solution may be responsible for the slow viscous relaxation of the resilin sample in our experiment. Other possible mechanisms of viscous energy loss in the wing hinge may be the interaction between resilin and chitin lamellae within the specimen. Typically for rubber-like materials, the storage modulus increases with decreasing relaxation times (Weng et al. 2007). The fact that the fast-decaying component has a smaller elasticity modulus (33% of that of the slowly decaying one) and viscosity (just 4% of that of the slowly decaying component) demonstrates the presence of a shear-thinning effect in resilin for time intervals in the range of seconds.

From the loss modulus (Gosline et al. 2002) utilizing, e.g., the Maxwell model, resilin viscosity in the frequency range 10–190 Hz might be estimated to be 1.8–0.4 kPa·s. Much higher resilin viscosities than those determined in our experiments have not been described before. This high viscosity may be important for the unfolding of complex and delicate organs, such as wings, in time intervals of seconds to minutes. The high resilin viscosity may prevent crack initiation in structures continuously exposed to strong bending loads, such as the locust wing hinge. The reason for this is that morphological structures containing resilin may strongly reduce stress concentrations under permanent loads.

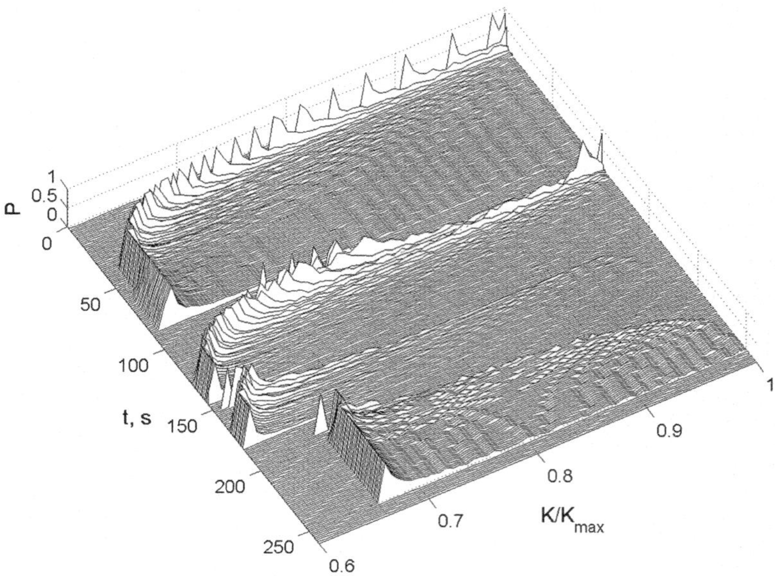

Fig. 7.18 Temporal development of the elasticity distribution $P\{K(y);t\}$, presented as a waterfall map. It is directly evident that every cycle of the circle's motion is accompanied by the evolution of $P\{K(y);t\}$ to $P\{K_{eq}(y)\}$, which each time starts from the new initial condition determined by the prior history of the process. (From Kovalev et al. 2018)

There is a standard model which can satisfactorily describe the viscoelastic properties of rubber-like materials. The Maxwell model presented above contains as many parameters (three Young's moduli and two viscosities) as are necessary to fit complex curves that demonstrate the behavior of a real material. However, the model does not physically explain the problem under consideration, but rather replaces it with a set of mathematical equations. In the case of our resilin experiment, discrete viscosities (relaxation times) do not have any relevant physical meaning.

The 1D model depends only on the system geometry. It allows testing a complex sphere-on-flat geometry, which would be optimal for a real experiment, but is highly complex for direct modeling. The model has only three physically reasonable parameters and clearly represents material properties in the contact region. The slow shift of the springs' stiffness distribution was not artificially introduced into the 1D model: rather, the shift corresponds to the slow effective spring relaxation. Although the shape of the force curve was not reproduced absolutely correctly using the 1D model, the model gives a good qualitative explanation of the mechanical behavior and coincides well with the experiment without the introduction of decay times, which are present in a generalized Maxwell model but are not related to any physically meaningful parameters.

The mechanical properties of materials demonstrating a strong viscoelastic component in their response to compression may also be characterized without the framework of the model suggested here. The dimension reduction approach provides a clear image of the contact of solid bodies for a complex (axisymmetric) geometry at low computational costs. A strong component of viscoelastic response to loads in the locust wing hinge suggests that, along with the flight support, wing folding is an equally important function of resilin, as previously hypothesized for earwigs and beetles (Haas et al. 2000a, b). The separation of these functions in structures containing resilin is possible because the characteristic times differ for different processes. The characteristic time, in turn, depends on the material's viscoelasticity and on the geometry of the structure. In the case of locust flight, the deformation frequency is high and the viscosity contribution is negligible. In the case of locust wing folding, the viscosity contribution is not negligible anymore, and it facilitates the stress relaxation of the deformed structure. Thus, the locust flight system demonstrates the co-evolution of material properties and morphological structure in the system of the wing hinge.

References

Ah-King M, Barron AB, Herberstein ME (2014) Genital evolution: why are females still understudied? PLoS Biol 12:e1001851

Andersen SO (1963) Characterization of a new type of cross-linkage in resilin, a rubber-like protein. Biochim Biophys Acta 69:249–262

Andersen SO, Weis-Fogh T (1964) Resilin. A rubberlike protein in arthropod cuticle. Adv Insect Physiol 2:1–65

Appel E, Gorb SN (2011) Resilin-bearing wing vein joints in the dragonfly *Epiophlebia superstes*. Bioinspir Biomim 6:046006. (11pp)

Arnqvist G (1998) Comparative evidence for the evolution of genitalia by sexual selection. Nature 393:784–786

Arnqvist G, Danielsson I (1999) Copulatory behavior, genital morphology and male fertilization success in water striders. Evolution 53:147–156

Arnqvist G, Rowe L (1995) Sexual conflict and arms races between the sexes: a morphological adaptation for control of mating in a female insect. Proc R Soc Lond B 261:123–127

Bailey K, Weis-Fogh T (1961) Amino acid composition of a new rubber-like protein, resilin. Biochim Biophys Acta 48:452–459

Bennet-Clark HC, Lucey ECA (1967) The jump of the flea: a study of the energetics and a model of the mechanism. J Exp Biol 47:59–76

Bennet-Clark H (2007) The first description of resilin. J Exp Biol 210:3879–3881

Brennan LR, Prum RO, McCracken KG, Sorenson MD, Wilson RE, Birkhead TR (2007) Co-evolution of male and female genital morphology in waterfowl. PLoS One 5:e418

Cheng L, Xia X, Scriven LE, Gerberich WW (2005) Spherical-tip indentation of viscoelastic material. Mech Mat 37:213–223

Christensen RM (1982) Theory of viscoelasticity, 2nd edn. Academic, London

Cordoba-Aguilar A (2005) Possible coevolution of male and female genital form and function in a calopterygid damselfly. J Evol Biol 18:132–137

Dougherty LR, Rahman IA, Burdfield-Steel ER, Greenway EV, Shuker DM (2015) Experimental reduction of intromittent organ length reduces male reproductive success in a bug. Proc R Soc Lond B 282:20150724

Eberhard WG (1985) Sexual selection and animal genitalia. Harvard University Press, Cambridge, MA

Elliott GF, Huxley AF, Weis-Fogh T (1965) On the structure of resilin. J Mol Biol 13:791–795

Filippov AE, Kovalev AE, Matsumura Y, Gorb SN (2015) Male penile propulsion into spiraled spermathecal ducts of female chrysomelid beetles: a numerical simulation approach. J Theor Biol 384:140–146
Filippov AE, Kovalev AE, Matsumura Y, Gorb SN (2016) Stiffness gradient of the beetle penis facilitates propulsion in the spiraled female spermathecal duct. Sci Rep 6:27608
Filippov AE, Popov VL (2007) Fractal Tomlinson model for mesoscopic friction: from microscopic velocity-dependent damping to macroscopic Coulomb friction. Phys Rev E 75:027103
Geike T, Popov VL (2007) Mapping of three-dimensional contact problems into one dimension. Phys Rev E 76:036710
Gorb SN (2004) The jumping mechanism of cicada *Cercopis vulnerata* (Auchenorrhyncha, Cercopidae): skeleton-muscle organisation, frictional surfaces, and inverse-kinematic model of leg movements. Arthr Str Dev 33:201–220
Gorb SN, Beutel RG (2001) Evolution of locomotory attachment pads of hexapods. Naturwissenschaften 88:530–534
Gorb SN, Filippov AE (2014) Fibrillar adhesion with no clusterisation: functional significance of material gradient along adhesive setae of insects. Beilstein J Nanotech 5:837–846
Gorb SN, Beutel RG, Gorb EV, Yuekan J, Kastner V, Niederegger S, Popov VL, Scherge M (2002) Structural design and biomechanics of friction-based releasable attachment devices in insects. Integr Comp Biol 42:1127–1139
Gosline J, Lillie M, Carrington E, Guerette P, Ortlepp C, Savage K (2002) Elastic proteins: biological roles and mechanical properties. Phil Trans Roy Soc B 357:121–132
Haas F, Gorb SN, Wootton RJ (2000a) Elastic joints in dermapteran hind wings: materials and wing folding. Arthr Str Dev 29:137–146
Haas F, Gorb SN, Blickhan R (2000b) The function of resilin in beetle wings. Proc R Soc Lond B 267:1375–1381
Holwell GI, Winnick C, Tregenza T, Herberstein ME (2009) Genital shape correlates with sperm transfer success in the praying mantis *Ciulfina klassi* (Insecta: Mantodea). Behav Ecol Sociobiol 64:617–625
House CM, Simmons LW (2003) Genital morphology and fertilization success in the dung beetle *Onthophagus taurus*: an example of sexually selected male genitalia. Proc R Soc Lond B 270:447–455
Hosken DJ, Stockley P (2004) Sexual selection and genital evolution. Trends Ecol Evol 19:87–93
Jensen M, Weis-Fogh T (1962) Biology and physics of locust flight. V. Strength and elasticity of locust cuticle. Philos Trans R Soc B 254:137–169
Kamimura Y (2013) Promiscuity and elongated sperm storage organs work cooperatively as a cryptic female choice mechanism in an earwig. Anim Behav 85:377–383
Kothari VK, Rajkhowa R, Gupta VB (2001) Stress relaxation and inverse stress relaxation in silk fibers. J Appl Polym Sci 82:1147–1154
van Lieshout E, Elgar MA (2010) Longer exaggerated male genitalia confer defensive sperm-competitive benefits in an earwig. Evol Ecol 25:351–362
Kovalev AE. Filippov AE, Gorb SN (2018) Slow viscoelastic response of resilin. J Comp Physiol A 204:409–417
Lombardi EC, Kaplan DL (1993) Preliminary characterization of resilin isolated from the cockroach, *Periplaneta americana*. Mater Res Soc Symp Proc 292:3–7
Lyons RE, Nairn KM, Huson MG, Kim M, Dumsday G, Elvin CM (2009) Comparisons of recombinant resilin-like proteins: repetitive domains are sufficient to confer resilin-like properties. Biomacromolecules 10:3009–3014
Macagno ALM, Pizzo A, Parzer HF, Palestrini C, Rolando A, Moczek AP (2011) Shape – but not size – codivergence between male and female copulatory structures in *Onthophagus* beetles. PLoS One 6:e28893
Matsumura Y, Machida R, Mashimo Y, Dallai R, Gottardo M, Kleinteich T, Michels J, Gorb SN, Beutel RG (2014) Two intromittent organs in *Zorotypus caudelli* (Insecta, Zoraptera): the

paradoxical coexistence of an extremely long tube and a large spermatophore. Biol J Linn Soc 112:40–54

Matushkina N, Gorb SN (2007) Mechanical properties of the endophytic ovipositor in damselflies (Zygoptera, Odonata) and their oviposition substrates. Zoology 110:167–175

Michels J, Appel E, Gorb SN (2016) Functional diversity of resilin in Arthropoda. Beilstein J Nanotech 7:1241–1259

Michels J, Gorb SN (2012) Detailed three-dimensional visualization of resilin in the exoskeleton of arthropods using confocal laser scanning microscopy. J Microsc 245:1–16

Michels J, Gorb SN, Reinhardt K (2015) Reduction of female copulatory damage by resilin represents evidence for tolerance in sexual conflict. J R Soc Interface 12:20141107

Miller GT, Pitnck S (2002) Sperm-female coevolution in *Drosophila*. Science 298:1230–1233

Misak HE, Sabelkin V, Miller L, Asmatulu R, Mall S (2013) Creep and inverse stress relaxation behaviors of carbon nanotube yarns. J Nanosci Nanotech 13:8331–8339

Peisker H, Michels J, Gorb SN (2013) Evidence for a material gradient in the adhesive tarsal setae of the ladybird beetle *Coccinella septempunctata*. Nat Commun 4:1661

Pohrt R, Popov VL, Filippov AE (2012) Normal contact stiffness of elastic solids with fractal rough surfaces for one- and three-dimensional systems. Phys Rev E 86:026710

Popov VL (2010) Kontaktmechanik und Reibung: von der Nanotribologie bis zur Erdbebendynamik. Springer, Berlin

Popov VL, Filippov AE (2012) Adhesive properties of contacts between elastic bodies with randomly rough self-affine surfaces: a simulation with the method of reduction of dimensionality. Phys Mesomech 15:324–329

Rodriguez V (1994) Function of the spermathecal muscle in *Chelymorpha alternans* Boheman (Coleoptera: Chrysomelidae: Cassidinae). Physiol Entomol 19:198–202

Rodriguez V (1995) Relation of flagellum length to reproductive success in male *Chelymorpha alternans* Boheman (Coleoptera: Chrysomelidae: Cassidinae). Coleopts Bull 49:201–205

Rodriguez V, Windsor DM, Eberhard WG (2004) Tortoise beetle genitalia and demonstrations of a sexually selected advantage for flagellum length in *Chelymorpha alternans* (Chrysomelidae, Cassidini, Stolaini). In: Jolivet P, Santiago-Blay JA, Schmitt M (eds) New developments in the biology of Chrysomelidae. Academic, The Hague, pp 739–748

Rowe L, Arnqvist G (2012) Sexual selection and the evolution of genital shape and complexity in water striders. Evolution 66:40–54

Sakurai G, Himuro C, Kasuya E (2012) Intra-specific variation in the morphology and the benefit of large genital sclerites of males in the adzuki bean beetle (*Callosobruchus chinensis*). J Evol Biol 25:1291–1297

Santangelo PG, Roland CM (1992) Chain ends and the Mullins effect in rubber. Rubber Chem Technol 65:965–972

Seber GAF, Wild CJ (1989) Nonlinear regression. Wiley, Hoboken

Sentenská L, Pekár S, Lipke E, Michalik P, Uhl G (2015) Female control of mate plugging in a female-cannibalistic spider (*Micaria sociabilis*). BMC Evol Biol 15:18

Simmons LW (2014) Sexual selection and genital evolution. Aust Entomol 53:1–17

Simmons LW, House CM, Hunt J, Garcia-Gonzalez F (2009) Evolutionary response to sexual selection in male genital morphology. Curr Biol 19:1442–1446

Tadler A (1999) Selection of a conspicuous male genitalic trait in the seedbug *Lygaeus simulans*. Proc R Soc Lond B 266:1773–1777

Tanabe T, Sota T (2013) Both male and female novel traits promote the correlated evolution of genitalia between the sexes in an arthropod. Evolution 68:441–452

Villavaso EJ (1975) Functions of the spermathecal muscle of the boll weevil, *Anthonomus grandis*. J Insect Physiol 21:1275–1278

Weis-Fogh T (1960) A rubber-like protein in insect cuticle. J Exp Biol 37:889–907

Weis-Fogh T (1961) Thermodynamic properties of resilin, a rubber-like protein. J Mol Biol 3:520–531

Weng L, Chen X, Chen W (2007) Rheological characterization of in situ crosslinkable hydrogels formulated from oxidized dextran and N-carboxyethyl chitosan. Biomacromolecules 8:1109–11115

Willkommen J, Michels J, Gorb SN (2015) Functional morphology of the male caudal appendages of the damselfly *Ischnura elegans* (Zygoptera: Coenagrionidae). Arthr Struct Dev 44:289–300

Chapter 8
Nanoscale Pattern Formation in Biological Surfaces

Abstract In the present chapter, three problems of nanoscale pattern formation will be discussed. (1) The particular symmetry violation in the dimple-like nano-pattern on the belly scales of the pythonid snake *Morelia viridis* is analyzed using correlation analysis of the distances between individual nanostructures. (2) Pattern formation in the multi-component colloidal secretion of whip-spiders (cerotegument) is numerically simulated and discussed. (3) Pattern formation of the springtail cuticle nanostructures.

Dimple-like nano-pattern on the belly scales of the snake skin is supposed to reduce both friction and abrasion. On the real snake skin surface, the pattern analysis revealed non-random, but very specific symmetry violation. The results of the analysis, performed on the snake were compared with nano-nipple pattern on the eye of the sphingid moth being well known reference of highly-ordered biological nanopatterns. In the case of the moth eye, the nano-nipple arrangement forms a set of domains, while, in the case of the snake skin, the nano-dimples arrangement resembles an ordering of molecules in amorphous state, which might provide friction isotropy to the skin. A simple model of such pattern formation is suggested, which almost perfectly reproduces the experimental results. Some other biological surfaces gain their super-hydrophobic properties by nano-structures on the surface. Some arachnids, such as the cryptic, large whip-spiders and some mites, exhibit a crust of dried secretion containing globular micro-structures covered with regularly arranged nano-particles built from a multi-phasic secretion. In order to gain a better understanding of the process of self-assembly of nanostructures on spherical microstructures, in the present chapter, we studied it from a theoretical point of view. It is demonstrated that slight changes of simple parameters lead to a variety of morphologies highly similar to the ones observed in the species specific cerotegument

---

Electronic supplementary material The online version of this chapter (https://doi.org/10.1007/978-3-030-41528-0_8) contains supplementary material, which is available to authorized users.

A. E. Filippov, S. N. Gorb, *Combined Discrete and Continual Approaches in Biological Modeling*, Biologically-Inspired Systems 16,
https://doi.org/10.1007/978-3-030-41528-0_8

structures. Also springtails have a complex hierarchically structured cuticle surface with even stronger repelling properties against water, low-surface-tension liquids, and sticky secrets of predatory insects. Non-wetting property of the collembolan cuticle is mainly based on the cuticle topography rather than on the surface chemistry. In material science, analogous surface coatings have been produced by colloidal lithography utilizing the self-assembly of nanoparticles on a substrate. We introduce here a numerical model to study the effect of different interactions between the substances on the morphology of the desired structure. In general the study of biological self-organising nanopatterns and their evolution seems to be a promising approach for generating new solutions for future industrial applications.

In contrast to the majority of inorganic or artificial materials, the surface structures of biological systems are usually not ordered or homogeneous. Local symmetry of the structures on biological surfaces is also often broken. In the present chapter, three problems along this line will be discussed:

1. First, the particular symmetry violation in the dimple-like nano-pattern on the belly scales of the pythonid snake *Morelia viridis* is analyzed using correlation analysis of the distances between individual nanostructures (Kovalev et al. 2016)
2. Second, pattern formation in the multi-component colloidal secretion of whip-spiders (cerotegument) is numerically simulated and discussed (Filippov et al. 2017)
3. Third, pattern formation of the springtail cuticle nanostructures (Filippov et al. 2018).

The examination of these biological examples of symmetry breaking in surface nanostructures provides strong evidence that the degree and type of disordering are determined by the presence of a specific molecular mechanism. We also hypothesize that the controlled disorder, at least in snake skin, is responsible for some additional functional features of the system.

8.1 Snake Skin Surface Nanostructures

The surface nanostructure of snake skin is supposed to reduce both abrasion and friction (Berthé et al. 2009; Klein et al. 2010; Klein and Gorb 2012, 2014; Baum et al. 2014) as has been previously shown for similar artificial nanostructures (Varenberg et al. 2002). On the ventral surface of snakes, this nanostructure typically consists of arrays of nano-dimples (Schmidt and Gorb 2012) (Fig. 8.1).

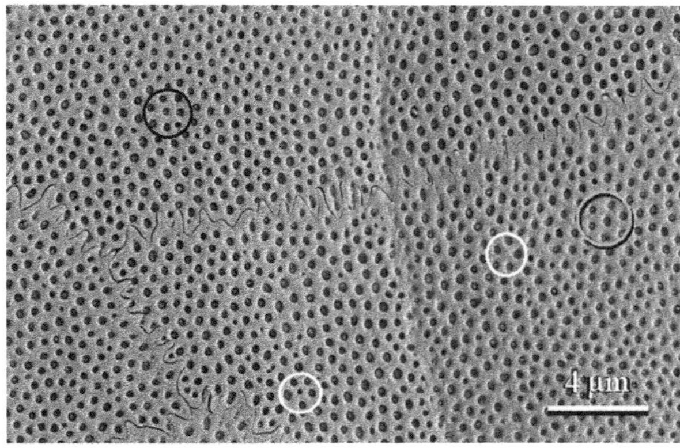

Fig. 8.1 Scanning electron microscopy image of a ventral scale from the tail region close to its cranial edge in the snake *Morelia viridis*. The black-shadowed gray circle indicates a typical hexagonal arrangement of dimples, while the white and black circles indicate five- and seven-fold symmetrical arrangements, respectively. (From Kovalev et al. 2016)

The arrangement of these nano-dimples in snake scales appears visually similar to the well known anti-reflective and anti-adhesive arrangement of nanostructures covering insect corneal surfaces (Stavenga et al. 2006; Peisker and Gorb 2010). Blagodatski et al. (2015) recently published a versatile classification of the insect corneal nanostructure arrangements based on the Turing pattern formation model, but without any further quantitative analysis of the patterns.

One of the most widespread arrangement types is the hexagonal arrangement (Fig. 8.1), which provides the highest structure density. However, it is well known that there is no perfectly regular arrangement in biological systems (Li and Bowerman 2010). Here, we will focus on the analysis of the arrangement imperfections using correlation analysis in a way similar to that previously used for the internal keratinous photonic nanostructures of bird feathers (Prum and Torres 2003).

The theoretical study below has the following structure. First, we will demonstrate on the well-known example of a lepidopteran eye, that of the Carolina sphinx moth (*Manduca sexta*, Insecta, Sphindidae) (Bernhard and Miller 1962; Miskimen and Rodriguez 1981; Stalleicken et al. 2006; Lee and Erb 2015), how the correlations in nanostructure arrangement can be analyzed numerically. Then we will apply this analysis to the nanostructure of snake skin. Further, a simple model of the mechanism of nanostructure formation on snake scales will be proposed. According to this model, the structure – which shows an arrangement similar to that in snake skin – appears due to the "freezing" of overdamped relaxation. It is caused by weak repulsion between the initially randomly distributed structures. Next, an analogous correlation analysis will be applied to this model and its results will be compared with the results of the correlation analysis of the real nanostructure arrangement in snake skin. Finally, two particular types of symmetry violation in biological surfaces will be characterized and the possible underlying developmental mechanisms of surface nanostructures will be discussed.

8.1.1 Correlation Analysis of the Nanostructures of Moth Eye and Snake Skin

Dry ventral tail scales from shed skin of the green tree python (*Morelia viridis*, Squamata, Pythonidae) and ommatidia of the complex eyes of the Carolina sphinx moth (*Manduca sexta*, Insecta, Sphindidae) were examined by scanning electron microscopy (SEM).

Grayscale SEM images of the nano-nipples were converted first to black and white ones by setting an appropriate grayscale threshold. This allowed us to apply a standard particle analysis method from ImageJ software. Using this method, a discrete array of coordinates of the centers of the nano-nipples (Fig. 8.2a) was produced to calculate a power spectrum of the coordinates (Fig. 8.2b) and to build a distribution of the distances between nearest neighbors (Fig. 8.2c).

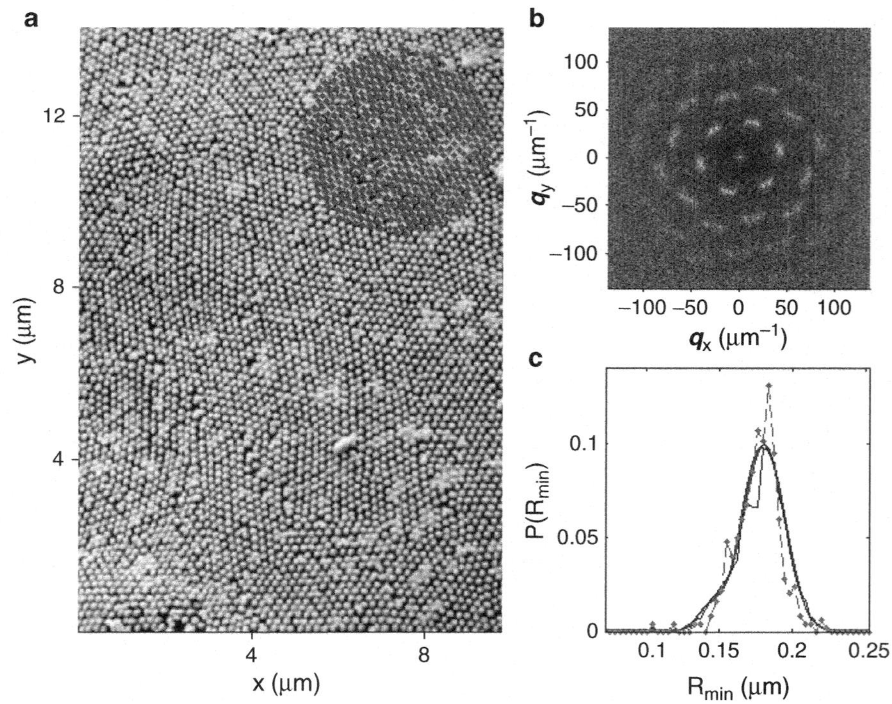

Fig. 8.2 Scanning electron microscopy image of a single ommatidium surface of an eye in the moth *Manduca sexta* (**a**). Fourier transformation (**b**) of the circle area highlighted in dark gray in (**a**). Histogram of the inter-nipple distances (**c**). Partial histogram built from the darkened area in (**a**) is shown as a dashed line. Histogram for the whole image is shown as a thin solid line. The fit of the whole image histogram with two Gaussian curves is shown as a thick solid line. (From Kovalev et al. 2016)

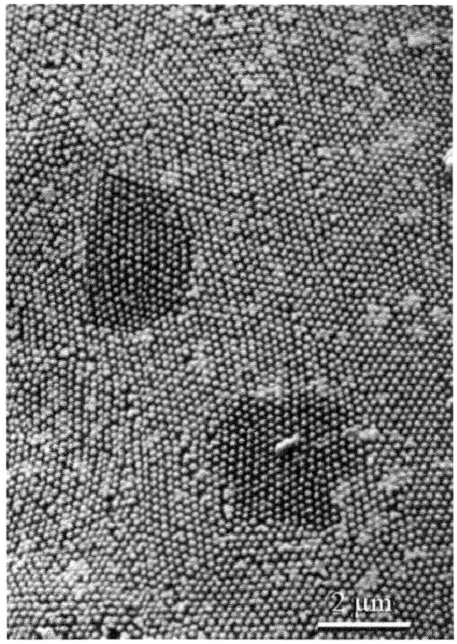

Fig. 8.3 Two domains with hexagonally ordered nipples indicated by enhanced contrast. (From Kovalev et al. 2016)

Eye Surface of the Moth *Manduca sexta* The hexagonal arrangement of ommatidial nano-nipples is obvious in both the SEM image (Fig. 8.2a) and an image in a frequency domain (2D Fourier transformation) (Fig. 8.2b).

The power spectrum indicates that the structure is highly ordered locally, but spatial correlation decreases with increasing distance between the structures. While the correlation between nano-nipples separated by three characteristic inter-nipple distances is weak, but still existent, the positions of structures separated by four characteristic inter-nipple distances show practically no correlation at all (Fig. 8.2b). The exact shape of the power spectrum essentially depends on the particular location and also varies for different individual specimens.

From the images in Figs. 8.2 and 8.3, it is directly evident that the nano-nipple pattern is actually a set of highly ordered domains, which is in good agreement with the findings of Sergeev et al. (2015). Besides, it can be seen from Fig. 8.3 that the relative orientation of neighboring domains can be essentially different. To elucidate this, we defined a circle and calculated the spectrum for the structures within the enclosed area. Then, we systematically moved the circle along the surface and calculated the spectra. This procedure is illustrated in Supplementary movie 8.1. In Fig. 8.2a, the area for which the spectrum is calculated is indicated as a darkened circle. The size of the circle was chosen to be comparable with the dimension of a typical domain.

Effectively, this procedure mimics an experimental observation where an observational window (in the shape of a circle) moves along the surface system and statistics obtained from differently oriented domains is accumulated. As a result, pronounced highly ordered and disordered spectra alternate while the circle moves over different locations. There is a simple reason why the power spectra seem disordered at some locations. Whenever the circle for which the spectrum is calculated covers two or more domains with different orientations, it combines differently oriented impacts which contribute different results to the overall spectrum. Therefore, angular histograms accumulated over areas bigger than a domain demonstrate no order, while the ordering in previously published images (Sergeev et al. 2015) is quite obvious.

In the vicinity of domain boundaries, some angles between structures and some distances between neighbors are distorted. In particular, some structures here have five- or seven-fold symmetries instead of the six-fold symmetry typical of perfectly ordered hexagonal domains. This spatial configuration leads to different distances between the neighbors in different positions of the system. To further characterize such an arrangement, we accumulated the distribution of the distances to the nearest neighbors in addition to the power spectrum calculation. The resulting histogram is shown in Fig. 8.2c.

Due to the relatively poor statistics, the histograms calculated over small local domains showed rather strong fluctuations. To illustrate this effect, one such histogram is shown as a dashed curve in Fig. 8.2c. Total distance distribution is independent of the structure orientation. One can easily accumulate such a histogram for the whole system as shown by the solid line in Fig. 8.2c.

The number of nanostructures in Fig. 8.2a exceeds one thousand. This allows us to analyze the shape of the distance histograms, which was not possible in previous studies because of the insufficient statistics. The obtained distance histogram can be perfectly fitted by two Gaussian curves $0.012 \exp\{-[(R - 0.121)/0.011]^2\} + 0.1 \exp\{-[(R - 0.165)/0.018]^2\}$. The smaller peak has an amplitude of ~6% smaller than the main one and distances that are 2 μm shorter. It corresponds to the population of smaller nano-nipples which can be found in distorted arrangements close to the boundaries between domains.

Mathematically, it is not possible to cover the curved ommatidia surface with a perfect hexagonal arrangement of nano-nipples; at least some elements with a five-fold symmetry are additionally required. Alternatively, the curved surface may be covered by a set of hexagonally arranged and locally flat domains, as observed in *M. sexta*.

A homogeneous model based on Turing pattern formation is unable to explain the formation of this poly-domain surface structure. To produce such a surface, the synthesis of nano-nipples in one domain needs to be sufficiently fast, e.g. vesicles should be released and appear at different moments in different domains. Thus, at least two hierarchy levels in time and in the corresponding molecular control are required in order to assemble this kind of structure in such a manner. There needs to be simultaneous growth of local domains which start from different positions and meet one another at the boundaries, as previously reported for non-biological systems, e.g. the structure evolution during the growth of polycrystalline films (Thompson 2000).

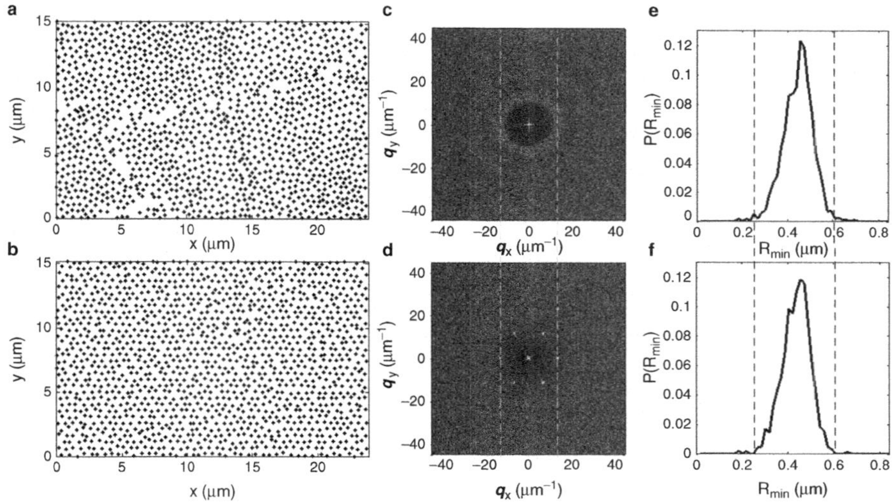

Fig. 8.4 (**a**) Natural nano-dimple pattern of snake skin and (**b**) a mathematically generated hexagonal arrangement where each element was slightly shifted in a random way. The amplitude of the random shift was chosen in such a way as to achieve a distribution of distances between nearest neighbors in a randomized hexagonal arrangement (**f**) which corresponds to that of the real system (**e**). (**c** and **d**) Fourier transformations for both systems. (From Kovalev et al. 2016)

Ventral Scale Surface of the Snake *Morelia viridis* The regular arrangement of nano-dimples on the scales of *M. viridis* looks a lot like the corneal nanostructure arrangement in *M. sexta* (Fig. 8.1). Neighboring elements are generally hexagonally ordered. Nanostructures arranged with five- and seven-fold symmetry (Fig. 8.1) are also present (like those found in *M. sexta*) and spread over the whole surface. Our study did not reveal any obvious domains with different orientations in the nanostructure arrangement.

The power spectrum and the distribution of distances between the nearest neighbors were calculated in the same manner as described above. In contrast to the arrangement observed in the nano-nipples of the moth eye (Fig. 8.2), where the pattern in Fig. 8.1 appears hexagonally arranged, the power spectrum here does not demonstrate any pronounced peaks (Fig. 8.4c). This is due to the absence of strongly oriented domains with a prominent arrangement. At the same time, there is a characteristic distance of around 0.4 μm between the nearest neighbors. The dark circle in the middle of Fig. 8.4c corresponds to the low probability of smaller distances. The first (gray) ring, which corresponds to the maximum probability of the distances, is well pronounced. A weak, second probability minimum and maximum (dark and light gray rings, respectively) follow the first ring and correspond to the much lower long-range order in the arrangement that was observed for the moth eye, as described above.

It can be assumed that the isotropy of the correlation function, which does not show specific directions in the dimple arrangement of the snake skin, is important for, e.g., maintaining isotropic tribological properties. The presence of a preferential direction would imply stress enhancement, and thus higher probability of material failure as well as stronger abrasion, in some direction. It is interesting to note that, despite the differences in the power spectra between these two biological nanostructures, the integral distribution of the distances to the nearest neighbor looks rather similar (Figs. 8.2c and 8.4e).

Comparison of the Two Nanostructures The similarity of the distance distributions and the domains of corneal and skin scale nanostructures raise an interesting general question: the two different surfaces are covered with regularly arranged nanostructures which seem to have practically the same "almost hexagonal" order and very similar distributions of distances between neighbors, but demonstrate different statistical properties, presumably corresponding to a different physical anisotropy of the structures. Understanding the differences between these structures is important from at least two points of view:

1. The different symmetries may be a result of the adaptation to different biological functions.
2. Knowledge about differences in biological functions may be important for the future numerical modeling of such systems.

8.1.2 Correlation Analysis of Numerically Generated Structure Arrangements

The standard modeling approach to weakly disordered hexagonal systems uses hexagonal arrangement which is slightly perturbed by random noise. In this approach, the particular amplitude of Gaussian random shifts may be appropriately adjusted to render the numerically found distribution of the distances close to those experimentally observed in real snake skin. This leads to practically identical distributions of distances (cf. Fig. 8.4e, f). Despite this, the real arrangement of the snake nano-dimples obviously differs from the simulated one with regard to the symmetry of its power spectrum (cf. Fig. 8.4c, d).

This means that the spatial distribution of the nano-dimples cannot be simulated by such a naive, slightly randomized hexagonal configuration. Rather, the real arrangement appears to be the result of some self-organization process caused by an interaction between the nanostructures. To prove this hypothesis we can perform the following simulation.

Let us generate, as an initial condition, an array of the randomly distributed dimples and apply short-range repulsion between them with the Gaussian potential $U\left(\vec{r}_{ij}\right) = U_0 \exp\left(-\left|\vec{r}_{ij}/r_0\right|^2\right)$, where $\vec{r}_{ij}$ is a vector connecting i and j elements, and the parameters r_0 and U_0 determine the characteristic distance and amplitude of the interaction, respectively. The system of the dimples evolves visco-elastically and moves in a strongly overdamped way. The evolution from their nanostructure distribution to the final configuration can be described by the equation of motion for their centers: $\partial \vec{r}_i / \partial t = -\gamma \sum_j \partial U / \partial \vec{r}_{ij}$, where γ is a damping constant which determines the characteristic time scale of the relaxation process.

It is important to note that freedom in the definition of particular values of the parameters γ, r_0, and U_0 does not introduce any uncertainty, at least not for the goals of the present study. Here we are actually interested only in the shape of the distribution of the distances and the symmetry of the correlation function. Besides, the total number of dimples in the numerical array is completely predefined by the dimple number in a given observation area of the real system. As a result, in the course of a kinetic process, the mean distance between the dimples (or their mean density) certainly remains fixed by the restricted size of the box in Fig. 8.4, and the only parameter that is changing during the relaxation process is the spatial configuration (symmetry) and, in turn, the shape of the distribution of distances.

Only one fitting parameter remains in the simulation. This is the width of the final distribution of distances. In other words, we have to continue the relaxation process up to the moment when its width coincides with the experimental one. This automatically defines the moment when the simulation has to be terminated, freezing the final configuration independently of the particular value of the damping constant γ. This procedure is illustrated in Supplementary movie 8.2.

We have demonstrated numerically that the spatial configuration and the power spectra obtained from the "frozen dynamics" (liquid- or glass-like state) look very similar to those calculated directly from SEM images of the snake skin (compare Figs. 8.4 and 8.5). Moreover, the distributions of distances numerically and experimentally found (dashed and solid lines in Fig. 8.4, respectively) coincide remarkably well. Please note that we use only one fitting parameter (width), but obtain the whole final function (shape of the distribution).

These results indicate that the nano-dimple arrangement in snake skin presumably appears in a visco-elastic overdamped system during kinetic relaxation, when the (pre-)dimples are initially randomly distributed and gradually repulse and move each other. Besides, at the moment of dimple motion, the high damping means that the system is in a highly viscous state and the interaction between single nanostructures is rather week.

8.2 3D Pattern Formation of Colloid Spheres in the Water-Repellent Cerotegument of Whip-Spiders

The exoskeleton of certain arachnids exhibits complex coatings consisting of globular structures with complex surface features. This so-called cerotegument is formed by a multi-component colloidal secretion that self-assembles and hardens on the body surface, leading to high water repellency. Previous ultrastructural studies revealed the involvement of different glandular cells that contribute different components to the secretion mixture, but the overall process of self-assembly into the observed complex structures has so far remained highly unclear. In this section we

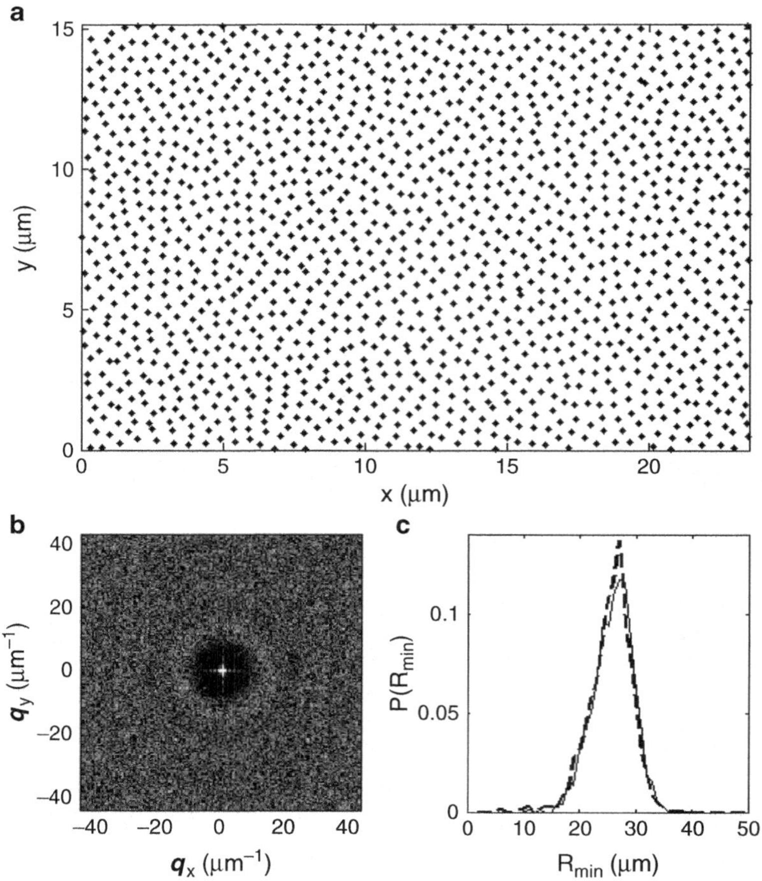

Fig. 8.5 "Frozen" arrangement obtained after relaxation with high damping (**a**) and its Fourier transformation (**b**). Histograms of the distances (**c**) calculated for the "frozen" system (dashed line) and for the natural snake skin (solid line) shown in Figs. 8.5a and 8.4a, respectively. (From Kovalev et al. 2016)

report on recent progress in this direction (Filippov et al. 2017) and will present a numerical model of the self-assembly process.

8.2.1 Water Repellence and Ultrastructure of Certain Granules in the Whip-Spider Cerotegument

For various materials, surface modification and functionalization through nanostructures is a versatile way of producing specific desirable properties. Biological systems have evolved such functional structures for a long time. Well-known examples of this kind are super-hydrophobic surfaces characterized by crystalline coatings or particles (Barthlott and Neinhuis 1997; Rakitov and Gorb 2013; Barthlott et al. 2016).

Some arachnids, such as the cryptic, large whip-spiders (Amblypygi) and some mites, are coated with a crust of dried secretion containing globular microstructures covered with regularly arranged nano-particles (Alberti et al. 1981; Raspotnig and Matischek 2010; Wolff et al. 2016, 2017). In whip-spiders, these structures are formed by a multi-phasic secretion that emerges from two distinct types of glandular apparatuses (Wolff et al. 2016). Within this relatively small arachnid group, a high diversity of structures has evolved that seems to be based on differences in the structure and properties of the basic colloidal particles (Wolff et al. 2016, 2017). Wolff et al. (2016) identified six clusters of related morphotypes, which correlate with the hypothesized inter-familiar relationships of whip-spiders (Weygoldt 2000).

In materials science, surface coatings with regularly arranged bi-hierarchical globular particles have been produced by colloidal lithography, a technique utilizing the self-assembly of nanoparticles on a substrate surface (Yang et al. 2006; Badge et al. 2013). However, the range of producible patterns and properties of these coatings is still restricted, as well as their durability. Therefore, the study of the versatile whip-spider cerotegument seems promising for generating new solutions for biomimetic novel surface coatings.

Water repellence and ultrastructure of granules are illustrated in Figs. 8.6, 8.7 and 8.8. All the figures are organized in the same manner and differ only by animal species. The figures for the three different groups of granules that are present in the system are separated for better magnification and clarity of the images. In each figure, the first row shows droplets of tap water directly ejected from a syringe onto the carapace, with a nearly spherical shape. The second row presents the fine structure of species-specific granules and the third row reproduces the ultrastructure of some colloid particles. Any correspondence between numerically found and natural structures is indicated by identical numbers in these figures, in the text below and in Fig. 8.9.

In order to gain a better understanding of the process of self-assembly and self-arrangement of nano-structures on spherical microstructures, we will now study this process from a theoretical point of view.

Our modeling experiment demonstrates that slight changes of simple parameters lead to a variety of morphologies that are highly similar to the ones observed in the species-specific cerotegument structures of whip-spiders. This can be explained by the effect of different interactions between the particles on the morphology of the final structure. This issue is related to the so-called Tammes problem, which refers to the optimal packing of a given number of pores or particles on a sphere so that the minimum distance between them is maximized (Tammes 1930; Tarnai and Gáspár 1987; Erber and Hockney 1991).

Our aim is to adjust the model parameters in such a way that structures similar to those observed in whip-spiders can be reproduced. In this way we want to test which alterations are necessary to transform the structures within a possible evolutionary scenario. However, these results are not only important for our understanding of the formation of globular hierarchical structures in nature, but also for the fabrication of novel surface coatings by colloidal lithography.

Fig. 8.6 Water repellence and ultrastructure of granules on the ceroteguments of *Charon* cf. *grayi* (left) and *Phrynichus ceylonicus* (right). First row: tap water droplets directly ejected onto the carapace form a nearly spherical shape. Second row: fine structure of species-specific granules. Numbered labels refer to the corresponding modeled structures shown in Fig. 8.9. Third row: ultrastructure of colloid particles. (Figure reproduced from Filippov et al. 2017)

8.2.2 Numerical Simulation of the Colloidal Self-assembly of Cerotegument Structures

The model of the system under consideration is organized as a combination of discrete and continuous approaches as introduced in Chap. 1. From a general (physical, chemical and biological) point of view one can expect that the observed structures

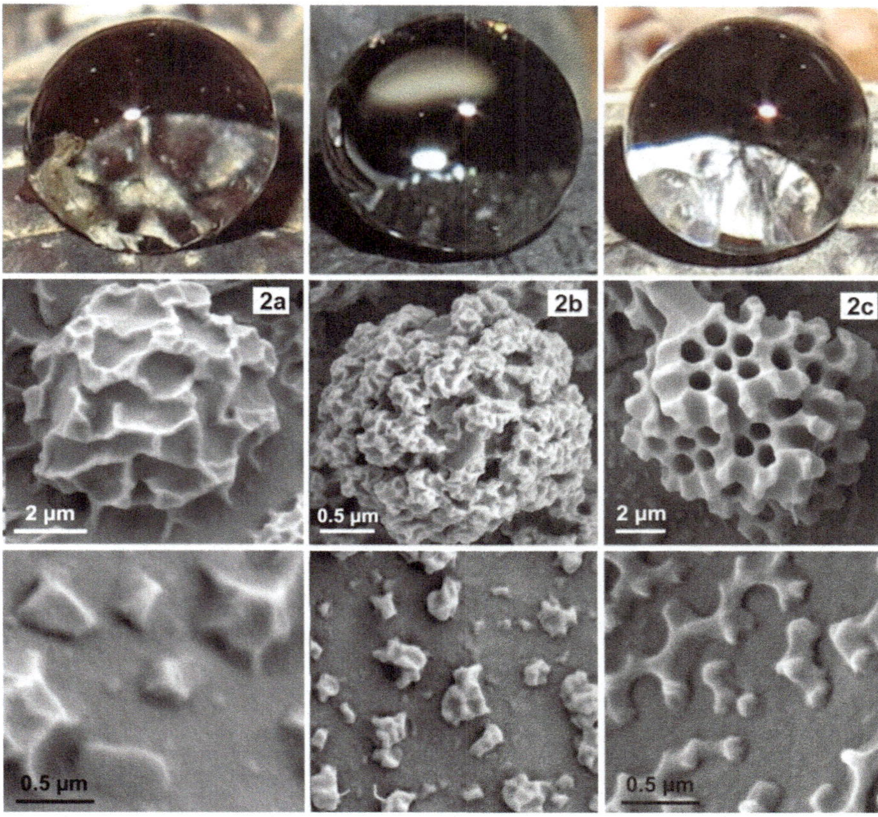

Fig. 8.7 Water repellence and ultrastructure of granules on the ceroteguments of *Phrynus longipes* (left), *Paraphrynus carolynae* (middle) and *Phrynus decoratus* (right). First row: tap water droplets directly ejected onto the carapace form a nearly spherical shape. Second row: fine structure of the species-specific granules. Numbered labels refer to the corresponding modeled structures shown in Fig. 8.9. Third row: ultrastructure of colloid particles. (Figure reproduced from Filippov et al. 2017)

appear in some kinetic process during which an initially more or less uniformly distributed substance redistributes and solidifies in a three-dimensional space.

Such a process can be observed in many different chemical and physical systems (and as underlying processes in biological systems). For example, it can occur during superconducting (Geim et al. 1997) or magnetic ordering (Filippov 1997) in quasi 2D systems, in the growth of surface structures (Filippov 1998; Kovalev et al. 2016), and so on. In all these cases, the process involves an interaction between many spatially distributed densities and its direct simulation can cause extremely time-consuming calculations.

In the particular case of the cerotegument, the problem is complicated by the specific topology of the surface on which the process unfolds, because the redistribution and solidification of the densities take place on a spherical object, so we are

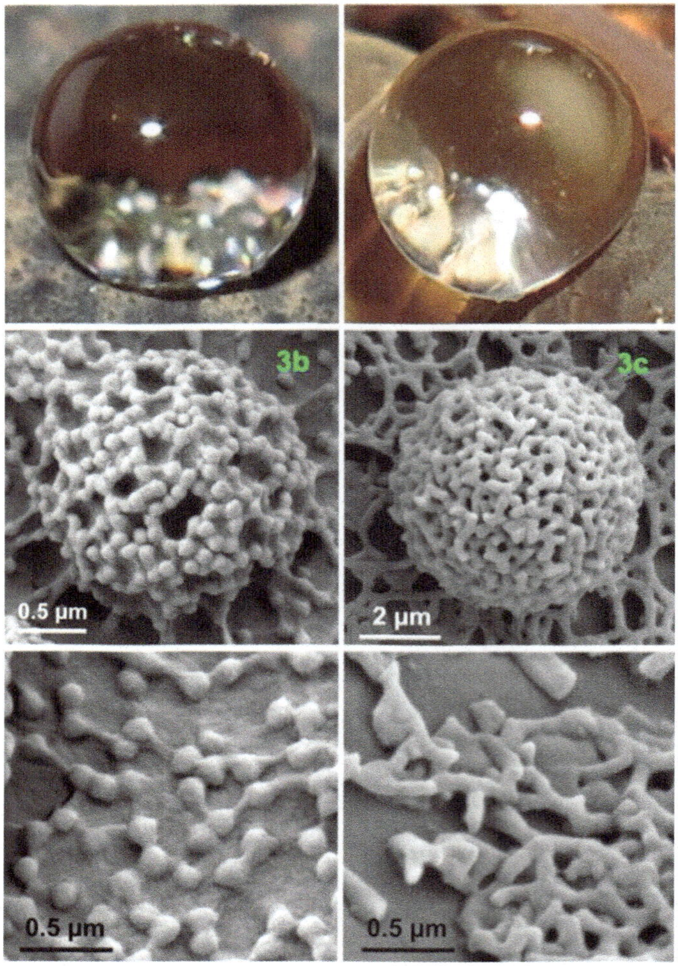

Fig. 8.8 Water repellence and ultrastructure of granules on the cerotegument of *Damon annulatipes* (left) and *Charinus acosta* (right). First row: tap water droplets directly ejected onto the carapace form a nearly spherical shape. Second row: fine structure of species-specific granules. Numbered labels refer to the corresponding modeled structures shown in Fig. 8.12. Third row: ultrastructure of colloid particles. (Figure reproduced from Filippov et al. 2017)

dealing with a phase transition (or phase separation) inside a spherical layer. As far as we know, such a topography in connection with this process has not previously been studied in conventional physics and chemistry. However, such a geometry is widespread in biological systems and therefore of particular interest.

The topological complexity of the problem forces us to reduce the numerical model as far as possible. Fortunately, there are good historical examples of such a simplification process for systems with complex topology. The main simplification

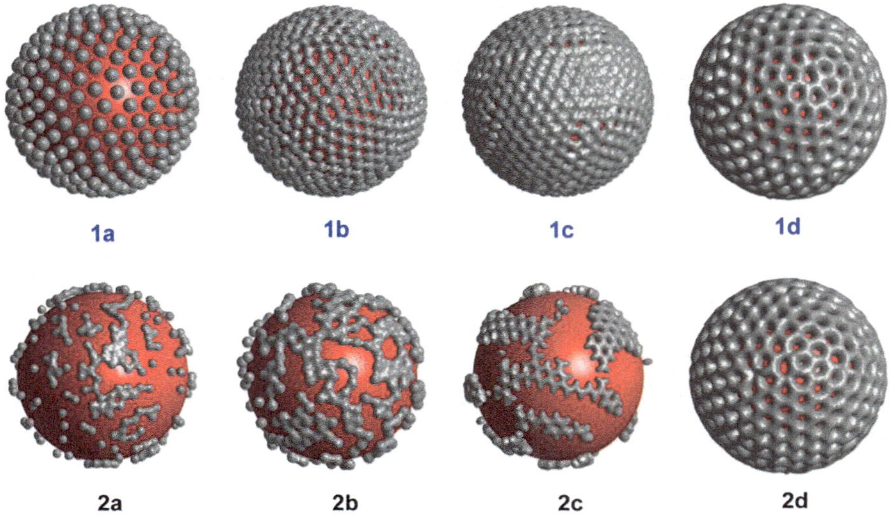

Fig. 8.9 Different structures obtained via different combinations of the interaction parameters described in the text. Structures with blue labels (upper row) were generated by repulsion only, structures with black labels (lower row) by combined interactions of attraction and repulsion. The red color visually represents the impact caused by the sphere, $\rho_S\left(\vec{r}\right)$. Rotating spatial distributions of the structures in 1a, 1c, 2a, 2b and 2d are shown in the Supplementary movies 8.5, 8.6, 8.7, 8.8, and 8.9, respectively. (Figure reproduced from Filippov et al. 2017)

is to combine the originally continuous problem with some preliminary discrete approach (which is used to achieve a more or less final or, at least, almost stationary, configuration).

A famous example of such an approach is the simulation of the topological phase transition in quasi 2D superconducting systems mentioned above (Kosterlitz and Thouless 1973), where the extremely complex evolution of the superconducting vortices was substituted to some extent by the motion of "charged particles". This approach proved extremely helpful and was even awarded the Nobel Prize in 2016. It is also applicable for the Tammes problem (Tammes 1930; Tarnai and Gáspár 1987; Erber and Hockney 1991), where the kinetic ordering on a sphere is modeled by finding an equilibrium ordering of "N equal charges" on the sphere. In all these cases, it is crucial that the approach leads to a discrete distribution of the maxima or minima of the densities; then, these distributions may be "dressed" with the continuous field with an expected density distribution.

It is important to note that in its classical form the Tammes problem is treated as an optimization problem: place a given number of n points (with n representing a natural number) on the surface of a sphere in a way that maximizes the shortest distance between any two points. This question normally dictates maximal simplification of the problem (e.g., by limiting the number of points to special cases which allow an analytical solution, or by limiting their interaction to the simple Coulomb one).

However, our goal here is almost the opposite: we want to examine which particular structures will appear in the course of different interactions and in which way they reproduce the observed structures. These simulated structures certainly are not optimal, and in many cases, the distribution of the final densities generated by frozen kinetics is extremely inhomogeneous. Depending on the relationships between the various interactions, it rather resembles the study of structural phase transitions between different phases in solid state physics. However, our simulation differs from the standard physical studies due to the newly introduced, atypical confinement of the particles on a spherical surface.

According to these considerations, we organize the procedure as follows. At the first stage, we randomly distribute some number of particles (henceforth called nucleation centers for the densities) on the surface of the sphere. At this stage, the most important parameters are the relationships between the sphere radius R_S, the number of particles N, and the characteristic distance of the repulsion interaction R_0 between them.

The total interaction can be represented by either pure short-range repulsion only:

$$U_1\left(\left|\vec{r}_j - \vec{r}_k\right|\right) = U_R \exp\left[-\left(\frac{\vec{r}_j - \vec{r}_k}{R_0}\right)^2\right]$$

(where $U_R > 0$ is the interaction intensity and $\vec{r}_j$ are the various positions, with $j = 1, 2, .., N$), or as a combination of both repulsion and attraction:

$$U_2\left(\left|\vec{r}_j - \vec{r}_k\right|\right) = U_A \exp\left[-\left(\frac{\vec{r}_j - \vec{r}_k}{R_A}\right)^2\right].$$

The second variant leads to a typical effective potential with a minimum, $U_{Interaction}\left(\left|\vec{r}_j - \vec{r}_k\right|\right) = U_1\left(\left|\vec{r}_j - \vec{r}_k\right|\right) + U_2\left(\left|\vec{r}_j - \vec{r}_k\right|\right)$, which normally determines an equilibrium distance between the particles in long time asymptotics.

In our particular problem, the movement of the particles is confined to the sphere. From a mathematical point of view, this means that we apply a strong potential which attracts them to the surface of the sphere with the given radius R_S:

$$U_{Sphere}\left(\left|\vec{r}_j - \vec{R}_S\right|\right) = U_S \exp\left[-\left(\frac{\vec{r}_j - \vec{R}_S}{\Delta_S}\right)^2\right].$$

If this potential is strong enough, the particles practically cannot leave the spherical surface despite the (relatively weak) repulsion $U_S \gg U_R$ between them. This statement can easily be verified visually and quantitatively. Here, we keep the relative deviation of the radius from the nominal radius smaller than 0.001: $|R_j - R_S|/R_S \leq 0.001$.

Taking into account that the original process is characterized by over-damped kinetics, the dynamic equations here should be over-damped as well:

$$\partial \vec{r}_j / \partial t = -\gamma \sum_k \partial \left[U_{Interaction} \left| \vec{r}_j - \vec{r}_k \right| + U_{Sphere} \left| \vec{r}_j - \vec{r}_k \right| \right] / \partial \vec{r}_k,$$

where the dissipative constant γ defines the characteristic time scale of the simulation. During the process, the particles dynamically rearrange with a rate which gradually decreases with time. One can control this slowing-down by calculating the mean velocity of the particles and stopping the procedure when it becomes negligible. In fact, the originally randomly distributed particles cannot reach a real "ground state" on the spherical surface. But in the long run, they will achieve some quite realistic distribution which resembles one of the realizations produced as a result of frozen kinetics in some real system in nature. Development of the frozen state on a sphere for monatomic and dumbbell-shaped particles is illustrated in Supplementary movies 8.3 and 8.4, respectively.

The final stage of the simulation consists of "dressing" the spherical surface $\rho_S\left(\vec{r}\right) = 1/\left[1 + \exp\left(\left|\vec{r} - \vec{r}_S\right|/R_0 - 1\right)\right]$ and the nucleation centers (particles) with continuous density distributions: $\rho_j\left(\vec{r}\right) = 1/\left[1 + \exp\left(\left|\vec{r} - \vec{r}_j\right|/R_0 - 1\right)\right]$, with the characteristic radii corresponding to the radius of the spherical layer in the potential (large sphere) and the effective size of the particles matching the radius of their interaction (small spheres). Total density around the nucleation centers should be accumulated by a summation of all particles: $\rho_{Total}\left(\vec{r}\right) = \sum_j \rho_j\left(\vec{r}\right) + \rho_S\left(\vec{r}\right)$.

It has to be admitted that such a model, which reproduces the kinetics via the interaction of discrete particles covered by the density distribution, completely ignores the hydrodynamics during the solidification process of the cerotegument. However, it may be assumed that at the scale of less than one micrometer (which applies to all the structures here) the system's behavior is strongly overdamped and all the dynamic effects such as inertia, vortices, or turbulences are relatively small.

Though we cannot directly visualize the obtained continuous density distributions as a projection on the 2D surface of the plots, we can easily plot the surfaces corresponding to any constant density distribution. Normally, such a constant should be chosen close to the mean value of the density typically distributed from 0 to the flat maxima around nucleation centers. The obtained results are summarized in Fig. 8.9.

These density distributions can be directly compared with the photographs of the real cerotegument structures presented in Figs. 8.6, 8.7 and 8.8. The numbers in the photographs correspond to the numbers in the theoretical sequences presented in Fig. 8.9. The red color in Fig. 8.9 visually separates the impact caused by the sphere, $\rho_S\left(\vec{r}\right)$, from that caused by the nucleation centers, $\rho_j\left(\vec{r}\right)$.

Each pattern is determined by the relationship between the area of the sphere and the size of the particles (i.e. their density on the surface) on the one hand, and the interaction between the particles on the other hand. This interaction, in turn, can follow two principles: (1) purely (short-range) repulsion and (2) a combination of (long-range) attraction and (short-range) repulsion. This causes two well pronounced branches of different structures depending on the characteristic distance of the interaction, shown as blue and black curves in Fig. 8.10.

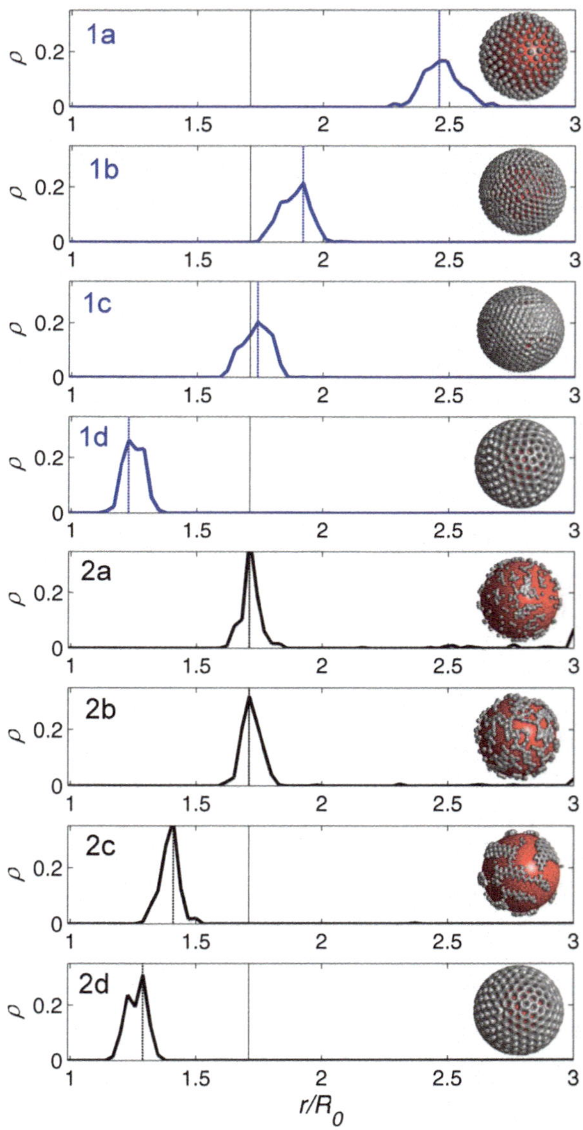

Fig. 8.10 Histograms of the distances between nearest neighbors calculated for the structures depicted on the right-hand side. The numbers in the panels refer to those in Fig. 8.9 and have the same meaning. (Figure reproduced from Filippov et al. 2017)

All the density distributions in Fig. 8.10 were calculated for $U_S = 100U_R$, with $U_R = 1$ serving as a unit of energy, and all the distances were normalized to R_0, with $R_0 = 1$ serving as a unit of the distances. The realizations numbered 1a–1d were obtained for pure repulsion $U_A = 0$ at the following values of the parameters:

(1a) $N = 300$, $R_S = R_0\sqrt{N/2}$;
(1b) $N = 600$, $R_S = 0.75R_0\sqrt{N/2}$;
(1c) $N = 600$, $R_S = 0.685R_0\sqrt{N/2}$;
(1d) $N = 300$, $R_S = 0.5R_0\sqrt{N/2}$.

The realizations numbered 2a–2d, were obtained when the repulsion core of interaction, $R_R = 0.75R_0$, was smaller than the distance of attraction interaction, $R_A = R_0$, the number of the particles (nucleation centers) was fixed ($N = 300$) and the radius of the sphere and the attraction constant were varied:

(2a) $R_S = R_0\sqrt{N/2}$, $U_A = -0.1U_R$;
(2b) $R_S = 1.25R_0\sqrt{N/2}$, $U_A = -0.1U_R$;
(2c) $R_S = 0.8R_0\sqrt{N/2}$, $U_A = -0.2U_R$;
(2d) $R_S = 0.5R_0\sqrt{N/2}$, $U_A = -0.2U_R$.

It can be intuitively predicted that in the case of pure repulsion the particles will tend to form uniform patterns with average spatial distributions and with a density depending on their total number, *N*, the radius of the large sphere, *R*, and the radius of each particle, r_p. In the simulation, the latter radius coincides with the radius of the density distribution surrounding every particle.

If the number of particles, *N*, increases at fixed values of both radii, $R_S = const.$ and $R_0 = const.$, the particles tend to be packed more densely and the system will ultimately reach the maximal packing of the spherical surface (1c in Fig. 8.9). If more particles are added, they will still be attracted to the sphere and must form some regular structure on the surface. But remember that the particles are only discrete representations of the maxima of their originally liquid and flexible density. For large numbers of particles, their density maximums will start to overlap. This causes a new kind of structure, which has visible holes between the particle centers (cf. 1d in Fig. 8.9).

Another behavior is observed if the total interaction combines repulsion and attraction. In this case, the interaction energy has a minimum and favors a particular equilibrium distance between the particles. If the number of particles, *N*, is not enough to cover the whole sphere with this equilibrium distance, the particles will "solve the contradiction" by forming some set of sufficiently dense clusters (see, e.g., 2a–2c in Fig. 8.9). The particular realization of the clusters is defined self-consistently by the reordering of a random initial configuration of the particles.

Inside every cluster (even small ones), the distance between the nearest particles remains practically the same for different total numbers of *N*, almost coinciding with the distance in large clusters. To prove this, one can calculate arrays of the distances between (formally defined) nearest neighbors for all numerically found structures

$r_k^{\min} = \min\left(\left|\vec{r}_k - \vec{r}_j\right|\right)$, where $j = 1, 2, \ldots N$, and accumulate their histograms $\rho = \rho\left(r_j^{\min}\right)$.

The results of these calculations are presented in Fig. 8.10, with the graphs numbered according to the numbers of the corresponding structures in Fig. 8.9. It is directly evident in this figure that for pure repulsion, as expected, the histograms continuously shift to shorter distances with increasing density. For combined repulsion and attraction in 2a and 2b, however, the position of the histogram does not move at all.

Physically, this means that when the number of particles grows, they will form more clusters, which will in turn connect with each other and grow bigger; but to some extent, the properties of these clusters will remain the same. If the attraction interaction U_A is strong enough (actually, $R_A > R_R$ is sufficient), however, the equilibrium distance between the particles will become so small that inside each cluster the particles will forms the same structure with gaps as observed for extremely strong packing at pure repulsion (2c in Fig. 8.9). If the number of clusters grows further, the ordered structure will cover the sphere completely and the final pattern (2d in Fig. 8.9) will practically coincide with the final pattern for repulsion only (1d in Fig. 8.9).

One can easily calculate the total densities for all mentioned types of structures and plot them together. The results are summarized in Fig. 8.11, where the densities (b) are compared with the histograms (a) corresponding to each case. Here, two separate branches associated with pure repulsion and with combined interactions may be distinguished. The most interesting region, where the distance between nearest neighbors $r \equiv r_k^{\min} = \min\left(\left|\vec{r}_k - \vec{r}_j\right|\right)$ for the densely packed uniform structure at pure repulsion coincides with the same distance in a set of clusters covering the spherical surface with a much smaller mean density, is indicated by the vertical line. That is why it is convenient to normalize all the densities C in the plot (Fig. 8.9) to the density at complete packing at pure repulsion ($C_{(1c)} = 1$).

Now let us turn to the remaining experimental cases with more complex local dumbbell-shaped substructures and labyrinth structures produced by chains of dumbbell structures. To model them, we have to create a more sophisticated interaction between the effective particles. We organize them as follows. The array of the particles becomes two-dimensional with the size $2 \times N$, where index $j = 1, \ldots, N$ numerates different dumbbells and $k = 1, 2$ corresponds to the individual particles inside each dumbbell. Two particles inside each dumbbell (for all indexes $j = 1, \ldots N$) are elastically connected by the strong two-valley potential $U_{Elastic}\left(\left|\vec{r}_{1,j} - \vec{r}_{2,j}\right|\right) = U_E\left(\vec{r}_{1,j} - \vec{r}_{2,j}\right)^2\left[R_{12}^2 - \left(\vec{r}_{1,j} - \vec{r}_{2,j}\right)^2\right]$ which naturally keeps the distance R_{12} between them almost fixed with relatively weak fluctuations around it. For definiteness, below we will always assume: $R_{12} = 4R_0$.

As usual, the particles are assumed to repulse each other at short distances $R_A > R_R$ corresponding to their dense cores. If there is no other interaction, this leads to the first experimental case with the surface covered by "simple" dumbbells. For a

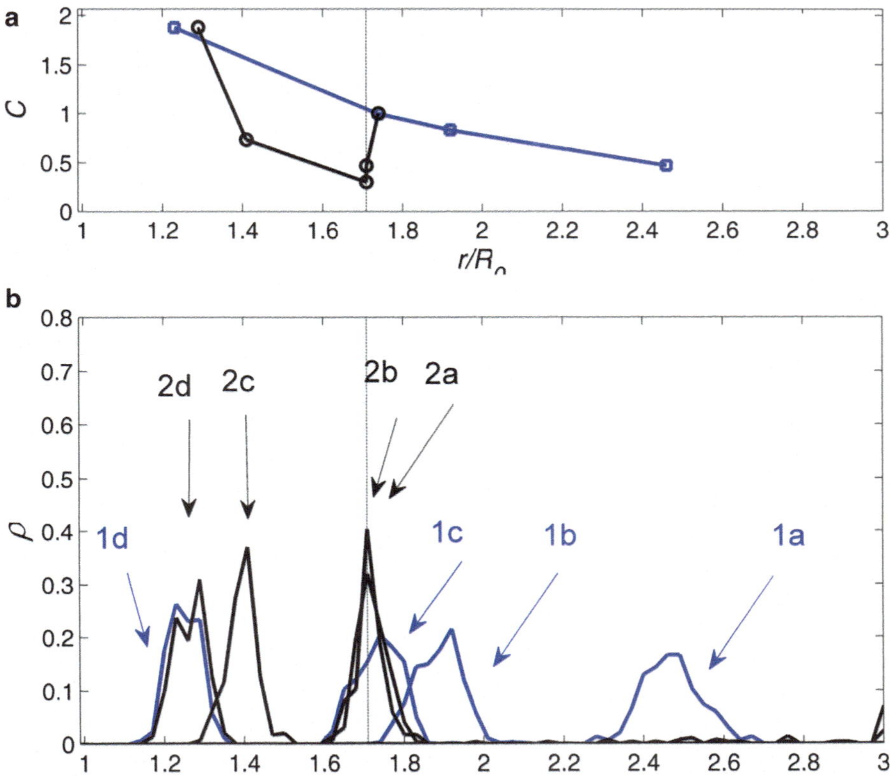

Fig. 8.11 Two branches of the density evolution with change of the parameters (**a**) accompanied by the corresponding histograms of the distances to the nearest neighbors accumulated in one figure (**b**). The colors and numbers of the curves in (**a**) have the same meaning as in Fig. 8.9. The point that marks equal distances for different structures with repulsion only and with combined interactions is indicated by the vertical line. (Figure reproduced from Filippov et al. 2017)

labyrinthine chain ordering one should add attraction between two different kinds of dumbbell ends and repulsion between equivalent ones. For both interactions, we will use exactly the same parameters as before: repulsion core of interaction $R_R = 0.75R_0$, distance of attraction interaction $R_A = R_0$ and energies $U_A = -0.2U_R$.

The interactions described here were implemented in the same numerical procedure as in the previous simulations, including the final "dressing" with densities. The results are presented in Fig. 8.12, with the graphs representing two different cases, which may also be compared with the observed natural structures (cf. 3b and 3c in Fig. 8.8). For a comparison with the previous numerical results, one can also artificially introduce an additional case where each side of the dumbbells can slightly attract each side of all other dumbbells, independently of their kind (3a in Fig. 8.12). As expected for such a case, the clusterization of the dumbbells is similar to that in the case of simple spherical nucleation centers (1a in Fig. 8.9).

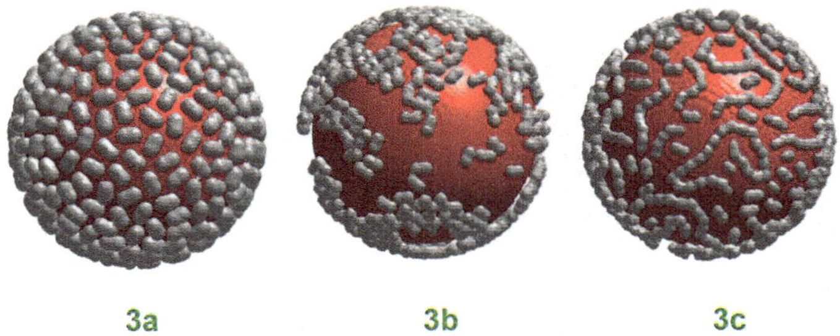

Fig. 8.12 Different structures of dumbbell-shaped building blocks, numerically obtained at different combinations of the interaction parameters as described in the text. (Figure reproduced from Filippov et al. 2017)

A quantitative analysis of the obtained structures can be carried out in the same manner as above. The only modification is that now we calculate the distances $r = r^{\min}_{k;k1,k2} = \min\left(\left|\vec{r}_{k;k1} - \vec{r}_{j,k2}\right|\right)$ between all the possible combinations of dumbbell ends: $(k_1, k_2) = (1, 1)$, $(k_1, k_2) = (2, 2)$, $(k_1, k_2) = (1, 2)$ and $(k_1, k_2) = (2, 1)$. The results are summarized in Fig. 8.13 (3a–3c). As expected, the histograms for the two last cases $(k_1, k_2) = (1, 2)$ and $(k_1, k_2) = (2, 1)$ are identical ($\rho_{12} = \rho_{21}$) and can thus be used to test the procedure.

The histograms for all the different combinations (ρ_{11}, ρ_{22} and ρ_{12}) found for the three different patterns (3a–3c) in Fig. 8.13 perfectly match the qualitative description mentioned above. In the first case (3a), the repulsion leads to maximal distances between all the dumbbells, and the distances do not statistically depend on the kinds of their ends. So in this case, all three independent histograms almost coincide. In the intermediate case (3b), the distances between adjacent dumbbells become shorter and tend to values close to the distances in a densely packed pattern. Again, the histograms almost coincide. The last case (3c) leads to a completely different picture. The maxima of two curves, ρ_{11} and ρ_{22}, corresponding to the combinations $(k_1, k_2) = (1, 1)$ and $(k_1, k_2) = (2, 2)$, remain at relatively long distances, but the bold curve ρ_{12} for $(k_1, k_2) = (1, 2)$ shifts to very short distances. This result obviously corresponds to the chain ordering where nearest neighbors of two different kinds almost touch each other along the chain.

One can also note that in this case (3c) all the curves for $(k_1, k_2) = (1, 1)$ and $(k_1, k_2) = (2, 2)$ as well as for $(k_1, k_2) = (1, 2)$ show two peaks. This corresponds to the possible ordering of the dumbbells according to two kinds of interaction. The first peak corresponds to the mentioned chain ordering and the second one appears due to localized pairs with differently oriented ends of the dumbbells inside each pair. Such a pair leads to the closest contact of the ends of different kind. Inside the chain, mutual attraction of the close ends is slightly "screened" by the mutual attraction of other ends along the chain, while in an isolated pair such "screening" is almost absent.

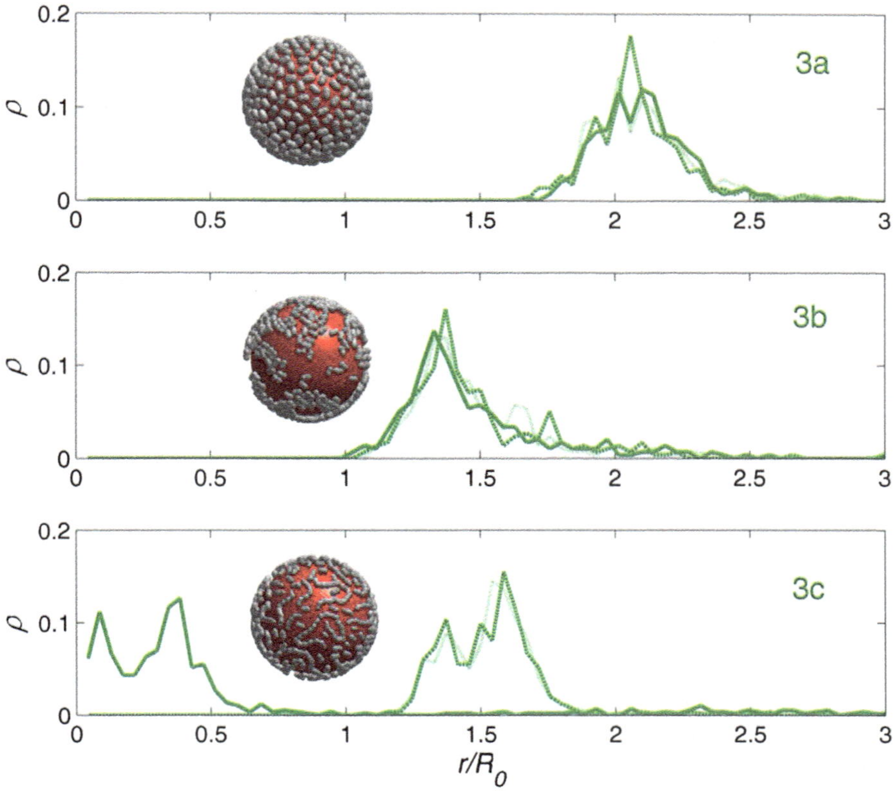

Fig. 8.13 Histograms of the distances between all combinations of dumbbell ends calculated for the globular structures depicted in the graphs. The bold lines correspond to the distances between different kinds of ends. (Figure reproduced from Filippov et al. 2017)

Obviously, this model can be developed in the direction of a more realistic position of the whole system in 3D space. Our previous simulation modeled an isolated sphere "levitating" in space. Strictly speaking, however, this is not correct. In fact, each sphere represents a granule that is glued to the substrate (i.e. the body surface of the animal) which is covered with interacting substructures of the same nature as the ones covering the spherical surface. Moreover, the geometry may not be exactly spherical, due to the usually flat contact area between the spheres and the substrate.

All these features can be accounted for by the appropriate choice of effective boundary conditions which also include the interaction of the "particles" with the substrate. The result of such a simulation for one particular combination of interactions is illustrated in Fig. 8.14. This particular case (corresponding to case 2c in Fig. 8.9) was chosen because it creates a pattern where the numerically obtained coverage of the substrate and the experimental one (shown for comparison in the

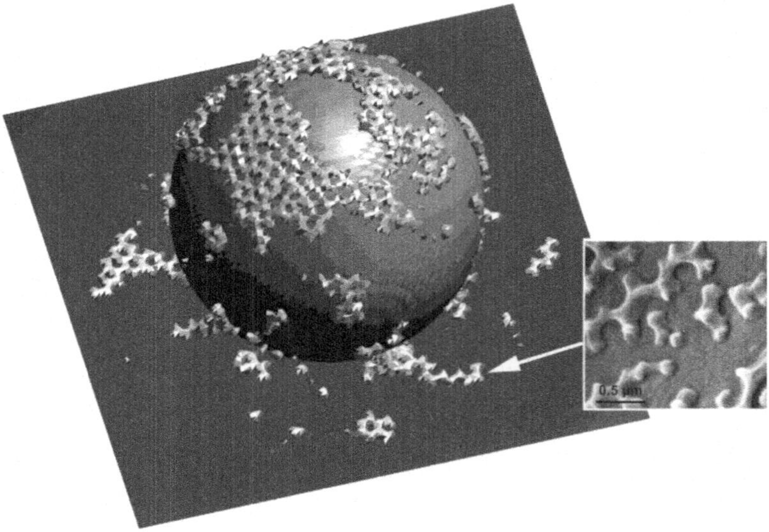

Fig. 8.14 Comparison between numerically obtained and experimentally found patterns. The small inset shows an SEM image of a real flat surface covered by nanostructures (see panel in the lower right-hand corner in Fig. 8.7). The arrow indicates an analogous model structure for comparison. (Figure reproduced from Filippov et al. 2017)

inset in Fig. 8.14) strongly coincide. Rotating spatial distributions analogous to the structures of the dumbbell building blocks shown in Figs. 8.13 and 8.14 are illustrated in the Supplementary movies 8.10 and 8.11, respectively.

8.2.3 Discussion of the Results and Their Biological Significance

It is important to note that both nature and our model can create a rather wide variety of structures, as observed in the very different cerotegument morphologies in related whip-spiders, by modifying the interaction properties of the colloid nano-particles that define their arrangement on the spherical microstructures. Normally it is expected that structures created by natural selection are more or less optimized for some biological function. In this particular case, the particular function may be its superhydrophobic nature. Certainly, all the structures found here have superhydrophobic properties due to their small scales and defined distribution density.

Of course, the superhydrophobic nature of the cerotegument may vary from case to case. Our model is limited and cannot answer these questions. But it provides a very good background for the study of this phenomenon. The densities obtained in the model may be used to compare them quantitatively with real densities, and to numerically analyze their spatial distribution, their correlation functions and the interaction with the surrounding liquids.

These interactions depend on the shape and the chemical properties of the (presumably proteinaceous) nano-particles, such as the distribution of "effective charges". From the biological point of view, it will be interesting to determine the genes producing these building blocks, and their changes throughout evolution. For taxonomists, it would be interesting to find some deeper relationships between cerotegument morphotypes and to derive some evolutionary hypotheses. Our results may also assist in the development of colloids for the functionalization of surfaces, by producing different structures with tailored mechanical and optical properties.

Our modeling experiment demonstrates that slight changes of simple parameters lead to a variety of morphologies that are highly similar to the ones observed in the species-specific cerotegument structures of whip-spiders. This can be explained by the effect of different interactions between the particles on the morphology of the final structure. This issue is related to the so-called Tammes problem, which refers to the optimal packing of a given number of pores or particles on a sphere so that the minimum distance between them is maximized (Tammes 1930; Tarnai and Gáspár 1987; Erber and Hockney 1991).

Our aim was to adjust the model parameters in such a way that structures similar to those observed in whip-spiders could be reproduced. In this way we wanted to test which alterations are necessary to transform the structures within a possible evolutionary scenario. However, these results are not only important for our understanding of the formation of globular hierarchical structures in nature, but also for the fabrication of novel surface coatings by colloidal lithography.

8.3 Numerical Simulation of the Pattern Formation of Springtail Cuticle Nanostructures

Springtails (Collembola) exhibit complex hierarchical nanostructures on their exoskeleton surface that repels water and other fluids with remarkable efficiency. These nanostructures have been widely studied with regard to their structure, chemistry, and fluid-repelling properties (see Sect. 8.3.1 below). These ultrastructural and chemical studies revealed the involvement of different components in different parts of the nanopattern, but the overall process of self-assembly remains unclear. Here we model this process from a theoretical point of view, partially using solutions related to the so-called Tammes problem. By using the densities of three different substances, we could obtain a typical morphology highly similar to the one observed on the cuticle of some springtail species. These results are not only important for our understanding of the formation of hierarchical nanoscale structures in nature, but also for the development of novel surface coatings.

In materials science, surface coatings with regularly arranged particles have been produced by colloidal lithography, a technique utilizing the self-assembly of nanoparticles on a substrate surface (Yang et al. 2006; Badge et al. 2013). However, the range of producible patterns and properties of these coatings is restricted, as well as their durability. Therefore, the study of biological self-organizing nanopatterns seems to be promising for generating new solutions (Kovalev et al. 2016; Filippov et al. 2017) for industrial applications.

8.3.1 Biological and Chemical Background of Pattern Formation in Springtail Cuticle

Springtails (Collembola) are soil-dwelling arthropods (Rusek 1998) with a complex, hierarchically structured cuticle surface (Nickerl et al. 2013) with strong repelling properties against water, liquids with low surface tension (Helbig et al. 2011; Hensel et al. 2013a), and sticky secretions of predatory insects (Körner et al. 2012). These properties reflect the collembolan cuticle adaptation to living in soil habitats (Hale and Smith 1966; Nickerl et al. 2013). Nanoscopic, comb-like structures are primarily responsible for the repellent properties, as has been previously demonstrated by polymer replication methods (Hensel et al. 2013b, 2014).

In general, the different layers of arthropod cuticle have different chemical compositions. Both the exocuticle and the endocuticle consist of chitin fibers and a proteinous matrix which may be more or less sclerotized (Richards 1951). The exocuticle and the endocuticle differ mainly in fiber orientation and in the degree of sclerotization (the endocuticle is less sclerotized). In the epicuticle, four layers with different compositions could be distinguished: the internal layer (polymerized lipoprotein stabilized by quinones; Dennell 1946); the cuticulin layer (quinone-tanned lipoproteins); the wax layer (mainly long-chain saturated alcohols esterified with acids; Wiggelsworth 1947); and the cement layer (proteins and lipids stabilized by various polyphenolic substances; Wiggelsworth 1976).

The chemical nature of the comb-like structure on top of the collembolan cuticle, however, was unknown until Nickerl et al. (2014) studied the chemical composition and architecture of the cuticle of *Tetrodontophora bielanensis*. These authors removed the different cuticle layers in a stepwise manner and separately analyzed their chemical composition. It was shown that in representatives of Collembola the endocuticle and the exocuticle consist mostly of chitin lamellae. Within the epicuticle, the comb-like nanostructures consist of glycine-rich structural proteins, and the outer wax layer consists of fatty acids (palmitic and stearic acid), fatty and steryl esters, terpenes (lycopaen), steroids (cholesterol and desmosterol), and triglycerides. The study also showed a regular distribution of pore channels within the chitin-containing deeper cuticle layers, which may be used for different functions, such as fluid transport from epidermal cells to the surface and respiration. The presence of pores with a possible respiratory function underlines the importance of the water-repellent properties of the collembolan cuticle surface. The main result of this study was that the non-wetting property of the collembolan cuticle is mainly based on its topography rather than its chemistry.

Even with this knowledge about the structure, chemistry and properties of a biological surface, however, it is difficult to trace the development of such elaborate hierarchical nanostructures as those of springtails. According to the dimensions of the structures, they are probably produced at the level of single cells, but even at the highest resolution of the SEM, the boundaries between single cells are difficult to discern, which supports the hypothesis that these structures appear due to a self-organization process based on the varying densities of materials synthesized by the underlying epidermal cells.

As a basis for the present modeling process, we examined the cuticle surface of the springtail *Orchesella cincta* in the scanning electron microscope (SEM). The images were obtained from dry specimens mounted on SEM aluminum stubs using double-sided carbon-containing sticky tape, sputter-coated with gold–palladium (thickness 10 nm) and examined in a SEM Hitachi S4800 at 3 kV accelerating voltage (Fig. 8.15). The cuticle of *O. cincta* is covered with nanoscale structures which form a complex hexagonal comb-like pattern. Since the comb elements were of slightly different sizes, structures with five or seven vertices occurred as well. Each comb element possesses a double-walled circular/oval circumference. At the sites where the boundaries of three comb elements meet, elevated triangle-like structures appear. That is why 5–7 triangles are located on each boundary of a comb element (heretoforth called circle). Each triangle contains a little pore in its center (Fig. 8.15b). Sometimes this pore is plugged by a nanoscopical droplet of an apparently amorphous solidified fluid. Additionally, within the circles, other slightly larger pores occasionally occur (Fig. 8.15a).

The mechanisms of nanostructure assembly are still not clearly understood, but some models already exist, in particular for the self-assembly processes in the ommatidia gratings of insect eyes, for snake skin nanodimples, and for the arachnid cerotegument (Blagodatski et al. 2015; Kovalev et al. 2016; Filippov et al. 2017), which are rather simple nanostructures in comparison with the collembolan cuticle pattern in Fig. 8.15). Several different approaches have previously been used in modeling these structures. Here, we introduce a numerical model that allows us to study the effect of different interactions between different substances on the morphology of the final structure. As above, this is related to the so-called Tammes problem, which searches for the optimal packing of a given number of pores or particles on a sphere with maximization of the minimum distance between them (Tammes 1930; Tarnai and Gáspár 1987; Erber and Hockney 1991). The aim was to alter the model parameters in such a way that 3D structures similar to those observed in springtails could be numerically reproduced based on the principle of frozen kinetics. Different sizes (polydispersity) disturb the optimal hexagonal distribution and generate other locally optimal distributions described in the framework of random packing (Mityushev 2014). Here, the optimal distribution corresponds to a minimum of energy.

In this way we want to test which minimal alterations are necessary to transform the structures within a possible evolutionary scenario, and likewise how colloids can be tailored to 3D functionalized surfaces by colloidal lithography.

8.3.2 Numerical Model of the Pattern Formation in Springtail Cuticle

From a physical point of view, one can expect that the surface nanostructures observed in springtail cuticle are formed by some kinetic process during which substances produced by some spatially distributed nucleation centers on the surface interact with each other, redistribute and solidify. Such a process can be observed in many different chemical and physical systems. That is why closely related patterns can appear convergently in very different systems. For example, patterns with hexagonal or honeycomb lattices can form during the ordering of superconducting vortices (Geim et al. 1997) or by magnetic ordering (Filippov 1997) in quasi 2D systems. The same structure may also be formed during the self-organization of surface structures in biological systems (Filippov 1998; Kovalev et al. 2016).

In all these cases, the process involves the interaction of many spatially distributed densities, and its direct simulation can cause extremely time-consuming calculations. In the following, we suggest a simple numerical model for a hexagonal comb-like pattern formation. Taking into account the complexity of the self-organization process under consideration, we will need several (at least 3) different kinds of interacting substances with densities ρ_k, corresponding to the fields inside the honeycomb-like formations (ρ_1) (dark gray in Fig. 8.15), to the boundaries between them (ρ_2) (medium gray in Fig. 8.15) and an additional density (ρ_3) which reproduces the "triangles" situated on top of their intersections (light gray in Fig. 8.15).

The main 'heuristic' hypothesis which we plan to use in the present study is that all of the densities are generated during a joint kinetic process and redistributed in space due to their mutual interactions. The second hypothesis, strongly related to the previous one, is that the material of the boundaries (ρ_2) is deposited due to an expansion of the dark gray fields of (ρ_1). In other words, it is synthesized and at the same time "pushed" by the expanding regions of (ρ_1).

Previous kinetic studies show that, if many individual sources of some density (ρ_1) repulse one another on a uniform surface, they will redistribute to form a slightly disturbed hexagonal lattice. This is close to the real pattern observed on the springtail cuticle surface. Such a lattice appears as a kind of "frozen" (or stopped) kinetic process (Kovalev et al. 2016; Filippov et al. 2017). Besides honeycomb cell domains, it normally contains some frozen "dislocations" with five- and sevenfold symmetry. This is also very close to what we see in our SEM images (Fig. 8.15).

Taking all the above into account, one can start with a set of naturally distributed sources of the first density (ρ_1) which grows from some initial value to a stationary

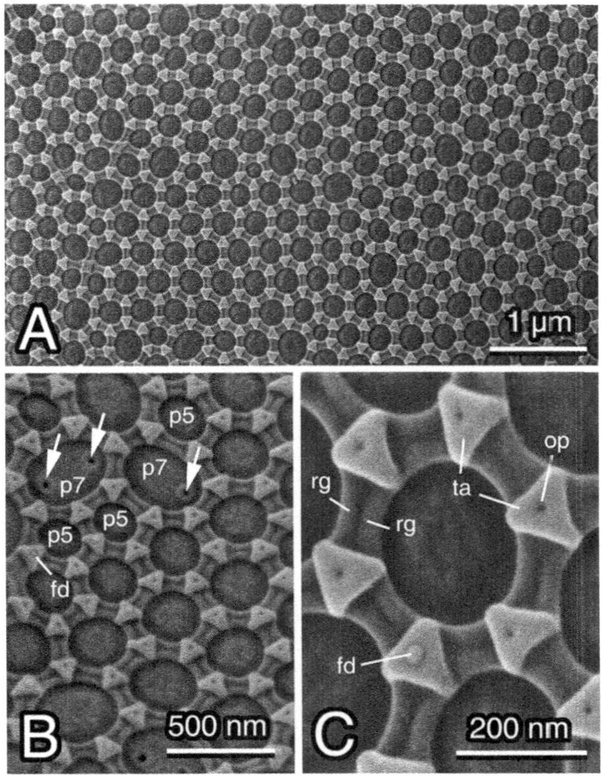

Fig. 8.15 Scanning electron images of the mainly hexagonal cuticle nanostructures on the abdomen of *Orchesella cincta*. The arrows indicate openings of circular structures. Abbreviations: *fd* fluid, *op* openings of triangle structures, *p5* pentagonal patterns, *p7* heptagonal patterns, *rg* ridges, *ta* triangles. (Figure reproduced from Filippov et al. 2018)

configuration (Fig. 8.16, ρ_{11}, ρ_{12}). During its growth and expansion, this density gradually creates and at the same time dislodges a second density (Fig. 8.16, ρ_2). Such a self-consistent process can naturally form the honeycomb-like boundaries (Supplementary movie 8.12), which are clearly visible in the SEM images (Fig. 8.15).

However, from a mathematical point of view, the following important problem appears when trying to realize such a process in a numerical experiment. If we use a common (unique) spatial distribution of the density ρ_1 for all the growing nucleation centers, the domains will collide without competition. As a result, the boundaries between them and the corresponding walls of the second density ρ_2, even though they are formed at intermediate stages of kinetics, will gradually wear off with time.

The only way to prevent such a "mathematical catastrophe" is to define all the sub-densities ρ_1 as formally independent particular densities $\rho_{1,j}$ interacting with all the other densities of the problem.

For example, if the number of the sources N is equal to $N = 100$ (the number we typically used for the calculations below), the number of the interacting densities of

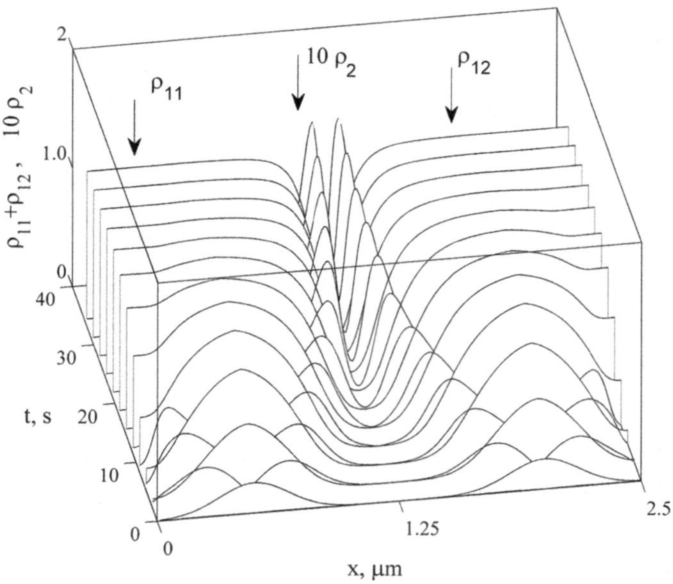

Fig. 8.16 Self-consistent space–time evolution in a 1D model of two domains ρ_{11} and ρ_{12} of increasing density which produce a stable double wall of the density ρ_2 in between

the first type (only!) also extends to $N = 100$, with $j = 1 : N$. The total (and experimentally measured) physical density ρ_1 of the substance is given by the sum of all the local ones:

$$\rho_1 = \sum_{k=1}^{N} \rho_{1,k} \tag{8.1}$$

Numerical experiments confirm that, because of the mutual repulsion, the local densities $\rho_{1,\ k}$ will be strongly localized inside their own domains. As a result, the difference $\rho_1 - \rho_{1,\ k}$ with a fixed number k reproduces the total density $\rho_1 = \sum_{m=1}^{N} \rho_{1,m}$ from which a particular subsystem $\rho_{1,\ k}$ is removed. The corresponding term $g_{11}(\rho_1 - \rho_{1k})$ being added below to the kinetic equations will describe an interaction between a given density $\rho_{1,\ k}$ and all the other densities with some intensity g_{11}.

According to our SEM images, the third density (ρ_3), corresponding to the light triangles in Fig. 8.15, seems to be the last one which enters the process at a very late stage. Thus, we will for now disregard this density and formulate the kinetic equations only for the densities $\rho_{1,\ k}$ and ρ_2.

Thus, the general set of the equations describing the ordering of these densities can be written as follows:

$$\begin{aligned} \partial\rho_{1k}/\partial t &= c_1\Delta\rho_{1k} + \rho_{1k}(1-\rho_{1k})(\rho_{1k}-1/2) - g_{11}(\rho_1-\rho_{1k}) - g_{12}\rho_2, \\ \partial\rho_2/\partial t &= c_2\Delta\rho_2 + \rho_2(1-\rho_2)(\rho_2-1/2) - g_{21}\rho_1 + c_{21}|\nabla\rho_1| \end{aligned} \tag{8.2}$$

with k = 1...N. Here, terms with a Laplacian operator, $c_j\Delta\rho_j = c_j(\partial^2\rho_j/\partial x^2 + \partial^2\rho_j/\partial y^2)$, simulate the surface tension with a strength controlled by the coefficient c_j. Surface tensions are different for different densities, but will be the same for all the variables $\rho_{1,\,k}$ of the same nature. That is why we have the terms $c_1\Delta\rho_{1k}$ and $c_2\Delta\rho_2$. The bigger the coefficient c_j, the larger its impact on the density gradients (mathematically corresponding to the surface tension), so the smoother density distributions will be energetically favorable. An analogous physical process would be the displacement of a solution on the cuticle surface by a solution excreted through the cuticle pores. Mutual repulsion of $\rho_{1,\,k}$ may be provided by the formation of a layer with high surface energy between interacting solutions/fluids.

Production of the second density is simulated by the term $c_{21}|\nabla\rho_1|$, which means that this density is created by the expanding boundaries of all the domains $\rho_1 = \sum_{k=1}^{N}\rho_{1,k}$. Mainly near the boundaries the absolute value of the gradient $|\nabla\rho_1| = \sqrt{(\partial\rho_1/\partial x)^2 + (\partial\rho_1/\partial y)^2}$ essentially deviates from zero and causes strong production of ρ_2. This is the case if ρ_2 for some reasons has a higher concentration close to the outer boundary of $\rho_{1,\,k}$.

The standard combination $\rho_j(1 - \rho_j)(\rho_j - 1/2)$ corresponds to the two-valley potential free energy which favors two equivalent equilibrium states and a barrier between them. In the trivial case when all other densities and gradients are absent, each solitary density ρ_j tends to approach one of the two values of the uniform equilibrium: $\rho_j = 0$ or $\rho_j = 1$. These terms could be alternatively assigned to dynamical processes with much shorter characteristic duration than in Eqs. 8.2. Other terms, which mix different densities in the equations, namely $-g_{11}(\rho_1 - \rho_{1k}) - g_{12}\rho_2$ and $-g_{21}\rho_1$, describe the repulsion between them, due to which every density tends to push another one from the same position in space.

A well-known difficulty of 2D simulations with a large number of spatially distributed densities is that they are relatively time-consuming and need a complex presentation of many consecutive time-dependent configurations of the surfaces or color density-maps. This makes it difficult to extract and analyze any reasonable amount of information about the space–time evolution of the system, especially with regard to the interacting densities which can overlap one another. So, before introducing the 2D model we will start out with a much simpler 1D model.

From a mathematical point of view, this means that we treat all the variables as depending on the coordinate x only and rewrite the Laplacian and gradient operators in their one-dimensional forms: $c_j\Delta\rho_j = c_j\partial^2\rho_j/\partial x^2$ and $|\nabla\rho_j| = \sqrt{(\partial\rho_j/\partial x)^2}$, respectively. This reduction gives us an opportunity to record the space–time evolution of the interacting densities and present time-dependent results statically. It allows us to visualize how the densities are generated, suppressed or overlapped by one another.

One-dimensional modeling also allows us to reduce the number of variables in Eqs. 8.2. In 2D space, where the expanding domains of the densities ρ_{1k} can grow in different directions, a relatively large number of neighboring domains of ρ_{1k} (normally 5 or 6) will be necessary to confine the density ρ_2. Due to the rigid confinement in 1D space, only two domains, ρ_{11} and ρ_{12}, are sufficient since they will inevitably meet one another when expanding along line x. This strongly simplifies calculations and allows us to perform an easy preliminary simulation.

The result of the simplified 1D simulations according to Eqs. 8.2 for all three variables ρ_{11}, ρ_{12} and ρ_2 in time–space coordinates is shown in Fig. 8.16. Here we can see the domains ρ_{11} and ρ_{12} growing and expanding from the left and the right side, respectively. Each one produces its own wall/boundary in the direction of ρ_2. With time, these boundaries will combine into a static double wall ρ_2 between the two mutually stabilized domains of ρ_{11} and ρ_{12}.

An important feature of the real structure in Fig. 8.15 is the existence of light triangles formed by a third material which is associated with the density ρ_3. These triangles form in the places where the double walls of ρ_2 meet one another on the 2D surface. This leads to the additional problem of defining these positions numerically. For the moment, however, we will postpone this question because the simplicity of the 1D model allows us to naturally include the third density ρ_3 into this preliminary study. Thus, the basic equation for the third density can be written as follows:

$$\partial\rho_3/\partial t = c_3\Delta\rho_3 + \rho_3(1-\rho_3)(\rho_3 - 1/2) - g_{32}\rho_2 + g_{31}\rho_1 + c_{32}|\nabla\rho_2|. \quad (8.3)$$

The structure of Eq. 8.3 reflects the general hypothesis that the density ρ_3 is created during the final stages of the process on the walls of the secondary density ρ_2. In other words, it is expected to be produced by a non-local term in the equation (gradient $c_{32}|\nabla\rho_2|$). In the trivial case of an energy barrier that is already produced and passed through, the density tends to an equilibrium in the two-valley free energy potential represented in the equation by the combination $\rho_3(1 - \rho_3)(\rho_3 - 1/2)$. The density has a surface tension $c_3\Delta\rho_3$ and interacts locally with all other densities $-g_{32}\rho_2 + g_{31}\rho_1$ in the same manner as in the two previous forms of Eqs. 8.2, so it has a high affinity to ρ_2 and a low affinity to ρ_1.

Surface tension, local interactions and influence of the gradient are regulated by the relations between all the coefficients of the problem, c_3, g_{32}, g_{31}, c_{32}. Direct observation of the natural pattern shows that the triangles of the third density ρ_3 demonstrate negative curvature and their vortices are turned into the direction of the first density, ρ_1. Very likely this combination of properties appears due to the low

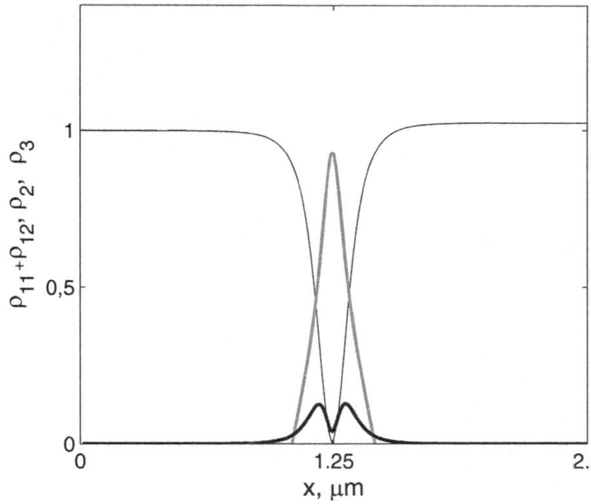

Fig. 8.17 Typical distribution of all the densities, $\rho_{11} + \rho_{12}$ (thin line), ρ_2 (bold line) and ρ_3 (gray line), in an asymptotic (stationary) configuration

surface tension, strong repulsion with regard to the second density ρ_2, attraction to ρ_1 and a sufficient rate of production regulated by the coefficients g_{31}, c_{32}.

This allows us to make some assumptions about the relationship between the phenomenological constants c_3, g_{32}, g_{31}, c_{21} and to adjust them preliminarily, using a simple 1D approximation. The typical distribution of all the densities in an asymptotic (stationary) configuration at a fixed set of the parameters is shown in Fig. 8.17. The kinetic process is illustrated in Supplementary movie 8.12.

Let us examine the structure of the curve of ρ_3 in Fig. 8.17. Its sharp central maximum is mainly due to the competition between the strong influence of the gradient term $|\nabla\rho_2|$ and repulsion by the maxima of ρ_2. At the same time, ρ_3 has relatively wide wings that penetrate into both domains of the first density, ρ_1, due to their mutual attraction. This configuration reflects, in the restricted 1D space, the specific distribution and orientation of the triangle-like density ρ_3 on a more realistic 2D surface.

Now we will return to the 2D configuration. The numerical complexity of the problem motivates us to introduce an additional reduction of the numerical model. The main idea of this simplification is to substitute an originally continuous problem by a combination with some preliminary discrete approach in order to achieve a final or at least almost stationary configuration. A well-known example of such an approach is the study of the topological phase transition in quasi 2D superconducting systems (Kosterlitz and Thouless 1973) where the extremely complex evolution of the superconducting vortices was substituted to some extent by the motion of charged particles. This approach was found to be extremely helpful and was even awarded the Nobel Prize in 2016. It is also applicable to the Tammes problem (Tammes 1930; Tarnai and Gáspár 1987; Erber and Hockney 1991), where the study of kinetic ordering on a sphere is substituted by the calculation of equilibrium ordering of N equal charges.

In all these cases and in the approach used here it is important to finalize the discrete distribution of the nucleation centers for the future densities before simulating the kinetic evolution of the interacting continuous fields. Here, we organize the discrete part of the procedure as follows.

Initially, we randomly distribute some number of particles (heretofore called nucleation centers for densities) on the limited rectangular planar surface.

In such a simulation, the most important parameters are the relationships between the area of the surface $L_x \times L_y$, the number of the particles N, and the characteristic distance of the repulsion interaction R_0 between them. To achieve the desired organisation of the particles, their interaction can be defined simply as a pure short-range (exponential) repulsion: $U_1\left(\left|\vec{r}_j - \vec{r}_k\right|\right) = U_R \exp\left[-\frac{\vec{r}_j - \vec{r}_k}{R_0}\right]$, where $U_R > 0$ is the interaction intensity and $\vec{r}_j$ represents the specific positions, $j = 1, 2, \ldots, N$.

We used the limited area $L_x \times L_y$ for our numerical simulation. In reality, however, the particles can move on a very large surface (compared to the distances between them). From the mathematical point of view, to study such a quasi-infinite system we must apply periodic (toroidal) boundary conditions in both directions. We have to take into account that the original process represents damped kinetics, so the dynamic equations here should involve damping as well: $\partial^2 \vec{r}_j / \partial t^2 = -\gamma \partial \vec{r}_j / \partial t + \sum_k \partial U_1 \left|\vec{r}_j - \vec{r}_k\right| / \partial \vec{r}_k$, where the dissipative constant γ defines the characteristic time scale and γ^{-1} can serve as a time unit for the simulation.

During the calculation routine, the particles dynamically rearrange with a gradually decreasing velocity. This slowdown can be controlled by calculating the mean velocity of the particles and stopping the procedure when their relative motion becomes negligible. In fact, originally randomly distributed particles practically cannot reach their real ground state in toroidal space. But in the long run, they demonstrate some quite realistic distribution. As we can see in the images of the real system, the same distribution actually appears in nature where the patterns are not perfect either. In both the numerical and the real configurations, various possible realizations are produced as a result of frozen kinetics.

One particular realization of almost (but not perfectly) periodically ordered particles is shown in Fig. 8.18 as the open circles represent nucleation centers for the individual domains $k = 1, 2, .., N$ of the first substance ρ_{1k}. Please note the dark circles in Fig. 8.18, which represent additional nucleation centers. They were added on purpose in order to reproduce the real configuration (see Fig. 8.15). Such configurations appear when two sources of a substance occasionally appear very close to each other at the beginning of the kinetic stage. In this case, the expanding domains of ρ_{1k} in this area are not able to form a double wall of ρ_2 in between and will remain connected until the very end of the process. We added these configurations in the simulation to take this observation into account.

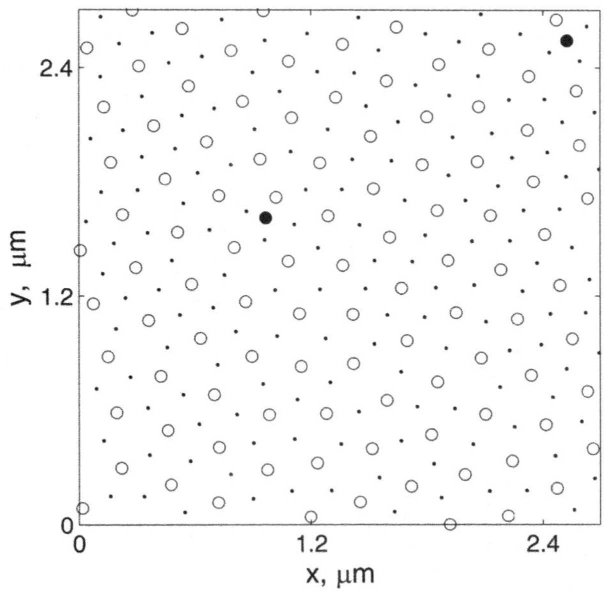

Fig. 8.18 Positions of the nucleation centers for the densities ρ_{1k} and ρ_3 represented by open circles and small black dots, respectively. Two additional nucleation centers of ρ_{1k} anomalously close to the normally distributed ones are represented by black circles. (Figure reproduced from Filippov et al. 2017)

Before taking up the continuous model, there is still one more question regarding the discrete model. When the double walls of the second density ρ_2 have met one another in the places of intersecting domain boundaries, the third density, ρ_3, starts to appear in these positions. To simplify the calculations, it is helpful to define the positions of the nucleation centers for ρ_3 in the discrete model as well. To do this, we create a new array of "particles" which repulse one another and of already fixed particles that define the positions of the growing domains of ρ_{1k}. With sufficient time, they will reach stationary positions which are as far as possible from all the previous ones. In Fig. 8.18, these positions of the already calculated configuration of open circles are represented by black dots.

Now everything is ready for the solution of our problem on a 2D surface. For this purpose, we distribute a set of the density nucleation centers, $\rho_{1k} = A\exp\left[-\frac{(x-x_k)^2+(y-y_k)^2}{b^2}\right]$, in the preliminary positions $\{x_k, y_k\}$, $k = 1, 2, \ldots N$ and solve the numerically connected kinetic Eqs. 8.2 for ρ_{1k} and ρ_2 with 2D Laplacians $c_j\Delta\rho_j = c_j(\partial^2\rho_j/\partial x^2 + \partial^2\rho_j/\partial y^2)$ and gradients $|\nabla\rho_1| = \sqrt{(\partial\rho_1/\partial x)^2 + (\partial\rho_1/\partial y)^2}$. As soon as the distributions of the densities ρ_{1k} and ρ_2 have been attracted to some state close to the equilibrium, we insert them into the 2D version of Eq. 8.3 and determine the third density distribution ρ_3 associated with ρ_{1k} and ρ_2. The result for a particular realization of the nucleation centers is shown in Fig. 8.18. The kinetics of this process is illustrated in Supplementary movie 8.13. Large-scale and small-scale representations are presented in Figs. 8.19 and 8.20, respectively.

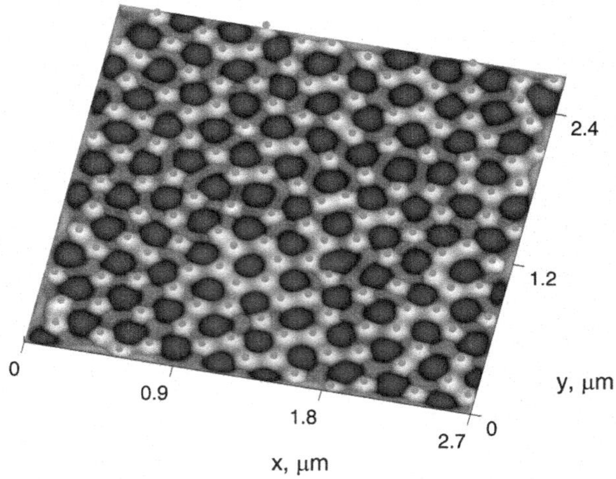

Fig. 8.19 Large-scale structure of the numerically generated surface of total density $\rho_{total} = \sum_k \rho_{1k} + \rho_2 + \rho_3$ formed by interacting densities ρ_{1k}, ρ_2 and ρ_3. The 3D structure of constant thickness was built up from the pseudo-2D density inhomogeneities $\rho_{1,\,2}$. (Figure reproduced from Filippov et al. 2017)

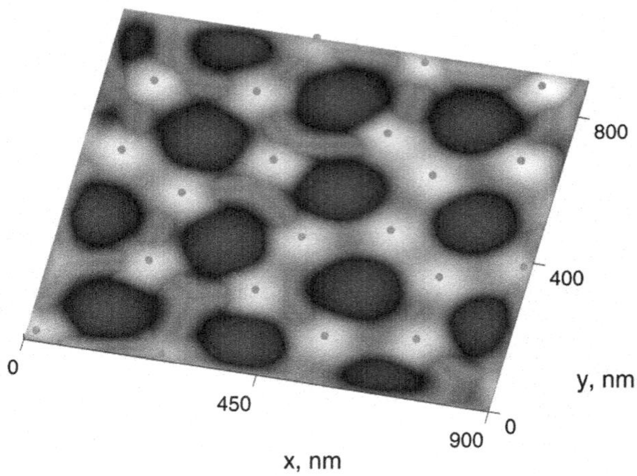

Fig. 8.20 Magnified fragment of the structure shown in Fig. 8.19. The double walls of ρ_2 and the different fine structures of ρ_3 (triangular and other possible shapes depending on the particular local configuration of the walls) are clearly visible. (Figure reproduced from Filippov et al. 2017)

8.3.3 Discussion of the Results and Biological Significance

The cuticle surface of the springtail *Orchesella cincta*, which shows the characteristic collembolan ornamentations, consists of a complex arrangement of nanostructures that interact during their formation. The three densities used in our model were chosen to correspond to the three chemically different layers recently found in collembolan cuticle (Nickerl et al. 2014). The thickest and deepest layer of the cuticle, the so-called procuticle, consisting of an exo- and endocuticle, is chitin-rich. The specifically arranged pore channels (Locke 1961; Gorb 1997) observed on the springtail cuticle surface and within its procuticle (Krzysztofowicz et al. 1972; Nickerl et al. 2014) enable material transport and are presumably involved in the complex pattern formation by depositing fluids that interact with their surroundings and with each other and, after solidifying, contribute to the formation of the surface pattern. Previous studies have demonstrated that the epicuticular structures contain proteins with high amounts of glycine, tyrosine and serine, with the amino acid composition resembling that of fibroin, collagen or resilin (Nickerl et al. 2014). The outermost layer consists of a lipid mixture of fatty acids, wax esters and terpenes, terminates the epicuticle and forms a protective barrier for the animal.

It was shown here that it is possible to explain the complex 3D nanoscale morphology found in springtails by the interaction of three different materials on the cuticle surface. Their interactions depend on the chemical and physical properties of the fluids and the presence/distribution of pore channels. Thus, according to the model described above, the complex 3D hierarchical surface structure of the collembolan cuticle with its anti-wetting properties might be assembled in a very simple self-organizing way. From the biological point of view, it would be interesting to determine the genes that produce the building blocks of this system and their alteration throughout evolution. For taxonomists it would be interesting to find some deeper relationships between different pattern morphotypes and derive some evolutionary hypotheses. These results may also assist in the engineering of colloids that may be used for the functionalization of surfaces by providing different 3D structures with tailored wetting, mechanical and optical properties.

References

Alberti G, Storch V, Renner H (1981) Über den feinstrukturellen Aufbau der Milbencuticula (Acari, Arachnida). Zool Jb Anat 105:183–236

Badge I, Bhawalkar SP, Jia L, Dhinojwala A (2013) Tuning surface wettability using single layered and hierarchically ordered arrays of spherical colloidal particles. Soft Matter 9:3032–3040

Barthlott W, Neinhuis C (1997) Purity of the sacred lotus, or escape from contamination in biological surfaces. Planta 202:1–8

Barthlott W, Mail M, Neinhuis C (2016) Superhydrophobic hierarchically structured surfaces in biology: evolution, structural principles and biomimetic applications. Phil Trans R Soc A 374:20160191

Baum M, Heepe L, Gorb SN (2014) Friction behavior of a microstructured polymer surface inspired by snake skin. Beilstein J Nanotechnol 5:83–97

Bernhard CG, Miller WH (1962) A corneal nipple pattern in insect compound eyes. Acta Physiol Scand 56:385–386

Berthé RA, Westhoff G, Bleckmann H, Gorb SN (2009) Surface structure and frictional properties of the skin of the Amazon tree boa *Corallus hortulanus* (Squamata, Boidae). J Comp Physiol A 195:311–318

Blagodatski A, Sergeev A, Kryuchkova M, Lopatinad Y, Katanaev VL (2015) Diverse set of Turing nanopatterns coat corneae across insect lineages. Proc Natl Acad Sci U S A 112:10750–10755

Dennell RA (1946) Study of an insect cuticle: the larval cuticle of *Sarcophaga faculata* Pand. (Diptera). Proc R Soc B 133:348–373

Erber T, Hockney GM (1991) Equilibrium configurations of *N* equal charges on a sphere. J Phys A Math Gen 24:L1369–L1377

Filippov AE (1997) Kinetics of vortex structure formation in magnetic materials. J Exp Theor Phys 84:971–977

Filippov AE (1998) Two-component model for the growth of porous subsurface layers. J Exp Theor Phys 87:814–822
Filippov AE, Wolff JO, Seiter M, Gorb SN (2017) Numerical simulation of colloidal self-assembly of super-hydrophobic arachnid cerotegument structures. J Theor Biol 430:1–8
Filippov AE, Kovalev A, Gorb SN (2018) Numerical simulation of the pattern formation of the springtail cuticle nanostructures. J R Soc Interface 15:20180217
Geim AK, Grigorieva IV, Dubonos SV, Lok JGS, Maan JC, Filippov AE, Peeters FM (1997) Phase transitions in individual sub-micrometre superconductors. Nature 390:259–262
Gorb SN (1997) Porous channels in the cuticle of the head-arrester system in dragon/damselflies (Insecta: Odonata). Microsc Res Tech 37:583–591
Hale WG, Smith AL (1966) Scanning electron microscope studies of cuticular structures in the genus *Onychiurus* (Collembola). Rev Ecol Biol Sol 3:343–354
Helbig R, Nickerl J, Neinhuis C, Werner C (2011) Smart skin patterns protect springtails. PLoS One 6:e25105
Hensel R, Helbig R, Aland S, Braun H-G, Voigt A, Neinhuis C, Werner C (2013a) Wetting resistance at its topographical limit: the benefit of mushroom and serif T structures. Langmuir 29:1100–1112
Hensel R, Helbig R, Aland S, Voigt A, Neinhuis C, Werner C (2013b) Tunable nano-replication to explore the omniphobic characteristics of springtail skin. NPG Asia Mater 5:e37
Hensel R, Finn A, Helbig R, Braun H-G, Neinhuis C, Fischer W-J, Werner C (2014) Biologically inspired omniphobic surfaces by reverse imprint lithography. Adv Mater 26:2029–2033
Klein M-CG, Gorb SN (2012) Epidermis architecture and material properties of the skin of four snake species. J R Soc Interface 9:3140–3155
Klein M-CG, Gorb SN (2014) Ultrastructure and wear patterns of the ventral epidermis of four snake species (Squamata, Serpentes). Zoology 117:295–314
Klein M-CG, Deuschle JK, Gorb SN (2010) Material properties of the skin of the Kenyan sand boa *Gongylophis colubrinus* (Squamata, Boidae). J Comp Physiol A 196:659–668
Koerner L, Gorb SN, Betz O (2012) Adhesive performance of the stick-capture apparatus of rove beetles of the genus *Stenus* (Coleoptera, Staphylinidae) toward various surfaces. J Insect Physiol 58:155–163
Kosterlitz JM, Thouless DJ (1973) Ordering, metastability and phase transitions in two-dimensional systems. J Physics C: Solid State Physics 6:1181–1203
Kovalev A, Filippov AE, Gorb SN (2016) Correlation analysis of symmetry breaking in the surface nanostructure ordering: case study of the ventral scale of the snake *Morelia viridis*. Applied Physics A 122:1–6
Krzysztofowicz A, Klag J, Komorowska B (1972) The fine structure of the cuticle in *Tetrodontophora bielanensis* (Waga), Collembola. Acta Biol Cracoviensia Ser Zool 15:113–119
Lee KC, Erb U (2015) Remarkable crystal and defect structures in butterfly eye nano-nipple arrays. Arthropod Struct Dev 44:587–594
Li R, Bowerman B (2010) Symmetry breaking in biology. Cold Spring Harb Perspect Biol 2: a003475
Locke M (1961) Pore canals and related structures in insect cuticle. J Biophys Biochem Cytol 10:589–618
Miskimen GW, Rodriguez NL (1981) Structure and functional aspects of the Scotopic compound eye of the sugarcane borer moth. J Morphol 168:73–84
Mityushev V (2016) Pattern formations and optimal packing. Math Biosci 274: 12–16 and Chapter 9 of Bressloff PC 2014 Stochastic processes in cell biology. Springer, Cham
Nickerl J, Helbig R, Schulz H-J, Werner C, Neinhuis C (2013) Diversity and potential correlations to the function of Collembola cuticle structures. Zoomorphology 132:183–195
Nickerl J, Tsurkan M, Hensel R, Neinhuis C, Werner C (2014) The multilayered protective cuticle of Collembola: a chemical analysis. J R Soc Interface 11:20140619

Peisker H, Gorb SN (2010) Always on the bright side of life: anti-adhesive properties of insect ommatidia grating. J Exp Biol 213:3457–3462
Prum RO, Torres RH (2003) Structural colouration of avian skin: convergent evolution of coherently scattering dermal collagen arrays. J Exp Biol 206:2409–2429
Rakitov R, Gorb SN (2013) Brochosomal coats turn leafhopper (Insecta, Hemiptera, Cicadellidae) integument to superhydrophobic state. Proc R Soc Lond B 280:20122391
Raspotnig G, Matischek T (2010) Anti-wetting strategies of soil-dwelling Oribatida (Acari). Acta Soc Zool Bohem 74:91–96
Richards AG (1951) The integument of arthropods. The chemical composition and their properties, the anatomy and development, and the permeability. University of Minnnesota Press, Minneapolis, London, Geoffrey Cumberlege, Oxford University Press
Rusek J (1998) Biodiversity of Collembola and their functional role in the ecosystem. Biodivers Conserv 7:1207–1219
Schmidt CV, Gorb SN (2012) Snake scale microstructure: phylogenetic significance and functional adaptations. Zoologica 157:1–106
Sergeev A, Timchenko AA, Kryuchkov M, Blagodatski A, Enin GA, Katanaev VL (2015) Origin of order in bionanostructures. RSC Adv 5:63521–63527
Stalleicken J, Labhart T, Mouritsen H (2006) Physiological characterization of the compound eye in monarch butterflies with focus on the dorsal rim area. J Comp Physiol A 192:321–331
Stavenga DG, Foletti S, Palasantzas G, Arikawa K (2006) Light on the moth-eye corneal nipple array of butterflies. Proc R Soc Lond B Biol Sci 273:661–667
Tammes PML (1930) On the origin of number and arrangement of the places of exit on pollen grains. Recl Trav Bot Néerl 27:1–84
Tarnai T, Gáspár Z (1987) Multi-symmetric close packings of equal spheres on the spherical surface. Acta Crystallogr 43:612–616
Thompson CV (2000) Structure evolution during processing of polycrystalline films. Ann Rev Mater Sci 30:159–190
Varenberg M, Halperin G, Etsion I (2002) Different aspects of the role of wear debris in fretting wear. Wear 252:902–910
Weygoldt P (2000) Whip spiders, (Chelicerata, Amblypygi), their biology, morphology and systematics. Apollo Books, Stenstrup, p 163
Wiggelsworth VB (1947) The epicuticle in an insect, *Rhodnius prolixus*. Proc R Ent Soc L B 134:163–181
Wiggelsworth VB (1976) The distribution of lipid in cuticle of *Rhodnius*. In: Hepburn HR (ed) The insect integument. Elsevier, Amsterdam, pp 89–106
Wolff JO, Schwaha T, Seiter M, Gorb SN (2016) Whip spiders (Amblypygi) become water-repellent by a colloidal secretion that self-assembles into hierarchical microstructures. Zool Letters 2(N23):1–10
Wolff JO, Seiter M, Gorb SN (2017) The water-repellent cerotegument of whip-spiders (Arachnida: Amblypygi). Arthropod Struct Dev 46:116–129
Yang SM, Jang SG, Choi DG, Kim S, Yu HK (2006) Nanomachining by colloidal lithography. Small 2:458–475

Chapter 9
Ecology and Evolution

Abstract Myrmecochory or plant seed dispersal by ants is a widely spread phenomenon. Seeds of such plants bear specialised lipid-rich appendages, elaiosomes, for attracting ants. Ant workers collect the seeds and usually carry them to their nests. The ant species complex in the ecosystem is continuously changing in time and space, and the question arises about the effect of the spatial distribution of different ant species in the ecosystem on the number and distribution of myrmecochorous plants with different dispersal strategies. In this chapter, we model the population dynamics of two myrmecochorous plants having various dispersal strategies in an ecosystem with two ant species differing in their seed preferences, colony territory size, and location of their waste piles. We find a correlation between the number of nests of different ant species and the stability of the ecosystem. In particular, if one ant species would partially or totally disappear from the system, this could cause dramatic changes in the plant populations as well. Another example treated in this chapter deals with animal aggregations, which are especially common in insects. The aggregations may result from an uneven distribution of resources or because an attraction of individuals to each other may be more efficient in defending the group against predators in general and each member of the group in particular. Tree trunks and other cylindrical objects, where aggregated insects live, represent a specific environment for predator-prey interactions, which is fundamentally different from the planar one. For a better understanding of the predator-prey interaction in a cylindrical space, we applied a numerical model that allows testing the effect of interactions between predator and aggregated prey on the plane and on the cylinder, taking into consideration different abilities of predators to visually detect the prey in these two types of space. It is shown that the aggregation in conjunction with a specific environment may bring additional advantages for the

---

Electronic supplementary material The online version of this chapter (https://doi.org/10.1007/978-3-030-41528-0_9) contains supplementary material, which is available to authorized users.

A. E. Filippov, S. N. Gorb, *Combined Discrete and Continual Approaches in Biological Modeling*, Biologically-Inspired Systems 16,
https://doi.org/10.1007/978-3-030-41528-0_9

prey. When one prey subgroup aggregates on the other side of the tree trunk and becomes invisible for the predator, it will survive with a higher probability. After all, the predator moving along one side of the tree will finally loose the major group of the prey completely.

9.1 Long-Term Dynamics of Ant-Species-Dependent Plant Seeds

Myrmecochory or seed dispersal by ants is a widely spread phenomenon and myrmecochorous plants constitute a large portion of species in many ecosystems. Since the ant species complex in an ecosystem is continuously changing in time and space, the long-term effects of such ant–plant interactions on the plant community remain unclear. The manifold information obtained from numerous previous studies of one ecosystem in the deciduous forests of Central Ukraine allowed us to simulate possible scenarios for plant survival and distribution in the ecosystem after a reduction or local disappearance of one of the ant species.

The results of this virtual long-term experiment show that the abundance and spatial distribution of myrmecochorous plants strongly depends on both the abundance of the ants and their species composition in the ecosystem. The results also demonstrate the positive role of ant species diversity for maintaining myrmecochorous plant species diversity. Competition between plant species for seed dispersers is influenced by the ant community in such a way that the disappearance of one ant species may lead to the reduction or even local disappearance of a particular plant population.

9.1.1 Myrmecochorous Plant Community

Myrmecochory or seed dispersal by ants is a widespread phenomenon found in representatives of nearly 80 families of plants (Beattie 1983, 1985; Giladi 2006). According to recent reports, it has evolved independently at least 100 times in angiosperms and is estimated to be present in over 11,000 species or more than 4.5% of all known angiosperm species (Lengyel et al. 2009). Myrmecochores are globally distributed and constitute large portions of the species in many ecosystems, e.g. accounting for up to 30–40% of spring-flowering herbs in temperate deciduous forests of North America (Beattie and Culver 1981; Handel et al. 1981; Beattie 1985) and 40–50% in Europe (Gorb and Gorb 2003).

Seeds of myrmecochorous plants bear specialized, lipid-rich appendages, so-called elaiosomes, for attracting ants (Bond and Stock 1989). Ant workers collect these seeds and usually carry them to their nests (Fig. 9.1). Some seeds reach the nests, whereas others are dropped during transport (Hughes and Westoby 1990,

Fig. 9.1 Ants (*Formica polyctena*) collecting elaiosome-bearing seeds of the violet (*Viola odorata*). (From Gorb et al. 2013)

1992). In the nests, the energy-laden elaiosomes are removed and consumed, whereas intact and viable seeds are commonly deposited either in underground nest chambers or in "waste piles" outside the nest (Buckley 1982; Beattie 1985; Keeler 1989; Gorb et al. 2000). The ants benefit by receiving high-quality food rich in fats, fatty acids, sugars, amino acids, and proteins (Bresinsky 1963; Thompson 1973; Beattie 1985; Soukup and Holman 1987; Lanza et al. 1992; Fischer et al. 2008).

Although mean dispersal distance is relatively short (Culver and Beattie 1978; Horvitz and Schemske 1986; Hughes and Westoby 1992; Gómez and Espadaler 1998; Ness et al. 2004), myrmecochory provides plants with several selective advantages, such as protection of the seeds against predators and fire, avoidance of interspecific competition and reduction of the competition between the parental plant and its seedlings as well as among seedlings, as well as transportation of the seeds to sites suitable for germination and seedling development (for review see Ulbrich 1928; Beattie 1985; Giladi 2006; Rico-Gray and Oliveira 2007).

For many plant species, seed dispersal by ants is the only dispersal method (*obligate myrmecochory*). Another part of the myrmecochorous group of plants is composed of diplochorous species using another dispersal method, mostly autochory, in addition to ant dispersal (*facultative myrmecochory*) (Sernander 1906; Ulbrich 1928; Berg 1966; Gorb and Gorb 2003). Depending on their dimensions and elaiosome size (or elaiosome-to-seed ratio), different seeds are attractive to different ant species and have different dropping rates during transport to/from the nest (Beattie and Lyons 1975; Culver and Beattie 1980; Davidson and Morton 1981a, b; Kjellsson 1985; Horvitz and Schemske 1986; Bond and Stock 1989; Higashi et al. 1989; Oostermejer 1989; Gorb and Gorb 1995, 1999a, b; Garrido et al. 2002; Mayer et al. 2005; Edwards et al. 2006). These distinctions, together with differences in the dimensions of the ant colonies' territories, sizes of ant individuals, locations of waste piles, and overall behavior (e.g. foraging strategy) of the ants, presumably influence dispersal success, abundance, distribution and survival of myrmecochorous plants in the ecosystem.

The ant species complex in any ecosystem is continuously changing in time and space, and the long-term effects of ant–plant interactions on the plant community are still not clear. Since the germination success of seeds, as well as the survival of seedlings, depend on the biology of the seed dispersal agents (Davidson and Morton 1981a, b; Hughes and Westoby 1992; Gorb and Gorb 2000, 2003; Gorb et al. 2000; Garrido et al. 2002; Cuautle et al. 2005; Giladi 2006; Rico-Gray and Oliveira 2007), this raises questions about the effect of the spatial distribution of different ant species in the ecosystem on the number and distribution of myrmecochorous plants with different dispersal strategies. Here, we aim at modeling the population dynamics of two myrmecochorous plants with different dispersal strategies in an ecosystem with two ant species that differ in their seed preferences, territory size, and location of their waste piles. The present model incorporates the interactions of these four species only, although ant and plant communities are usually a lot more complex.

We used our observations and experimental data on ant–seed interactions obtained in the deciduous forest of the Central Ukraine (Gorb and Gorb 2003) as a basis for the model. This set of data contains information on the removal and dropping rates of ant-dispersed seeds (Gorb and Gorb 1995, 2000, 2003; Gorb 1998), seed dispersal distance (Gorb and Gorb 1999b), the soil seed bank (Gorb and Gorb 2003) and seed-related ant behavior (Gorb et al. 2000). This manifold information from one ecosystem allowed us to simulate possible scenarios of plant distribution within the ecosystem when one of the ant species is reduced in abundance or goes locally extinct.

The main goal of this study is to determine a correlation between the number of nests of different ant species and the stability of the ecosystem under consideration. Taking into account that the survival of different plant species strongly depends on the location and number of ant nests, one could expect that the partial or total disappearance of one ant species from the system would cause dramatic changes in the plant populations as well. To elucidate this, we performed a set of numerical experiments with different numbers of ant nests.

9.1.2 Temporal Development of the Forest Ecosystem

To simulate the temporal development of the forest ecosystem under consideration, we apply the following combination of a discrete and a continuous model (Gorb et al. 2013): A subsystem of ant nests is treated as two separate arrays of discrete points $\{x_{kj}, y_{kj}\}$. The nests are assumed to be randomly placed inside a fixed area in 2D space $[x, y]$. Here, the indices $\{k\}$ and $\{j\}$ identify two different species of ants, $k = 1, 2$, and a particular ant nest $j = 1, \ldots, N_k$ in each set of ant nests.

Index $k = 1$ corresponds to the nest of the larger ants with a larger foraging territory, and $k = 2$ corresponds to smaller ants with a smaller territory. Initial numbers of the nests at $t = 0$ are always fixed and equal to $N_1 = 10$ and $N_2 = 50$, respectively. For simplicity in further simulations, we selected a square forest area of size $L_x \times L_y$ with constant equal sides L_x and L_y: $L_x = L_y = 50$ meters.

Spatial distributions of the ant nests and the corresponding plant densities are developed by the system self-consistently as a result of a self-organized process and in general are unknown to us '*ab initio*'. For appropriate initial conditions, we have to start from some more or less natural random initial conditions and wait for some transient time to get a self-organized structure. Here, we apply this approach and determine the initial positions of the ant nests at $t = 0$ by the following formulae:

$$x_{kj} = L_x \zeta_{kj}, \quad y_{kj} = L_y \zeta_{kj}, \tag{9.1}$$

where $\zeta_{kj}(x, y)$ are δ-correlated random numbers $< \zeta_{kj}\zeta_{k'j'} >= \delta_{kk'}\delta_{jj'}$ uniformly distributed in the interval $[-1/2 \div 1/2]$ along both spatial coordinates, x and y. Briefly said, the original nests are distributed uniformly and independently in the square of the forest domain. Initially, they do not necessarily lie outside the territories of each other, but this naturally happens after some transient time for subsequent generations.

One of the most important sources for the dynamic behaviour of the populations presented here is the ants' ability to change the position of their nests from time to time, relocating (almost randomly) from one place to another within the forest. According to biological observations (Smallwood 1982; Hölldobler and Wilson 1990), each nest exists for a limited time in a particular position $\{x_{kj}, y_{kj}\}$. This time is defined by the ant species, the prevalent environmental conditions and other factors and can be estimated at 15–20 years. During the main part of this period (hereafter called "lifetime" τ_L), the probability of finding the ant nest in the same place, $P_k = P_k(t)$; $k = 1, 2$, is very close to $P_k = 1$ and is almost constant. For simplicity, we will assume that the nests of both ant species have existed in fixed positions approximately for the same lifetime τ_L, which was determined to be 20 years. In the frame of the model, this means that nests live, on average, for 20 years, with the individual lifetime chosen from the given distribution. The locations for new nests are then chosen uniformly from the region, but excluding areas within existing territories.

To simulate this process numerically, we apply the following procedure. For each time step t for both arrays of ant nests, we generate arrays of random numbers $\zeta_{kj}(t)$, uniformly distributed in the interval [0, 1], and compare them with a smoothed step-like function $P_k(\delta t_j) = 1/\left(\exp \frac{\delta t_j - \tau_L^k}{0.1\tau_L^k} + 1\right)$, which varies with regulated width from 1 to 0 and is shown in Fig. 9.2.

The procedure is organized as follows. Individual "living time" δt_{kj} is calculated for every ant nest, starting from the moment of its appearance in a given place. When a randomly generated number $\zeta_{kj}(t)$ exceeds the function $P_k(\delta t_{kj})$, a nest from the array k with the index j leaves its current place with the coordinates (x_{kj}, y_{kj}) and randomly moves to a new place. At this time, we also reset its age ("living time") δt_{kj} to zero and its "life" begins in a new position from $\delta t_{kj} = 0$.

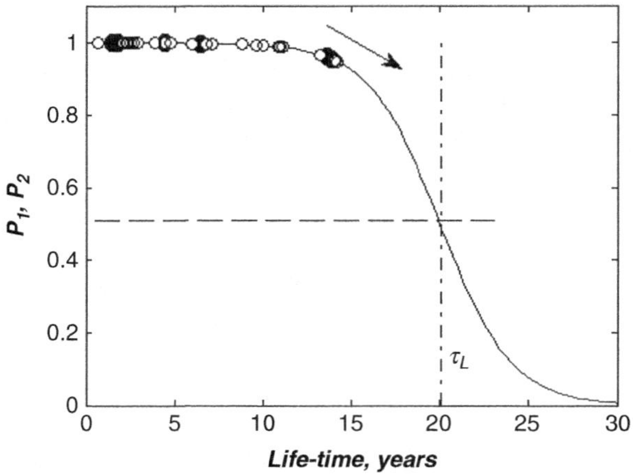

Fig. 9.2 Dependence of the probability of an ant nest to remain ("survive") in the same position in a forest on the length of the 'lifetime', for which the nest already exists in this place. Small and large circles mark the positions of the specific 'lifetimes' of two different ant species. The arrow indicates the direction along the curve in which these positions move with time. When the current ant nest is left and moved to another position in the forest, the corresponding circle is transferred back to its initial point at zero time. (From Gorb et al. 2013)

It is assumed that each ant colony has its own territory (circle with radius R_k) that differs for large and small ants: $R_1 = 5$; $R_2 = 1$. Formally, when an ant nest moves to a new place, its position can be randomly chosen within the entire forest area $L_x \times L_y$. Regions (circles) already occupied by a colony of the same ant species, however, have to be excluded (Gorb et al. 2000). If, occasionally, a new ant nest position appears within an already existing circle, the program repeats the process of randomly choosing a new position until a place in an empty region is found.

To simulate the ability of the ants to disperse plant seeds, each ant nest is assumed to have, around its current position, a region with a positive impact on the probability $z_k^{k'}(x, y)$ of the plants to survive and produce new seeds for the next generation. Here, we use the subscripts $k = 1, 2$ to designate the nests of the two different ant species, and the superscripts $k' = 1, 2$ to designate different species of plants (large-seeded and small-seeded, respectively). In general, both large and small ants can take up and transport both kinds of seeds, but with different probabilities $B_k^{k'}$. This was previously shown experimentally for a series of plant species and ant species (Gorb and Gorb 1995, 1999a, b; Gorb 1998). The probabilities can strongly correlate with the presence or absence of other ant species (Gorb and Gorb 1999a, b).

Large and small ants collect the differently sized seeds of different plant species and dispose them either in a ring close to the border of their territory (large ants) or in small circles within the entire territory of the ant colony (small ants). The deposition of seeds close to the territory border is characteristic for large *Formica polyctena* ants (Gorb et al. 2000). Seed distribution over the entire territory results from a high dropping rate during transport (Gorb and Gorb 1999b): collected seeds do not reach the nest and are dispersed along the foraging trails. In the model, this is accounted for by a different spatial structure of the "preference coefficients":

$$z_1(x,y) = \sum_{k=1}^{2} B_k^1 \sum_{j=1}^{N_1} \exp\left[-\left(\frac{r-R_1}{R_2}\right)^2\right]; \quad z_2(x,y) = \sum_{k=1}^{2} B_k^2 \sum_{j=1}^{N_1} \exp\left[-\left(\frac{r}{R_2}\right)^2\right], \tag{9.2}$$

where $r = \sqrt{(x - x_{kj})^2 + (y - y_{kj})^2}$ is the distance from an arbitrary point in the area $L_x \times L_y$ to the position of the ant nest $\{x_{kj}, y_{kj}\}$. The values of the coefficients $B_k^{k'}$ should be chosen to reflect experimentally observed preferences in the choice of different seeds by the different ant species (Kjellsson 1985; Lanza et al. 1992; Mark and Olesen 1996; Rico-Gray and Oliveira 2007). Here, they were assumed to be $B_1^1 = 4$, $B_1^2 = 0.1B_1^1$, $B_2^2 = 1$, $B_2^1 = 0.1B_2^2$.

The space/time evolution of the seed densities $f_{1,\,2}(x, y)$ is determined by the following equations:

$$\begin{cases} \dfrac{\partial f_1}{\partial t} = c_1 \Delta f_1 + f_1(G_1 - g + u z_1) \\ \dfrac{\partial f_2}{\partial t} = c_2 \Delta f_2 + f_2(G_2 - g + u z_2) \\ \dfrac{\partial u}{\partial t} = c_u \Delta u + u(G_u - f_1 - f_2) \end{cases}. \tag{9.3}$$

The last equation in the system of Eq. (9.3) describes time- and space-distributed resources (soil fertility, light, water supply), which are not directly specified within the framework of this model, but define necessary conditions for seed (plant) survival. As usual for all populations (including density $u(x, y)$ of the resources), the evolutionary equations include the "birth-rate" constants $G_{1,\,2}$ and G_u that describe the fertility of the particular species $G_{1,\,2}$ or the monotonous growth of the resources G_u in the absence of their consumption and ant activity. Activity- and resource-dependent impacts to the density evolution are incorporated into the model by terms proportional to the products $f_{1,\,2}uz_{1,\,2} \equiv f_{1,\,2}(x, y)u(x, y)z_{1,\,2}(x, y)$. The biological meaning of these terms is as follows: the activity of the ants has an impact on the local growth of the plants, which is proportional to the density $f_{1,\,2}$ itself, the amount of resources available in this territory and the "preference coefficients" defined by Eq. (9.2). The products $f_{1,\,2}uz_{1,\,2} \equiv f_{1,\,2}(x, y)u(x, y)z_{1,\,2}(x, y)$ in the equations are written in a dimensionless form and all conversion factors for the transformation of the resource u into seed densities and vice versa are included in the coefficients connecting these values.

The system of Eq. (9.3) also contains non-local terms with Laplace operators:

$$\Delta f(x, y) \equiv \frac{\partial^2 f}{\partial x^2} + \frac{\partial^2 f}{\partial y^2}, \tag{9.4}$$

which describe the spatial dispersion of the seeds and resources in each generation due to various physical factors (like wind or random deposition, for example) that are independent of the activity of the ants. The coefficients $c_1 = 0.1$; $c_2 = 0.3$ and $c_u = 0.1$ near non-local terms of the system of Eq. (9.3) regulate the intensity of the spatial dispersion $f_{1,\ 2}$ for both plants and resources, according to the biological meaning of each density. The death rate $g = 0.1$ is taken to be equal for both species of plants.

As mentioned above, the initial conditions for solving the equations in Eq. (9.3) self-consistently with variable positions of the ant nests are unknown. However, if the dynamics of the system described by these equations is correct, it must be stable and, under stationary external conditions, must tend, at $t \to \infty$, to some attracting scenario (some stationary, but generally non-static behavior). In this way, we can "naturally" produce an appropriate distribution starting from an arbitrary one. The quickest and simplest way to get there is to start from uniformly distributed, random initial densities $f_{1,\ 2}$ and u. One can easily demonstrate that within the time scale of this particular problem, it takes around 40–60 years to reach almost stationary system behavior with deviations of the mean values smaller than 10%. Below, we always used a period of 50 years as a transient run to create natural initial distributions from the randomly generated ones. Random initial conditions, together with the randomness in the motion of the ant nests during the process, including their possible appearance and disappearance in given positions, lead to different realizations of the stationary state. However, if the total number of ant nests is fixed, the process is stable and will lead to the same total plant populations, $F_{1,\ 2} = \int_{Area} f_{1,\ 2}(x, y)dxdy$, with deviations in the mean number of plant individuals <10%.

9.1.3 Integral Values of Time-Depending Behavior and Their Biological Interpretation

Typical density distributions $f_{1,\ 2}$ and u at some point in time obtained at $N_1 = 10$ and $N_2 = 50$ are shown in Fig. 9.3 by grey-scale maps with brighter regions corresponding to higher densities. Nests of larger and smaller ants are represented by large and small dots, respectively. Mutual correlation between the space distributions of two plant species is presented in Fig. 9.4 by the difference of the densities $f_1 - f_2$.

Generally, each plant species avoids places already occupied by another species. As a result, different regions are found to be occupied by different plant species. This

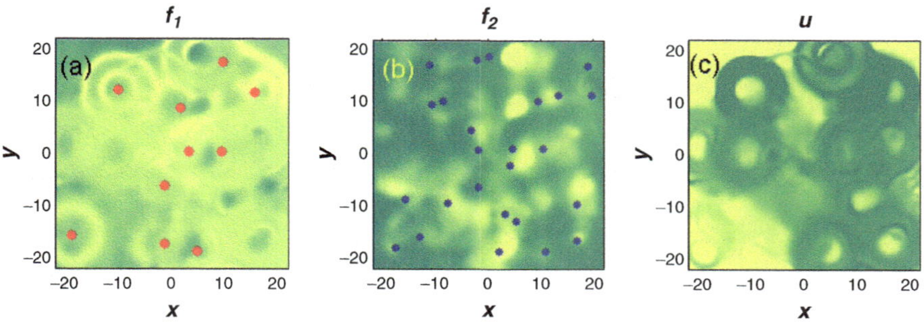

Fig. 9.3 Typical instant configurations at some point in time. Positions of the ant nests (large and small ants) are shown by large and small dots in panels (**a**) and (**b**), respectively. Spatial configurations of the corresponding densities of plants $f_{1,\,2}$ are presented in the same panels by color maps, with bright regions corresponding to higher densities. Distribution of the resources u at the same moment in time is shown in panel (**c**). (From Gorb et al. 2013)

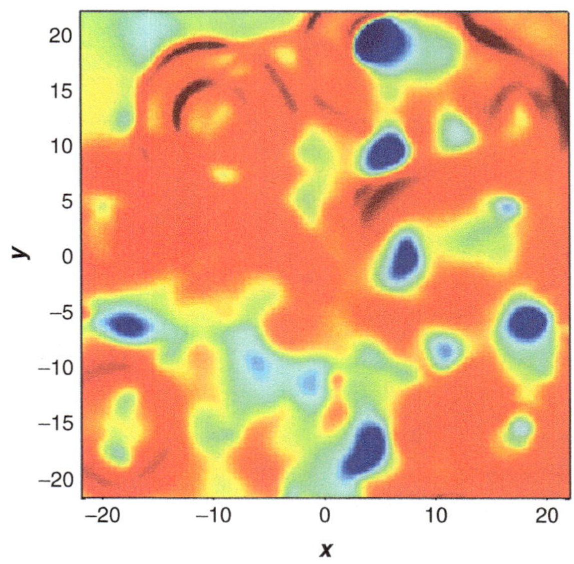

Fig. 9.4 Spatial correlation of the distributions shown in Fig. 9.3. Each density distribution $f_{1,\,2}$ avoids places occupied by other plant species. As a result, the regions occupied by different species are visualized here by a color map for the difference $f_1 - f_2$, with contrast regions corresponding to higher densities (red and blue correspond to the densities f_1 and f_2, respectively). (From Gorb et al. 2013)

can be visualized by the color map in Fig. 9.4 where red regions are occupied by f_1 and blue ones by f_2 densities, respectively.

A typical scenario of the development is presented in supplementary videos 9.1, 9.2 and 9.3. These movies present the dynamics of particular realizations of the process. Supplementary video 9.1 corresponds to Fig. 9.3, supplementary video 9.2 summarizes the information presented in Figs. 9.2 and 9.8, and supplementary video 9.3 reproduces a configuration analogous to that shown in Fig. 9.4.

The results of the numerical experiments with a decreasing number of ant nests are summarized in Figs. 9.5, 9.6, 9.7, 9.8, 9.9 and 9.10. For the first 50 years, we generated different stationary realizations of all the subsystems (ants and plants) with

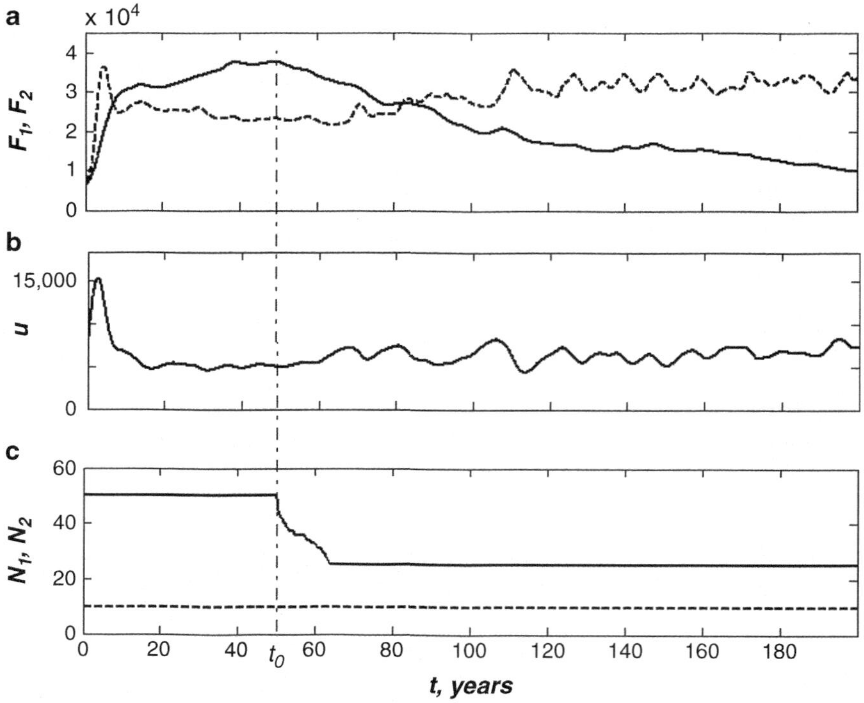

Fig. 9.5 Temporal development of the total populations. The values $F_{1,\,2} = \int_{Area} f_{1,\,2}(x, y)dxdy$ are shown in panel (**a**) by a dashed (F_1) and a solid line (F_2), respectively. Panel (**b**) presents the corresponding temporal evolution of the resources u. Time-dependent numbers of the ant nests N_1 and N_2 are shown by the dashed and the solid line in panel (**c**), respectively. The vertical dash-dotted line indicates the point in time when N_2 begins to decrease from its initial value, $N_2 = 50$, to the final one, $N_2 = 30$. (From Gorb et al. 2013)

fixed numbers $N_1 = 10$ and $N_2 = 50$, and then gradually decreased the total number of one type of ant nests down to various final values $N_{1,\,2final}$.

The numerical procedure of changing the numbers N_1 or N_2 resembles the one we used for the relocation of the nests from one place to another. For this purpose, we generate uniformly distributed random numbers $\zeta_{kj}(t)$ and compare them with a preliminary specified threshold. If $\zeta_{kj}(t)$ exceeds the threshold, the ant nest disappears. All continuous variables corresponding to the vanishing ant nest become *sparse* arrays, and the corresponding numbers $N_{1,\,2} = N_{1,\,2}(t)$ decrease as well. The process continues until each of the numbers $N_{1,\,2}$ are larger than their final values$N_{1,\,2}>N_{1,\,2final}$. Both plant species compete for the resources. A change in the numbers $N_{1,\,2}$ modifies competition conditions and shifts the equilibrium between the populations.

In the presence of large ants, the smaller ants usually avoid confronting them and are not able to collect the large seeds. When large ants are absent, however, the small ants do collect large seeds (Lanza et al. 1992). One can account for such a possibility

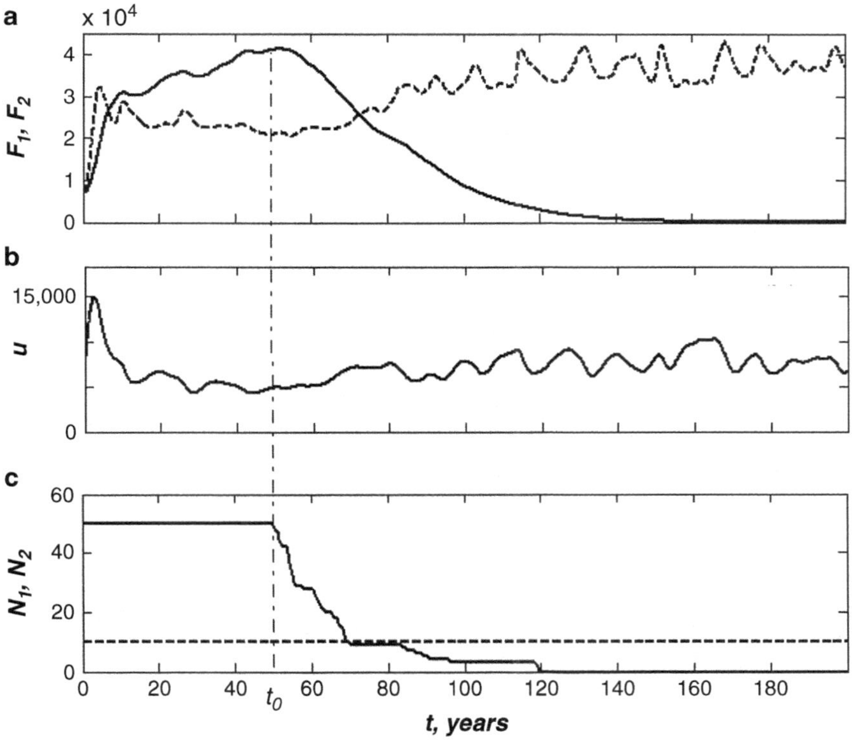

Fig. 9.6 The same graphs as in Fig. 9.5 for a scenario in which N_2 varies from $N_2 = 50$ to $N_2 = 0$. (From Gorb et al. 2013)

in the model by varying the coefficient B_2^1. Instead of a 'naive' constant value $B_2^1 = 0.1B_2^2$, B_2^1 is assumed to depend on a number $N_1(t)$: $B_2^1(t) = \frac{2B_2^2}{1+1.9N_1(t)}$. This value tends to the limit $B_2^1 = 0.1B_2^2$ at $N_1 = 10$ and moves to a new relation $B_2^1 = 2B_2^2$ when $N_1 \rightarrow 0$, which corresponds to the above-mentioned preference $B_2^1 > B_2^2$.

Figures 9.7 and 9.8 show how the total populations $F_{1,\ 2}$ vary for two different final numbers of ant nests, $N_{2final} = 25$ and $N_{2final} = 0$, at a fixed number of $N_1 = 10$. It can be clearly demonstrated that a decrease of N_{2final} leads to a strong reduction of the corresponding plant species and can even cause its local disappearance if N_{2final} is smaller than some threshold. When N_2 is fixed and N_1 decreases, however, this can lead to a wider variety of scenarios. This is due to the ants' preferences for the seeds of different plant species with different sizes. Experimental observations showed that smaller ants also prefer larger seeds (Lanza et al. 1992) and only resort to smaller seeds in the presence of larger ants.

For a general picture of the phenomenon, one can vary each final value, N_{1final} or N_{2final}, in the entire interval $0 \leq N_{1final} \leq 10$ and $0 \leq N_{2final} \leq 50$, with the other value fixed ($N_1 = const$ for varied N_2, and vice versa). The corresponding results are

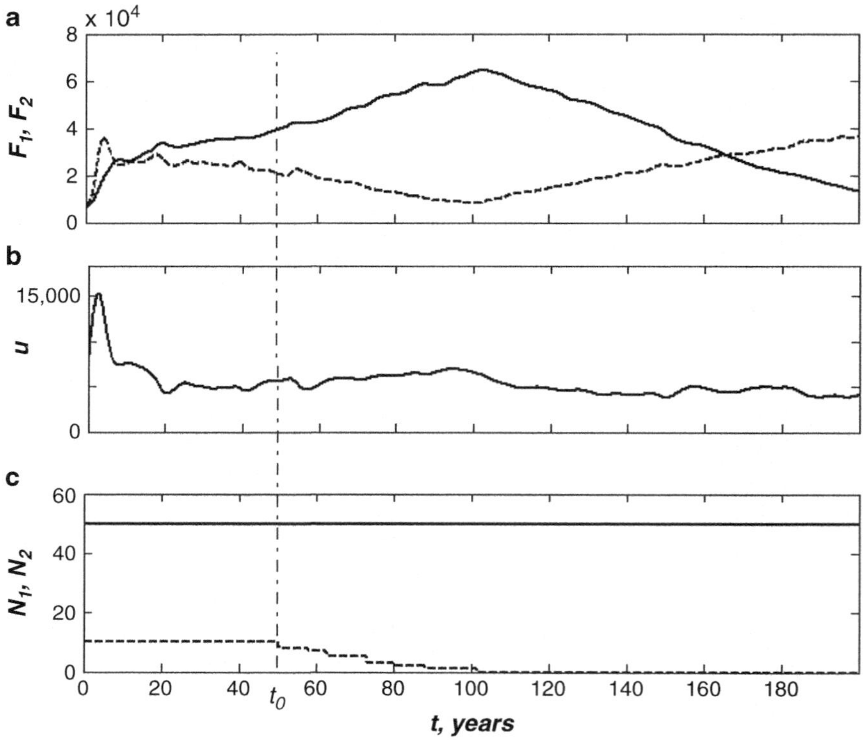

Fig. 9.7 The same graphs as in Fig. 9.5 for N_1 varying from $N_1 = 10$ to $N_1 = 0$. (From Gorb et al. 2013)

summarized in Figs. 9.9 and 9.10, which present the final total plant populations $F_{1,\,2}$ at $t \to 200$ years for both cases. Some representative final configurations of the plant densities $f_1 - f_2$ are shown in the insets in Fig. 9.9.

9.1.4 Discussion of the Modeling Results

Seed dispersal has an important effect on plant communities: it influences the dynamics and persistence of populations and the abundance and distribution of species, it determines the spatial structure of populations, and it affects their adaptation to environmental changes and survival (Nathan and Muller-Landau 2000; Rico-Gray and Oliveira 2007).

Our study demonstrates the dynamics of the spatial pattern of two myrmecochorous plant species that vary in their dispersal strategies in the forest ecosystem in the presence of two ant species with different seed-related behaviors. We found that in a stable ecosystem with a constant number of nests of two ant

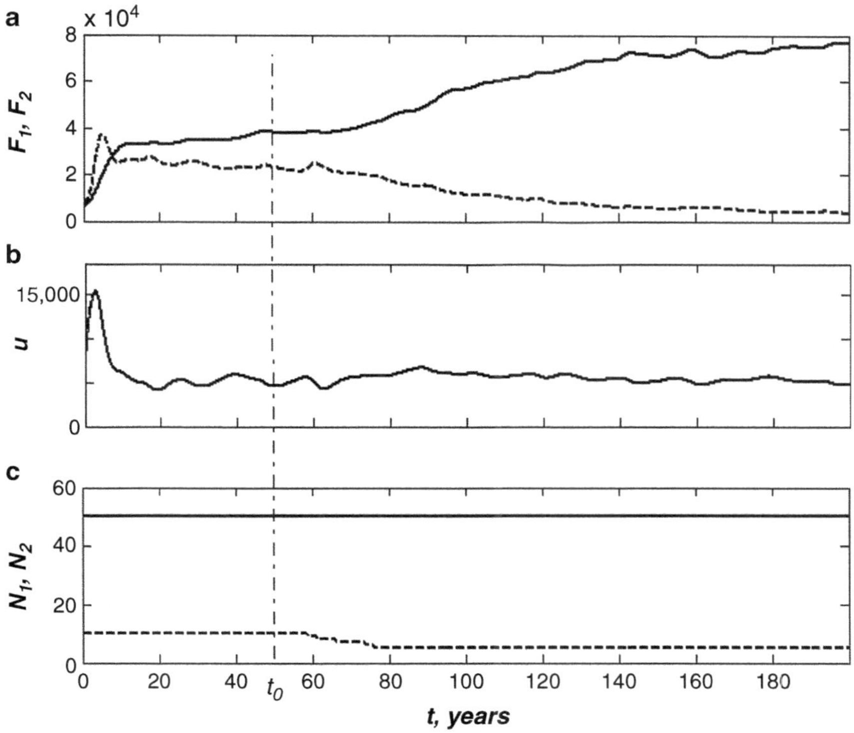

Fig. 9.8 The same graphs as in Fig. 9.5 for N_1 varying from $N_1 = 10$ to $N_1 = 5$. (From Gorb et al. 2013).

species, different plants tend to occupy different sites (Fig. 9.3a, b). Differences in the distribution and population densities between the plant species (Fig. 9.4) result from different seed preferences of the ants, different deposition sites of seeds within the ant territories, and competition for resources (Fig. 9.3c).

A decrease in the nest number of small ants causes, over time, a gradual reduction in the total population of the small-seeded plant species, whereas the population of the large-seeded plant species continuously grows and occupies more sites, profiting from decreased competition for resources (Fig. 9.5). These effects become more pronounced with a decreasing number of surviving nests of the small ants (Fig. 9.9). In the case of local extinction of these ants, the small-seeded plant species also disappears ($N_{2final} = 0$), and the entire territory will be occupied by a highly (up to two-fold) increased population of the other plant species (Figs. 9.6 and 9.9).

A reduction in the nest number of the large ants leads to various scenarios, depending on the number of remaining nests. For a given ant species complex, when the nest number of large ants decreases down to less than one third of the initial number ($3 < N_{1final} < 10$), a continuous strong reduction in the large-seeded plant population occurs simultaneously with an increase in the small-seeded plant

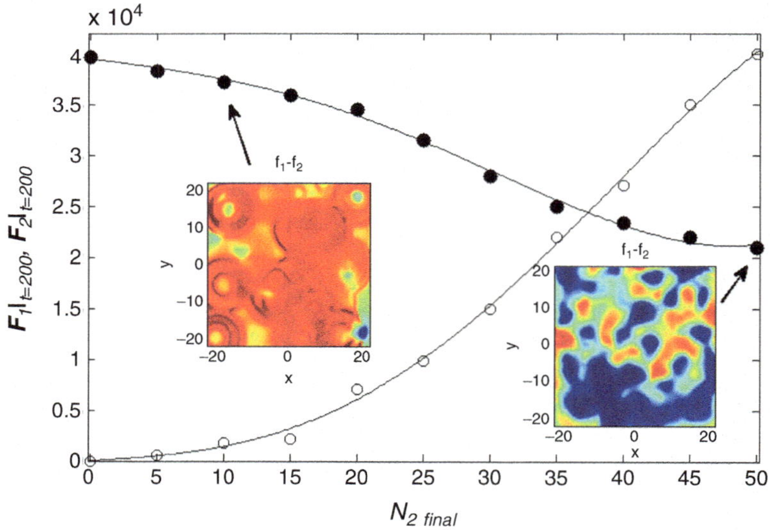

Fig. 9.9 Final total populations. The values of $F_{1,\,2}$ at $t = 200$ years obtained for different N_{2final} are presented by black and empty circles, respectively. The curves connecting the circles are inserted as guiding lines only. Typical density maps of the difference $f_1 - f_2$ at large $N_{2final} = 50$ and small $N_{2final} = 10$ values are shown in the insets. The arrows indicate the total amounts of F_1, to which these distributions are related. In the insets, red and blue colors correspond to the densities f_1 and f_2, respectively, and contrast (blue and red) regions correspond to higher densities. (From Gorb et al. 2013)

population (Figs. 9.8 and 9.10). At a certain (critical) nest number of large ants ($N_{1final} = 3$), the large-seeded plant species will go almost extinct in this local area, while the small-seeded plant strongly dominates (Fig. 9.10).

The situation will differ fundamentally, however, when the large ants disappear completely. Since large seeds are also attractive to small ants, they will readily collect these seeds when large ants are absent (at $0 \leq N_{1final} \leq 2$). Ultimately, the small ants will prefer large seeds and refuse to disperse the small seeds of the respective plants. This situation results in an extremely rapid increase in the large-seeded plant population, even above its initial density, whereas the population density of the small-seeded plant, initially associated with small ants, will drastically drop to a level much lower than the initial one (Figs. 9.7 and 9.10).

This study shows that the abundance and spatial distribution of myrmecochorous plants strongly depend on both ant abundance and ant species composition in the ecosystem. The long-term modeling experiment revealed the importance of ant species diversity for maintaining myrmecochorous plant species diversity in the ecosystem. Competition for resources between the plant species is tuned by the ant community in such a way that it may lead to the reduction or even local extinction of a particular plant population when one ant species disappears. The present model can be potentially extended to a higher number of ant and plant species and adapted to various seed-related behavioral features of different ant species.

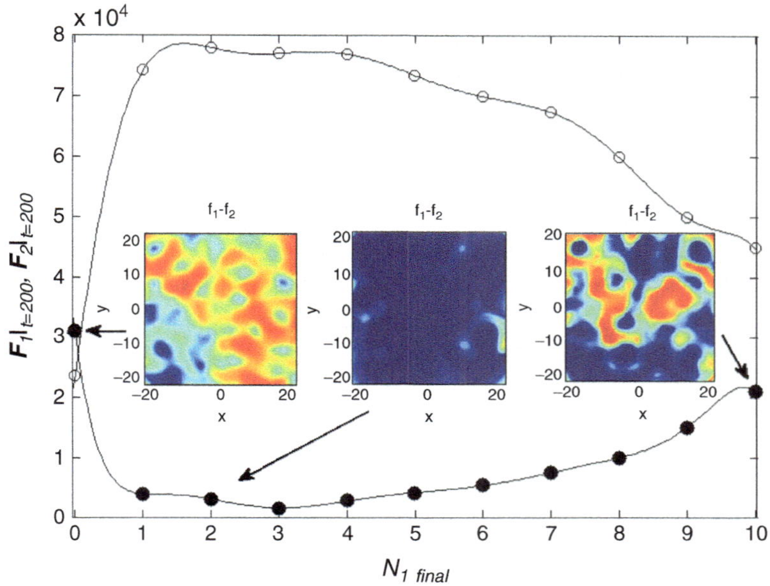

Fig. 9.10 The same graphs as in Fig. 9.9, obtained for varying $N_{1\,final}$. Typical density maps of the difference $f_1 - f_2$ at large ($N_{1\,final} = 10$), small ($N_{1\,final} = 2$) and zero ($N_{1\,final} = 0$) final numbers of $N_{1\,final}$ are shown in the insets. The arrows indicate the total amounts of F_1, to which these distributions are related. In the insets, red and blue colors correspond to the densities f_1 and f_2, respectively, and contrast (blue and red) regions correspond to higher densities. (From Gorb et al. 2013)

9.2 Influence of Aggregation Behavior on Predator–Prey Interactions

A vast number of literature sources reports on animal aggregations, which are especially common in insects, including non-social ones. Such aggregations are formed for a number of reasons and by a number of mechanisms. Aggregations may result from an uneven distribution of resources or from the attraction of individuals to each other due to the stimuli of conspecifics (Bengtsson 2008).

Aggregation behavior must have some selective advantages or at least be neutral in comparison to alternative kinds of behavior. There are some indications that individual advantages are positively correlated with the density of individuals in the group (Stephens and Sutherland 1999): for example, for various reasons, a group may be more efficient in defending itself in general and each member of the group in particular (Alcock 1982; Raffa 2001). The mechanisms of aggregation formation can be rather different. Sensory systems, such as olfaction and taste, are often used by insects for this purpose (Durieux et al. 2015). Some insects can directly produce aggregation pheromones (Vite and Pitman 1969; Klein et al. 1973; O'Ceallachain and Ryan 1977; Lockwood and Story 1985; Hodges et al. 2002), use non-specific compounds produced by conspecifics (Lorenzo Figueiras and Lazzari 1998) or react to some compounds released by a damaged plant that specifically attract herbivore conspecifics (Schlyter and Birgesson 1999). Additionally, some insect species use visual cues and signals for maintaining aggregations. For example, larvae of the bug *Nezara viridula* (Hemiptera, Pentatomidae) aggregate due to visual and tactile stimuli (Lockwood and Story 1985), similar to those of some other hemipterans (Minoli et al. 2007). Later larvae of *N. viridula* produce their own aggregation pheromone that is attractive at low to intermediate concentrations, but repellent at high concentrations (Bengtsson 2008).

The advantages of living in a group have been examined in earlier publications (Handegard et al. 2012). A widely accepted benefit of aggregation behavior is enhanced protection against predators (Cocroft 2002), e.g., by the "dilution of risk" (Treherne and Foster 1982; Watt et al. 1997; Aukema and Raffa 2004; Morrell and James 2007), since an individual is less exposed to potential predators when in an aggregation. Another benefit could arise in the presence of a predator that is less efficient in utilizing large amounts of aggregated prey (Johannesen et al. 2014). For example, it was demonstrated that in aphids (*Aphis varians*, Homoptera, Aphididae) the rate of population growth in colonies exposed to ladybird beetles (*Hippodamia convergens*, Coleoptera, Coccinellidae) increases with increasing colony size, whereas for colonies protected from predators, population growth rates are generally higher, but decline with increasing colony size (Turchin and Kareiva 1989; Bengtsson 2008). Thus, even though the evolutionary scenarios and functional mechanisms are not entirely clear (Ruxton and Sherratt 2006), the data supports the hypothesis that the individual risk for an aphid to be captured by a ladybird beetle decreases as colony size increases. Also, theoretical models predict that the probability that feeding individuals will detect approaching predators increases with group size (Miller 1922; Pulliam 1973).

Some insects, such as water striders (*Halobates robustus*, Hemiptera, Gerridae), aggregate on a 2D plane (Foster and Treherne 1980; Treherne and Foster 1980, 1981, 1982). In insects living in vegetation (which is a 3D terrain), however, the aggregation strategy may interact with the geometry of the space they live in. Tree trunks and other cylindrical objects where aggregated insects live represent a very specific environment for predator–prey interactions, which will necessarily differ from the situation on a 2D plane, as can be observed, for example, in aggregations of juvenile *Cerastipsocus sivorii* (Psocoptera, Psocidae) on tree trunks in Brazil (Fig. 9.11a). Nearly 50% of the aggregations studied so far had up to 90 individuals, but large groups presenting 240 individuals or more were also frequent, comprising 10% of all aggregations found in the field (Requena et al. 2007). The group is maintained by some chemical cues (Buzatto et al. 2009), but psocopterans also use their antennae to monitor/control the distance to the nearest neighbor by tactile cues (Fig. 9.11b). The *C. sivorii* group in Fig. 9.11 was repeatedly and successfully attacked by an assassin bug (*Graptocleptes* sp., Hemiptera, Reduviidae) (Fig. 9.11c). The group used various escape strategies, from splitting apart into smaller groups up to hiding on the other side of the trunk.

Fig. 9.11 Aggregation of juvenile *Cerastipsocus sivorii* (Psocoptera: Psocidae) on a tree trunk, Brazil. (**a**) A group of *C. sivorii* escaping from a predator along the tree trunk. (**b**) Arrangement of individuals within the group. Note the positions of long antennae overlapping with neighboring individuals. (**c**) Reduviid bug – a typical predator – pursuing the group. Note the skin of the sucked-out *C. sivorii* on the bug's mouthparts. (From Filippov et al. 2019)

9.2.1 Numerical Model of Interactions Between a Predator and Aggregated Prey

In order to model the interactions between a predator and aggregated prey in an environment which realistically reproduces the conditions in which aggregated insects usually live, we will introduce a numerical model that allows us to test the effect of different interactions between predator and aggregated prey on a plane (2D) and on a cylinder (3D), taking into consideration the differing ability of predators to visually detect their prey in these two types of space.

The model is organized on a two-tier basis. First, we formulate a model which reproduces predator–prey behavior in ordinary (2D) space, similar to some other publications (Strömbom et al. 2014). Next, the model is generalized to include cylindrical spaces, in accordance with the biological systems described above. Many existing models (Lotka 1925; Volterra 1928; Thompson 1924; Nicholson and Bailey 1935; Watt 1959; Hassell and Varley 1969) consider random search only, which implies one predator or an even distribution of predators throughout the whole prey area and makes the particular type of prey distribution irrelevant to the model outcome (Hassell and May 1974). Here, for definiteness, we will limit the model to only one predator. Conceptually, the interaction between a predator and prey items is modeled by the specific spatial dependence of attraction–repulsion interactions between the predator and the prey and between the prey items themselves (Schellinck and White 2011). The predator "sees" the prey and is attracted to the current positions of the prey items with a force that depends on the distance to each one of them. The simplest way to simulate an interaction with a certain characteristic distance is to use a Gaussian effective potential with given width R_0:

$$f_j^{pred} = A_0\left(\vec{R} - \vec{r}_j\right)\exp\left[-\left(\frac{\vec{R} - \vec{r}_j}{R_0}\right)^2\right],$$

where factor $A < 0$ defines the amplitude of the attraction, vector $\vec{R} = \{X, Y\}$ defines the position of the predator and the array of the vectors $\vec{r}_j = \{x_j, y_j\}$ describes the positions of the prey items $j = 1, .., N$, with N representing the total number of the population. The total attraction force acting on the predator is given by the sum of all these forces:

$$F^{pred}\left(\vec{R}\right) = \sum_{j=1}^{N} f_j^{pred}\left(\vec{R}, \left\{\vec{r}_j\right\}\right).$$

The prey animals are repulsed from the predator with a force that has the same structure as the attraction force acting on the predator, but with a different amplitude B_{0j} and radius R_{0j}:

$$f_j = B_{0j}\left(\vec{R} - \vec{r}_j\right)\exp\left[-\left(\frac{\vec{R} - \vec{r}_j}{R_{0j}}\right)^2\right],$$

which can slightly vary from one animal to another around some mean value typical for the particular species. Additionally, the prey items interact with each other. Their tendency to aggregate is simulated by their mutual attraction at longer distances:

$$f_{jk}^{attract}\left(\vec{r}_j, \vec{r}_k\right) = B_{jk}^{attract}\left(\vec{r}_j - \vec{r}_k\right)\exp\left[-\left(\frac{\vec{r}_j - \vec{r}_k}{R_{jk}^{attract}}\right)^2\right].$$

This attraction, however, is limited by the characteristic maximum density of the animals. Mathematically, this means that they cannot penetrate into the "individual domain" of another animal. This should be represented by a strong repulsion force $B_{jk}^{repuls} \gg \left|B_{jk}^{attract}\right|$ at shorter distances $R_{jk}^{repuls} \ll R_{jk}^{attract}$, which represents the characteristic scale of the "individual domain" of the animal:

$$f_{jk}^{repuls}\left(\vec{r}_j, \vec{r}_k\right) = B_{jk}^{repuls}\left(\vec{r}_j - \vec{r}_k\right)\exp\left[-\left(\frac{\vec{r}_j - \vec{r}_k}{R_{jk}^{repuls}}\right)^2\right].$$

For different species, the radius of such a domain R_{jk}^{repuls} can be very different and can vary from a minimal length (coinciding with the physical size of the animal) to much greater lengths (up to dozens of meters or more).

For simplicity and definiteness, we here define each animal by two different points, "head" and "tail", with different radiuses of repulsion and attraction, $R_{jk,1}^{repuls} > R_{jk,2}^{repuls}$:

$$R_{jk}^{repuls} = \left\{ R_{jk,1}^{repuls}, R_{jk,2}^{repuls} \right\}.$$

This means that the complete array of the points that describe the potential positions of the prey is split into two arrays:

$$\vec{r}_j = \left\{ \vec{r}_{j1}, \vec{r}_{j2} \right\},$$

with fixed distances between the points in each pair $l = \left| \vec{r}_{j1} - \vec{r}_{j2} \right| = const$, which corresponds to the length of the animal body. Below, we assume that the minimal possible domain of the animal is equal to this length, so corresponding short-range repulsion is comparable to it: $R_{jk}^{repuls} \approx l$. The nonzero length $l = \left| \vec{r}_{j1} - \vec{r}_{j2} \right| = const$ as well as the difference between the "head" and "tail" interaction parameters, $R_{jk,1}^{repuls} > R_{jk,2}^{repuls}$, allow us to reproduce naturally the anisotropy of the real bodies and their tendency to sort themselves out and move inside the groups in parallel or one after another.

Naturally, the predator can catch prey from some limited distance R^{catch}, which is normally bigger than, or at least equal to, the size of the animals: $R^{catch} \geq l$. In numerical simulations, such an event can happen with a given probability p^{catch} at every discrete time step Δt, depending on the interval of discretization, $p^{catch} \propto \Delta t$.

In the framework of the present numerical procedure, this means that a member of the array corresponding to the captured/eaten prey must be omitted. As a result, the index describing the total number of eaten animals increases:

$$N_{eaten} \rightarrow N_{eaten} + 1.$$

To characterize the intensity of the process of eating, we can also introduce some time window τ a couple of orders longer ($n_\tau = 100$) than the minimal time step of the calculation $\tau = n_\tau \Delta t$ and calculate a variable number of eaten prey items for every interval equal to such a time window:

$$N_{eaten}^{\tau} = \sum_{k=1}^{n_\tau} N_{eaten}(k).$$

Certainly, this value can strongly vary with time, depending on the number of prey items in close proximity to the predator N^{τ}_{proxy} during the observation interval τ. Thus, in order to understand the process, it is important to calculate this number as well:

$$N^{\tau}_{proxy} = \sum_{k=1}^{n_{\tau}} N_{proxy}(k).$$

According to a general understanding of the system under consideration, it must have inertia in both "subsystems". When a real predator loses sight of a potential prey, it will continue to move in the same direction for some time. In part, this happens because of the physical inertia of its body, and in part because there is also an "inertia of thinking". For the same reasons, the prey items will continue their motion, even if the predator has visually disappeared. Their inertia involves simple physical inertia of the body, but also involves some other biological reasons, such as some kind of "collective behavior"; when each animal sees that others are still moving, it will in turn continue its own run and thus stimulate the motion of others.

From a mathematical point of view, this means that the equations of motion must contain terms with second-order time derivatives responsible for inertia. But with repulsion present between the prey and the predator, pure inertia would cause infinite motion, and the system would move infinitely, in contrast to reality. Both subsystems, predator and prey, must gradually "forget" the immediate past and stop their motion after some characteristic time τ_{γ}. This means that the equations of motion must include an effective "dissipation term" γv proportional to the velocity v and with a dissipation constant $\gamma = 1/\tau_{\gamma}$ which gradually decreases motion when the predator is not present any more.

In general, the equations should also incorporate uncertainties in the behavior of all the "participants of the game", such as wrong guesses about the actions of other participants or about the entire situation, as well as influences from other species in the surrounding ecological niche also acting on the open system. This uncertainty can be simulated by a random source on the right side of the equations of motion.

Under all these assumptions, the equations of motion will take the following form:

$$\frac{\partial^2}{\partial t^2} = F^{pred}\left(\vec{R}\right) - \gamma_{pred}\frac{\partial \vec{R}}{\partial t} + \zeta\left(\vec{R}\right);$$

$$\frac{\partial^2 \vec{r}_j}{\partial t^2} = \sum_{k=1}^{N}\left[f^{pred}\left(\vec{R}, \left\{\vec{r}_j\right\}\right) + f^{attract}_{jk}\left(\vec{r}_j, \vec{r}_k\right) + f^{repuls}_{jk}\left(\vec{r}_j, \vec{r}_k\right)\right]$$

$$- \gamma_{prey}\frac{\partial \vec{r}_j}{\partial t} + \zeta\left(\vec{r}_j\right), \quad j = 1, \ldots N$$

where all the forces mentioned above are included, with effective "damping constants" which define a characteristic time of "forgetting" and are different for the predator (γ_{pred}) and the prey (γ_{prey}), and with $\zeta\left(\vec{R}\right)$, $\zeta\left(\vec{r}_j\right)$ representing δ-correlated random sources for both sides of the action:

$$< \zeta\left(\vec{R};t\right) \geq 0; \quad < \zeta\left(\vec{R};t\right)\zeta\left(\vec{R};t'\right) \geq D^{pred}\delta(t-t');$$

$$< \zeta\left(\vec{r}_j;t\right) \geq 0; \quad < \zeta\left(\vec{r}_j;t\right)\zeta\left(\vec{r}_k;t'\right) \geq D^{prey}\delta_{jk}\delta\left(\vec{r}_j - \vec{r}_k\right)\delta(t-t').$$

Here,

$$\delta_{jk} = \begin{cases} 1 & j=k \\ 0 & j \neq k \end{cases} \quad \text{and} \quad \delta(\mathrm{x}-x') = \begin{cases} 1 & x=x' \\ 0 & x \neq x' \end{cases}$$

are Kronecker delta-symbol and Dirac delta-function, respectively.

9.2.2 Model Behavior in a "Flat" World

One can generally predict the typical behavior of the system. Due to the long-range attraction, the randomly placed population of the prey aggregates into some number of groups, where the typical distance between the nearest neighbors inside each group is defined by the characteristic scale of their short-range repulsion.

The predator is preferably attracted to the closest prey group with the biggest number of individuals. In many spatial realizations, this causes a "saddle point" configuration where the predator has to choose between two or more close groups. After some extremely slow transient motion it definitely chooses one of the local groups, starts slowly in its direction and moves into this direction with some acceleration.

When the predator appears inside the group, the prey items are repulsed from it in different directions. Their trajectories diverge, but after a moment, they either join already existing groups or accumulate to form new ones. This scenario generally happens again and again with modifications that may be quantitatively large but are not very important qualitatively. In this sense, one can call the motion "*quasi-periodic*". At the same time, such a motion follows a so-called "*strange-attractor*".

Due to such a random (*quasi-periodic* or *strange-attractor*) behavior of the system, the effectiveness of different aggregation strategies must be characterized by some statistical value. In this context, one of the most convenient values would be the total number of eaten animals accumulated during periods much longer than the typical time of the particular interventions of the predator in the groups (quasi-periods of the motion).

This time must be chosen from the numerical observations and can strongly vary, depending on the mutual relationships between all the parameters. Therefore, in

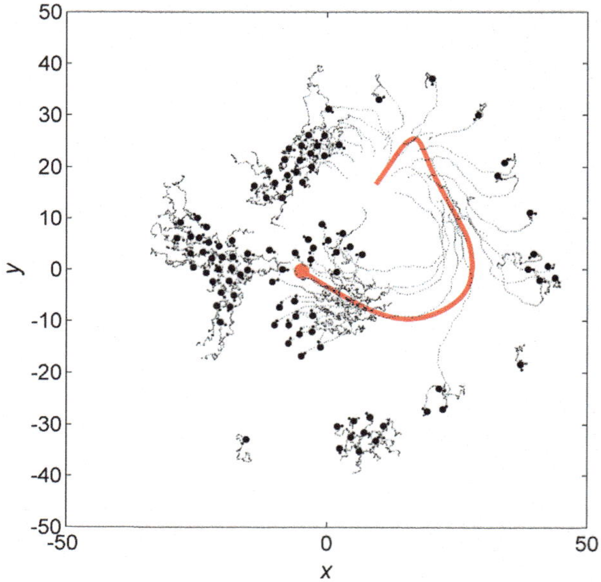

Fig. 9.12 Some momentary configuration of the system. The prey items are represented by pairs of big and small dots, the predator is represented by a large red dot. The red curve reproduces a part of the predator's trajectory preceding this configuration; the trajectories of the prey items are shown by the dotted lines. Saddle configurations produced by the divergence of the prey trajectories are directly evident. (From Filippov et al. 2019)

order to compare the effectiveness of different strategies one needs to perform an observation of the long-term, relatively stationary behavior. In other words, one needs a long time of observation with a quasi-infinite reservoir of prey with a density and distribution that remain practically the same although some of them are eaten and removed. In practical numerical simulations, with limited array sizes, one can simulate such a stationary behavior only by keeping the total number of prey items fixed. Technically, in our simulations this is organized as follows: when one animal is eaten by the predator, and the corresponding member of the array has disappeared, a new one is created with a coordinate randomly placed between the maximal and minimal prey positions. This guarantees the conservation of the total number of prey during the entire simulations.

Typical results of the simulation are presented in Figs. 9.12, 9.13 and 9.14. Figure 9.12 depicts some momentary configuration of the system. The prey items are represented by pairs of big and small dots, the predator is shown by a larger red dot with a curve which reproduces a part of its trajectory that preceded this particular configuration. Every particular scenario can be characterized by two important time-dependent variables: the number of prey items which appear inside a certain "time window" (as described above) in the proximity of the predator where it can catch them (N_{proxy}), the number of actually eaten prey items during the same interval (N_{eaten}) and the same values integrated over the complete time elapsed from the beginning of the procedure ($TotalN_{proxy}$ and $Total\,N_{eaten}$). These values are shown in the four panels of Fig. 9.13. It is directly evident that the maxima of N_{proxy} are associated with a greater time/density of N_{eaten} and as a result their accumulated values correlate with each other.

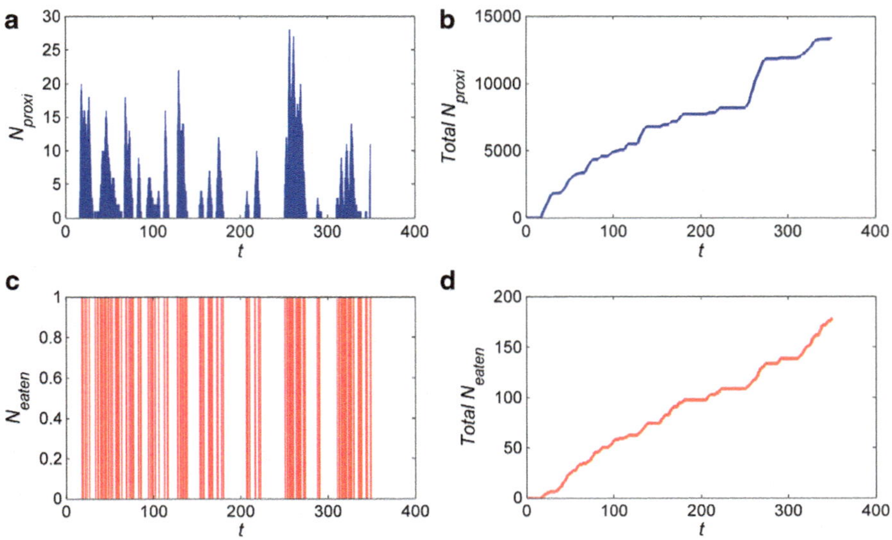

Fig. 9.13 (**a**) Histogram of the number of prey items appearing in the proximity of the predator inside the time window in which it can catch them. (**c**) Number of actually eaten prey items during this interval. On the right-hand side, panels (**b**) and (**d**) present both these values integrated over the entire time from the beginning to the end of the procedure. (From Filippov et al. 2019)

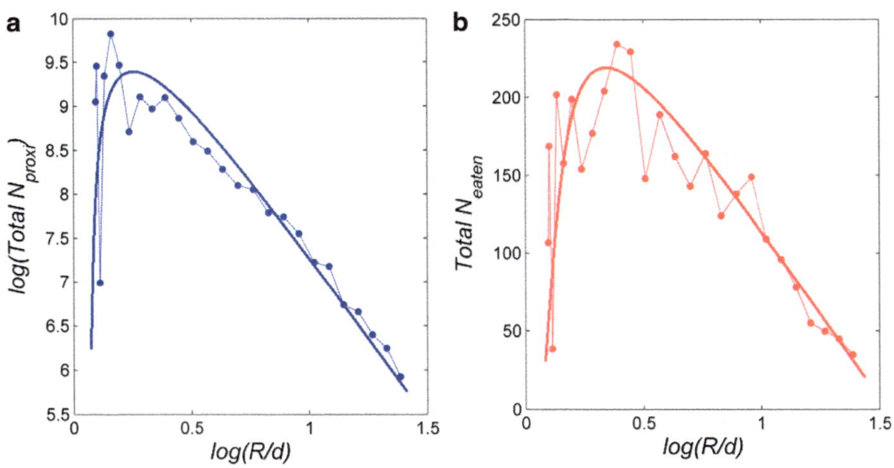

Fig. 9.14 The total number of prey items in close proximity to the predator (**a**) and the total number of captured/eaten prey items (**b**) during long-run simulations for different relationships between the interaction radii. The bold line represents the curve numerically smoothed to even out the fluctuations. (From Filippov et al. 2019)

It is quite obvious that the more animals have accumulated in a group, the more of them will be close to the predator when it arrives at the center of the group. Thus, however, there will be a greater distance between the predator and the other groups when the prey is scattered and either forms another group or joins some group distant from the predator. These scenarios are clearly evident in the simulations in supplementary movie 9.4.

Intuitively, one can expect two different survival strategies: (1) very dense packing of animals restricted only by their own size, with large distances between the groups, and (2) very scattered distribution of animals placed as far as possible from each other. In the first case, the predator spends much time for reaching the various groups, in the second case it wastes a lot of time chasing each single animal.

To verify this supposition, we calculated the total numbers of prey items in close proximity to the predator and eaten animals during long-run simulations at different relations between the defined above radii of attraction and repulsion interactions between the prey items (this relation mainly regulates different aggregation strategies). Accumulated plots are presented in Fig. 9.14.

The stronger the tendency to aggregation, the longer the distance between two quasi-periods of the motion, and, in turn, the smaller the number of such periods available for the statistical averaging of the results. The strategy of maintaining very short distances between prey items causes relatively big fluctuations on averaged curves. The greater the distance between prey items, the more regular motion is normally observed in the numerical experiments, and the more regular statistics and smoother curves will be obtained. At this limit, the distance between the animals can be much greater than their size. To make both limits (of extremely long and extremely short mean distances between the prey items) visible in the same plots in Fig. 9.14, we use a logarithmic scale on the horizontal axis, which visually compresses long distances and extends short ones.

Good correspondence between the results obtained and preliminary qualitative expectations for two limiting cases is directly evident in the plots in Fig. 9.14. It is important, however, to note that aggregation is found to be a losing strategy for a wide range of intermediate distances between the animals comparable to few animal sizes.

9.2.3 Model Behavior in a "Cylindrical World"

The above results are strongly related to the important assumption of the model that all the events happen in empty (mathematically flat) space, where the predator can see the prey, in principle, from a far-off (formally infinite) distance. In reality, of course, this distance is always limited by some physical "horizon of events", e.g., a real horizon on the spherical surface of the Earth, mountains and so on. In some cases, these distances will be sufficiently long to justify the model's assumption of a practically infinite planar space. In other cases, the model can be modified to adapt to the real surface by adding a relief inside which the animals actually move.

Here we will not develop the model in this direction, however, but will concentrate on another possible geometry of the system, which widely exists in the biological world. This is the cylindrical symmetry of surfaces (tree trunks, branches of a bush, grass etc.). In many cases, animals are strongly attached (virtually "glued") to such cylindrical surfaces. From the mathematical point of view, these animals move within a so-called topological confinement on effective "cylindrical world".

The standard technique to modify the model for this case is to apply periodic boundary conditions in one of the directions. It is convenient to parametrize motion around the cylinder by the angle $0 \leq \varphi < 2\pi$ of winding and to calculate – at the given radius of the cylinder, $R_{cylinder}$, the physical distances as the corresponding product:

$$y = \varphi R_{cylinder}.$$

A typical configuration of a system with such a geometry is presented in Fig. 9.15. It is important to note that despite the formally flat map "angle vs. coordinate" in Fig. 9.15a, the conditions of motion along both the vertical and the horizontal axes are different, because here the real restriction by a "horizon of events" along the angle-coordinate φ is present. Both participants of the "game", the prey and the predator, can see only a very limited strip along this coordinate (in some interval $|\varphi - \varphi_j| < \delta\varphi_{horizon}$ only). Depending on the relationship between their physical size and the radius of the cylinder, the strip can be wider or smaller, but, generally speaking, it is limited. This limitation can cause an additional advantage for the aggregation strategy as well as for the survival strategy, because the prey items can simultaneously and completely aggregate on the "other side of the tree" (Fig. 9.15b).

Dynamic simulation shows that the prey develops a specific survival strategy in this cylindrical space. When a local group of animals incidentally appears on the side of the cylinder opposite to the current location of the predator, this group is invisible to the predator. As a result, the predator will follow other groups which may in fact be much less dense and have a smaller total number of individuals than the group on the other side of the cylinder. Of course, from time to time the predator may switch to the other side of the cylinder, e.g., when following a group of prey that is trying to escape from the predator. In this case the predator will reach the other side and will have the chance to catch prey animals accumulated there. Such invasions, however, are relatively rare and the long time intervals in between will provide a competitive advantage to the prey animals surviving on "the other side" of the world. These scenarios are clearly evident in the simulations presented in supplementary movies 9.5–9.7).

However, how to describe the corresponding behavior quantitatively and/or in static pictures? Figures 9.16 and 9.17 propose two variants of such a representation. Both of them are based on the idea that the prey animals will preferably survive on the other side of the "tree trunk", and as a result the distribution of the trajectories will appear inhomogeneous along the angular coordinate $0 \leq \varphi < 2\pi$.

Yet, taking into account the possibility of the predator moving of its own accord along the same coordinate $0 \leq \varphi < 2\pi$, this distribution should become redistributed

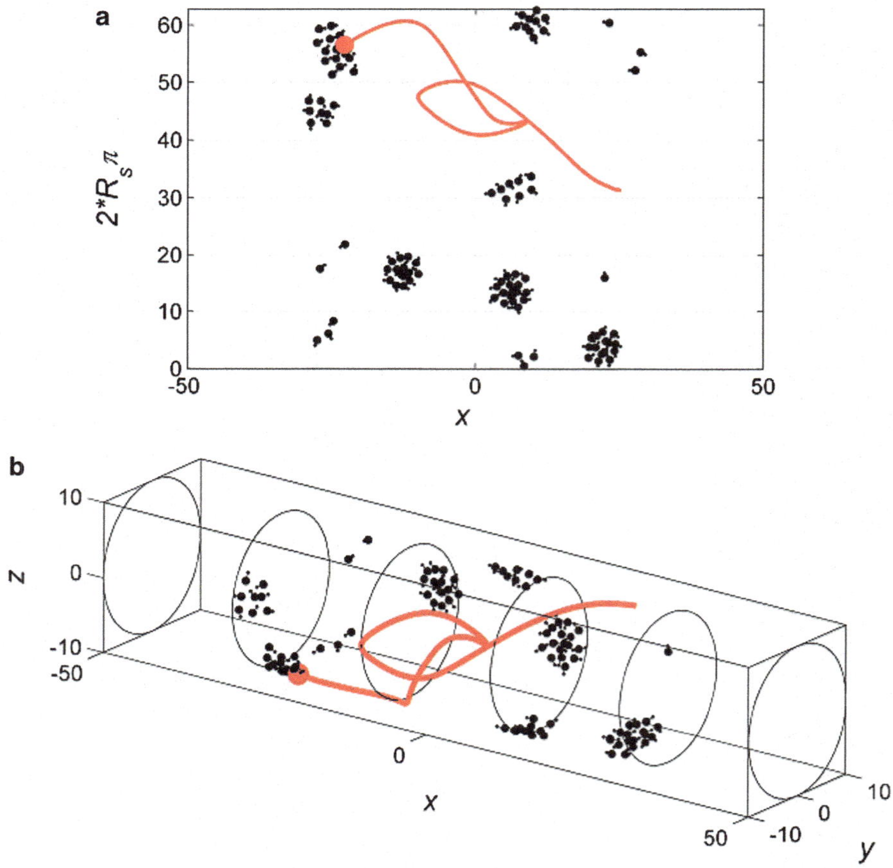

Fig. 9.15 Momentary configuration of a cylindrical system with trajectory of the predator (**a**) in angle-coordinate φ projection and (**b**) in 3D space. The symbols are the same as in Fig. 9.12. (From Filippov et al. 2019)

with time t, ultimately covering all the possible angles. As a result, at $t \to \infty$, the angle distribution of the prey items should also tend to a uniform configuration. To avoid this self-averaging, one can shift the coordinate φ_j to the movable system with the beginning of the instant position of the predator: $\varphi_j(t) \to \delta\varphi_j(t) = [\varphi_j(t) - \varphi_{pred}(t)]$. The sequence of this operation is illustrated in supplementary movie 9.7. In a static form, it may be represented by a plot of the complete array of trajectories $\delta\varphi_j(t)$ versus time (Fig. 9.16c). The corresponding histogram of the probability $P(\delta\varphi)$ of finding a trajectory with a given $\delta\varphi_j$ is accumulated during a sufficiently long run (Fig. 9.16d).

It is directly evident from this plot that, despite the well-pronounced maxima, this histogram still has a high plateau corresponding to the uniform impact from all possible angles . Moreover, the difference between this plateau and the maxima of the histogram gradually disappears with time, especially for extremely long runs

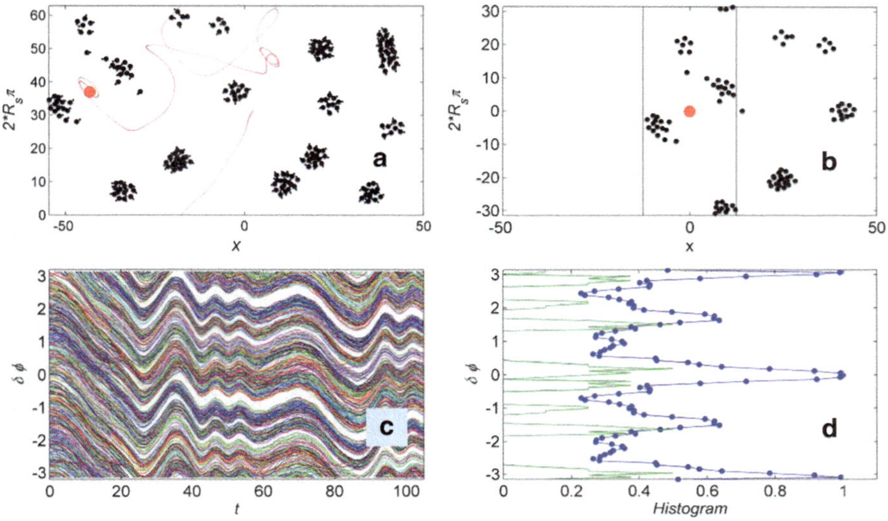

Fig. 9.16 Snapshot of a particular realization of the angle φ – space x configuration (**a**), the same configuration in a movable system of the coordinate (**b**), the complete array of all the trajectories $\delta\varphi_j(t) = [\varphi_j(t) - \varphi_{pred}(t)]$ over time (**c**) and the corresponding histogram for the probability $P(\delta\varphi)$ to find a trajectory with given $\delta\varphi_j$ accumulated during a sufficiently long run (**d**). (From Filippov et al. 2019)

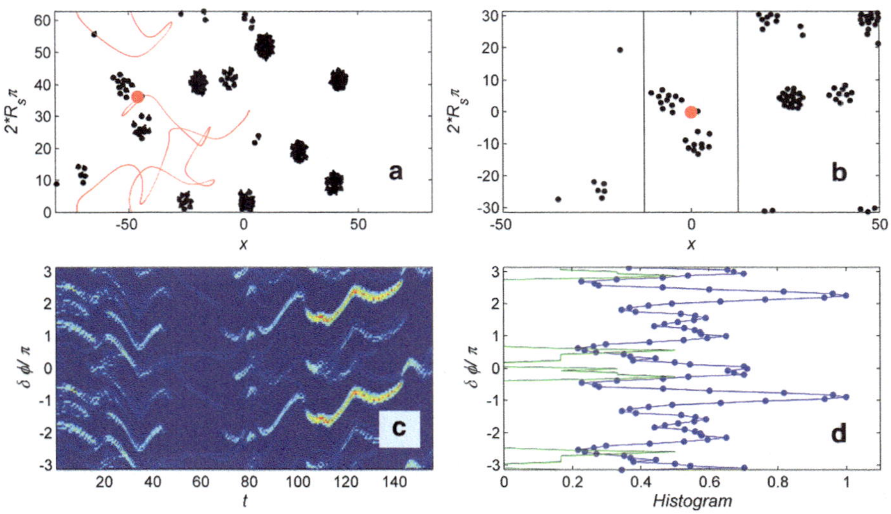

Fig. 9.17 Snapshot of a particular realization of the angle φ – space x configuration (**a**), the same configuration in a movable system of the coordinate (**b**), the density map of $p(\delta\varphi; t)$ (**c**) and the integral distribution $P(\delta\varphi)$ (**d**). Vertical lines in panel (**b**) enclose an interval along the x-coordinate, over which the histogram is calculated. Green and blue lines in the panel (**d**) represent the momentary distribution of the probability and the accumulated histogram for the complete interval of the simulation. (From Filippov et al. 2019)

($t \to \infty$). The contradiction between the originally expected and the observed results appears due to the fact that the redistribution of the prey items along the angular coordinate $\delta\varphi_j$ appears only for a relatively short distance along the cylinder in the x-direction. When $|x_j - X| \gg R_0$, the prey animals will practically ignore the predator and their distribution along the $\delta\varphi_j$-coordinate remains practically unaffected.

To avoid this problem and correct the calculation of the histogram, one can restrict the interval for which the histogram is calculated by the condition $|x_j - X| < R*$, where $R* \geq R_0$. In this case, at each particular moment in time the only prey which strongly interacts with the predator will be accounted for in the distribution. The number of the prey items which satisfy this condition will, however, vary with time, and the corresponding array will have different lengths as well. Therefore, one cannot plot all the trajectories together. To avoid this problem, we can calculate a partial histogram of the probabilities for every moment $p(\delta\varphi; t)$, plot it as colored "density map" and accumulate the final histogram $P(\delta\varphi)$ by integrating $p(\delta\varphi; t)$ over the complete run from $t = 0$ to t_{max}:

$$P(\delta\varphi) = \int_{t=0}^{t=t_{max}} p(\delta\varphi; t)dt.$$

The results of these calculations are presented in supplementary movie 9.7 and in Fig. 9.17 for a snapshot of the particular realization, the density map of $p(\delta\varphi; t)$ and the integral distribution $P(\delta\varphi)$, respectively.

9.2.4 *Biological Consequences of Motion in Worlds of Different Topologies*

Aggregations in psocopterans may be a defensive strategy against bark gleaning visually oriented predators, such as spiders, ants, bugs, and some birds. Previous authors suggested that psocopteran aggregations may provide some selective advantages for the individual insects in three different ways resulting in a higher probability of survival (New and Collins 1987):

- The aggregated larvae look cryptic to predators, because the group resembles a piece of bark or a patch of lichens (Fig. 9.1a). As soon as a visually oriented predator approaches, the group might be alarmed by the jerky movements of some individuals. This signal can be quickly transferred due to the almost continuous contact between neighboring individuals through their antennae. On the one hand, the predator might mistake the whole aggregation for a single organism, but, on the other hand, the contact may aid in quickly transferring the alerting signal to other group members.

- The sudden scattering of the prey in all directions after the predator strikes, as clearly shown by our modeling, confuses the predator (due to the effect of "Buridan's ass").
- Some subgroups of the split large group may use the predator's confusion and escape to the other side of the "cylindrical world" of the tree trunk.

Aggregation strategies, in general, might increase the defense efficiency (Lockwood and Story 1985), because natural enemies are more efficiently repelled by the active defense of an aggregation of potential prey than by single individuals. As mentioned above, in an aggregation, individuals may also benefit from a greater vigilance due to the stronger collective sensory performance. For example, pine aphids (*Schizolachnus pineti*) are warned against potential attacks of syrphid larvae by swinging abdomens and kicking hind legs sending an alert to the aggregated neighbors that may then walk away or drop off the plant (Kidd 1982; Bengtsson 2008). Aggregation has an additional benefit for insects that use aposematic coloration to warn predators with bright colors or contrasting patterns (Riipi et al. 2001). Here, we showed that aggregation in conjunction with a specific environment may convey additional advantages. Our predator–prey model for a cylindrical world demonstrated some advantages for individual subgroups of aggregated prey compared to aggregation on a 2D plane. In a cylindrical world these groups may potentially remain hidden to predators mainly relying on visual cues for prey detection, even in the vicinity of the predator, whereas uniform distribution of the prey would lead to an enhanced effectiveness of prey capture by the predator. When one subgroup aggregates on the back side of the tree trunk and becomes invisible to the predator, this subgroup will survive with an extremely high probability. In time, this might lead to the evolution of the specific survival strategy of anisotropic repulsion. Anisotropic repulsion manifests in running around, rather than along, the tree trunk. We are planning to extend the existing model to this particular feature in future.

Aggregation behavior has not only benefits, but also some costs, such as (1) increased competition due to overheard aggregation signals, (2) increased risk of infection by pathogens, (3) increased competition, (4) deterioration of the local microclimate, and, most importantly, (5) increased detectability by predators (Bengtsson 2008). In *Drosophila melanogaster* (Diptera, Drosophilidae) larvae, aggregation pheromones make them more noticeable to the parasitoid *Leptopilina heterotoma* (Hymenoptera, Figitidae) (Wertheim et al. 2003). The predatory clerid beetle *Thanasimus undatulus* (Coleoptera, Cleridae) is attracted more intensively to the higher concentration of aggregation pheromones of the spruce beetle, *Dendroctonus rufipennis* (Coleoptera, Curculionidae) (Poland and Borden 1997). In the case of a cylindrical world, smell produced by a prey group on the back side of a trunk might be misjudged by a predator, which, seeing a small subgroup of prey on its side of the tree may regard it as the major one and pursue it without exploring alternative possibilities. Ultimately, a predator moving only along one side of the tree may completely miss out on the major group.

References

Alcock J (1982) Natural selection and communication among bark beetles. Fla Entomol 65:17–32

Aukema BH, Raffa KF (2004) Does aggregation benefit bark beetles by diluting predation? Links between a group-colonisation strategy and the absence of emergent multiple predator effects. Ecol Entomol 29:129–138

Beattie AJ (1983) Distribution of ant-dispersed plants. Sonderbl Naturwiss Ver Hamburg 7:249–270

Beattie AJ (1985) The evolutionary ecology of ant-plant mutualisms. Cambridge University Press, Cambridge

Beattie AJ, Culver DC (1981) The guild of myrmecochores in the herbaceous flora of West Virginia forests. Ecology 62:107–115

Beattie A, Lyons N (1975) Seed dispersal in *Viola*: adaptations and strategies. Am J Bot 62:714–722

Bengtsson J (2008) Aggregation in non-social insects: an evolutionary analysis. Introductory paper at the Faculty of Landscape Planning, Horticulture and Agricultural Science, Swedish University of Agricultural Sciences, Alnarp, Sweden, vol 2, pp 1–18

Berg RY (1966) Seed dispersal of *Dendromecon*: its ecologic, evolutionary, and taxonomic significance. Am J Bot 53:61–73

Bond WJ, Stock WD (1989) The costs of leaving home: ants disperse myrmecochorous seeds to low nutrient sites. Oecologia 81:412–417

Bresinsky A (1963) Bau, Entwicklungsgeschichte und Inhaltsstoffe der Elaiosomen. Studien zur myrmekochoren Verbreitung von Samen und Fruechten. Bibl Bot 126:1–54

Buckley RC (1982) Ant-plant interactions: a world review. In: Buckley RC (ed) Ant-plant interactions in Australia. Dr. W. Junk, The Hague, pp 111–141

Buzatto BA, Requena GS, Machado G (2009) Chemical communication in the gregarious psocid *Cerastipsocus sivorii* (Psocoptera: Psocidae). J Insect Behav 22(5):388–398

Cocroft RB (2002) Antipredator defense as a limited resource: unequal predation risk in broods of an insect with maternal care. Behav Ecol 13(1):125–133

Cuautle M, Rico-Grey V, Díaz-Castelazo C (2005) Effects of ant behaviour and extrafloral necteries presence on seed dispersal of the neotropical myrmecochore *Turnera ulmifolia* L. (Turneraceae), in a sand dune matorral. Biol J Linn Soc 86:67–77

Culver D, Beattie A (1978) Myrmecochory in *Viola*: dynamics of seed-ant interactions in some West Virginia species. J Ecol 66:53–72

Culver D, Beattie A (1980) The fate of *Viola* seeds dispersed by ants. Am J Bot 67:710–714

Davidson DW, Morton SR (1981a) Competition for dispersal in ant-dispersed plants. Science 213:1259–1261

Davidson DW, Morton SR (1981b) Myrmecochory in some plants (F. Chenopodiaceae) of the Australian arid zone. Oecologia 50:357–366

Durieux D, Fassotte B, Deneubourg J-L, Brostaux Y, Vandereycken A, Joie E, Haubruge E, Verheggen FJ (2015) Aggregation behavior of *Harmonia axyridis* under non-wintering conditions. Insect Sci 22(5):670–678

Edwards W, Dunlop M, Rodgerson L (2006) The evolution of rewards: seed dispersal, seed size and elaiosome size. J Ecol 94:687–694

Filippov AE, Guillermo-Ferreira R, Gorb SN (2019) "Cylindrical worlds" in biology: does the aggregation strategy give a selective advantage? Biosystems 175:39–46

Fischer R, Richter A, Hadacek F, Mayer V (2008) Chemical differences between seeds and elaiosomes indicate an adaptation to nutritional needs of ants. Oecologia 155:539–547

Foster WA, Treherne JE (1980) Feeding, predation and aggregation behaviour in a marine insect, *Halobates robustus* barber (Hemiptera: Gerridae), in the Galapagos Islands. Proc Roy Soc B 209:539–553

Garrido JL, Rey PJ, Cerda X, Herrera CM (2002) Geographical variation in diaspore traits of an ant-dispersed plant (*Helleborus foetidus*): are ant community composition and diaspore traits correlated? J Ecol 90:446–455

Giladi I (2006) Choosing benefits or partners: a review of the evidence for the evolution of myrmecochory. Oikos 112:481–492

Gómez C, Espadaler X (1998) Myrmecochorous dispersal distances: a world survey. J Biogeogr 25:573–580

Gorb O (1998) Seed morphology and seed dispersal in two *Corydalis* species. Ukrainian Bot J 55:62–66

Gorb SN, Gorb EV (1995) Removal rates of seeds of five myrmecochorous plants by the ant *Formica polyctena* (Hymenoptera: Formicidae). Oikos 73:367–374

Gorb SN, Gorb EV (1999a) Effects of ant species composition on seed removal in deciduous forest in Eastern Europe. Oikos 84:110–118

Gorb SN, Gorb EV (1999b) Dropping rates of elaiosome-bearing during transport by ants (*Formica polyctena* Foerst.): implications for distance dispersal. Acta Oecol 20:47–53

Gorb E, Gorb S (2000) Effects of seed aggregation on the removal rates of elaiosome-bearing *Chelidonium majus* and *Viola odorata* seeds carried by *Formica polyctena* ants. Ecol Res 15:187–192

Gorb E, Gorb S (2003) Seed dispersal by ants in a deciduous forest ecosystem. Kluwer Academic Publishers, Dordrecht

Gorb SN, Gorb EV, Puntilla P (2000) Effects of redispersal of seeds by ants on the vegetation pattern in a deciduous forest: a case study. Acta Oecol 21:293–301

Gorb EV, Filippov AE, Gorb SN (2013) Long-term ant-species-dependent dynamics of a myrmecochorous plant community. Arthr-Plant Interact 7(3):277–286

Handegard NO, Boswell KM, Ioannou CC, Leblanc SP, Tjøstheim DB, Couzin ID (2012) The dynamics of coordinated group hunting and collective information transfer among schooling prey. Curr Biol 22:1213–1217

Handel SN, Fisch B, Schatz GE (1981) Ants disperse a majority of herbs in the Mesic forest community in New York state. Bull Torrey Bot Club 108:430–437

Hassell MP, May RM (1974) Aggregation of predators and insect parasites and its effect on stability. J Anim Ecol 43(2):567–594

Hassell MP, Varley GC (1969) New inductive population model for insect parasites and its bearing on biological control. Nature 223:1133–1137

Higashi S, Tsuyuzaki S, Ohara M, Ito F (1989) Adaptive advantages of ant-dispersed seeds in the myrmecochorous plant *Trillium tschonoskii* (Liliaceae). Oikos 54:383–394

Hodges RJ, Birkinshaw LA, Farman DI, Hall DR (2002) Intermale variation in aggregation pheromone release in *Prostephanus truncatus*. J Chem Ecol 28:1665–1674

Hölldobler B, Wilson EO (1990) The ants. Harvard University Press, Cambridge

Horvitz CC, Schemske DW (1986) Seed dispersal of a neotropical myrmecochore: variation in removal rates and dispersal distance. Biotropica 18:319–323

Hughes L, Westoby M (1990) Removal rates of seeds adapted for dispersal by ants. Ecology 71:138–148

Hughes L, Westoby M (1992) Fate of seeds adapted for dispersal by ants in Australian sclerophyll vegetation. Ecology 73:1285–1299

Johannesen A, Dunn AM, Morrell LJ (2014) Prey aggregation is an effective olfactory predator avoidance strategy. Peer J 2:e408

Keeler KH (1989) Ant-plant interactions. In: Abrahamson WG (ed) Plant-animal interactions. McGraw-Hill, New York, pp 207–242

Kidd NAC (1982) Predator avoidance as a result of aggregation in the Grey pine aphid, *Schizolachnus pineti*. J Animal Ecol 51:397–412

Kjellsson G (1985) Seed fate in a population of *Carex pilulifera* L. I Seed dispersal and ant seed mutualism. *Oecologia* 67:416–423

Klein M, Ladd T, Lawrence K (1973) Simultaneous exposure of phenethyl propionate-eugenol (7:3) and virgin female japanese beetle as a lure. J Econ Entomol 66:373–374
Lanza J, Schmitt MA, Awad AB (1992) Comparative chemistry of elaiosomes of 3 species of *Trillium*. J Chem Ecol 18:209–221
Lengyel S, Gove AD, Latimer AM, Majer JD, Dunn RR (2009) Ants sow the seeds of global diversification in flowering plants. *PLoS One* 4(5):e5480
Lockwood JA, Story RN (1985) Bifunctional pheromone in the first instar of the southern green stink bug, *Nezara viridula* (L.) (Hemiptera: Pentatomidae): its characterization and interaction with other stimuli. Ann Entomol Soc Am 78:474–479
Lorenzo Figueiras AN, Lazzari CR (1998) Aggregation behaviour and interspecific responses in three species of Triatominae. Mem Inst Oswaldo Cruz 93(1):133–137
Lotka AJ (1925) Elements of physical biology. Williams & Wilkins, Baltimore
Mark S, Olesen JM (1996) Importance of elaiosome size to removal of ant-dispersed seeds. Oecologia 107:95–101
Mayer V, Ölzant S, Fischer RC (2005) Myrmecochorous seed dispersal in temperate regions. In: Forget P.-M, Lambert JE, Hulme PE, Vander Wall, SB (eds) Seed fate: predation, dispersal and seedling establishment. CABI Publishing, Wallingford, 176–195
Miller RC (1922) The significance of the gregarious habit. Ecology 3:122–126
Minoli SA, Baraballe S, Lorenzo Figueiras AN (2007) Daily rhythm of aggregation in the haematophagous bug *Triatoma infestans* (Heteroptera: Reduviidae). Mem Inst Oswaldo Cruz 102(4):449–454
Morrell LJ, James R (2007) Mechanisms for aggregation in animals: rule success depends on ecological variables. Behav Ecol 19:193–201
Nathan R, Muller-Landau HC (2000) Spatial patterns of seed dispersal, their determinants, and consequences for recruitment. Trends Ecol Evol 15:278–285
Ness JHJ, Bronstein L, Andersen AN, Holland JN (2004) Ant body size predicts the dispersal distance of ant-adapted seeds: implications of small-ant invasions. Ecology 85:1244–1250
New TR, Collins NM (1987) "Herd-grazing" in tropical Psocoptera. Entomol Monthl Mag 123:229–230
Nicholson AJ, Bailey VA (1935) The balance of animal populations. Part I. Proc Zool Soc Lond B 3:551–598
O'Ceallachain DP, Ryan MF (1977) Production and perception of pheromones by the beetle *Tribolium confusum*. J Insect Physiol 23:1303–1309
Oostermejer JGB (1989) Myrmecochory in *Polygala vulgaris* L., *Luzula campestris* (L.) DC and *Viola curtisii* Forster in a Dutch dune area. Oecologia 78:302–311
Poland TM, Borden JH (1997) Attraction of a bark beetle predator, *Thanasimus undatulus* (Coleoptera: Cleridae), to pheromones of the spruce beetles and two secondary bark beetles (Coleoptera: Scolytidae). J Entomol Soc Br Columbia 94:35–41
Pulliam HR (1973) On the advantages of flocking. J Theor Biol 38:419–422
Raffa KF (2001) Mixed messages across multiple trophic levels: the ecology of bark beetle chemical communication systems. Chemoecology 11:49–65
Requena GS, Buzatto BA, Machado G (2007) Habitat use, phenology, and gregariousness of the Neotropical psocopteran *Cerastipsocus sivorii* (Psocoptera: Psocidae). Sociobiology 49 (2):1–19
Rico-Gray V, Oliveira PS (2007) The ecology and evolution of ant-plant interactions. University of Chicago Press, Chicago
Riipi M, Alatalo RV, Lindstrom L, Mappes J (2001) Multiple benefits of gregariousness cover detectability costs in aposematic aggregations. Nature 413:512–514
Ruxton GD, Sherratt TN (2006) Aggregation, defence and warning signals: the evolutionary relationship. Proc Roy Soc B 273:2417–2424
Schellinck J, White T (2011) A review of attraction and repulsion models of aggregation: methods, findings and a discussion of model validation. Ecol Model 222:1897–1911

Schlyter F, Birgesson G (1999) Forest beetles. In: Hardie J, Minks AK (eds) Pheromones in non-Lepidopteran insects associated with agricultural plans. CAB International, Oxford
Sernander R (1906) Entwurf einer Monographie der europäischen Myrmekochoren. K Sven Vetenskapsacad Handl 41:1–410
Smallwood J (1982) Nest relocation in ants. Insect Soc 29:138–147
Soukup VG, Holman RT (1987) Fatty acids of seeds of North American pedicillate *Trillium*-species. Phytochemistry 26:105–108
Stephens PA, Sutherland WJ (1999) Consequences of the Allee effect for behaviour, ecology and conservation. Trends Ecol Evol 14:401–405
Strömbom D, Mann RP, Wilson AM, Hailes S, Morton AJ, Sumpter DJT, King AJ (2014) Solving the shepherding problem: heuristics for herding autonomous, interacting agents. J R Soc Interface 11:20140719
Thompson WR (1924) La theorie mathematique de l'action des parasites entomophages et le facteur du hasard. Ann Fae Sci Marseille 2:69–89
Thompson SN (1973) A review and comparative characterization of the fatty acid compositions of seven insect orders. Comp Biochem Physiol A 45:467–482
Treherne JE, Foster WA (1980) The effects of group size on predator avoidance in a marine insect. Anim Behav 28:1119–1122
Treherne JE, Foster WA (1981) Group transmission of predator avoidance in a marine insect: the Trafalgar effect. Anim Behav 29:911–917
Treherne JE, Foster WA (1982) Group size and anti-predator strategies in a marine insect. Anim Behav 32:536–542
Turchin P, Kareiva P (1989) Aggregation in *Aphis varians*: an effective strategy for reducing predation risk. Ecology 70:1008–1016
Ulbrich E (1928) Biologie der Fruechte und Samen (Karpobiologie). Springer, Berlin
Vite JP, Pitman GB (1969) Aggregation behaviour of *Dendroctonus brevicomis* in response to synthetic pheromones. J Insect Physiol 15:1617–1622
Volterra V (1928) Variations and fluctuations of the number of individuals in animal species living together. J Cons Perm Int Explor Mer 3:3–51
Watt KEF (1959) A mathematical model for the effect of densities of attacked and attacking species on the number attacked. Can Ent 91:129–144
Watt PJ, Nottimgham SF, Young S (1997) Toad tadpole aggregation behaviour: evidence for a predator avoidance function. Anim Behav 54:865–872
Wertheim B, Vet LEM, Dicke M (2003) Increased risk of parasitism as ecological costs of using aggregation pheromones: laboratory and field study of *Drosophila-Leptopilina* interaction. Oikos 100:269–282

Index

A. E. Filippov, S. N. Gorb, *Combined Discrete and Continual Approaches in Biological Modeling*, Biologically-Inspired Systems 16,
https://doi.org/10.1007/978-3-030-41528-0

G

H

T

Printed by Printforce, the Netherlands